Multidisciplinary Approaches to Cholinesterase Functions

Multidisciplinary Approaches to Cholinesterase Functions

Edited by

Avigdor Shafferman and Baruch Velan

Israel Institute for Biological Research
Ness-Ziona, Israel

Springer Science+Business Media, LLC

Library of Congress Cataloging-in-Publication Data

Multidisciplinary approaches to cholinesterase functions / edited by
  Avigdor Shafferman and Baruch Velan.
      p.   cm.
    "Proceedings of the Thirty--sixth OHOLO Conference on
  Multidisciplinary Approaches to Cholinesterase Functions, held April
  6-10, 1992, in Eilat, Israel"--T.p. verso.
    Includes bibliographical references and index.
    ISBN 978-1-4613-6328-6   ISBN 978-1-4615-3046-6 (eBook)
    DOI 10.1007/978-1-4615-3046-6
    1. Acetylcholinesterase--Congresses.   I. Shafferman, A. (Avigdor)
  II. Velan, Baruch. III. OHOLO Conference on Multidisciplinary
  Approaches to Cholinesterase Functions (1992 : Eilat, Israel)
    [DNLM: 1. Cholinesterases--congresses. 2. Structure-Activity
  Relationship--congresses.   QU 136 M961 1992]
  QP609.A25M85 1992
  591.19'253--dc20
  DNLM/DLC
  for Library of Congress   92-49360
                                                                  CIP

36th OHOLO Conference Organizing Committee

Shafferman, A. (Co-Chairman)	Israel Institute for Bioligical Research, Ness-Ziona, Israel
Velan, B. (Co-Chairman)	Israel Institute for Biological Research, Ness-Ziona, Israel
Taylor, P.	University California San Diego, California, U.S.A.
Massoulie, J.	Centre National de la Recherche Scientific, Paris, France
Amitai, G.	Israel Institute for Biological Research, Ness-Ziona, Israel
Ashani, Y.	Israel Institute for Biological Research, Ness-Ziona, Israel
Silman, I.	Weizman Institute of Science, Rehovot, Israel
Soreq, H.	Hebrew University, Jerusalem, Israel

Proceedings of the Thirty-Sixth OHOLO Conference on
Multidisciplinary Approaches to Cholinesterase Functions,
held April 6-10, 1992, in Eilat, Israel

ISBN 978-1-4613-6328-6

© 1992 Springer Springer Science+Business Media New York
Originally published by Plenum Press, New York in 1992
Softcover reprint of the hardcover 1st edition 1992

CONTRIBUTORS

Arpagaus, Martine
   INRA Différenciation Cellulaire et Croissance, 2 Place Viala
   34060 Montpellier Cedex, FRANCE

Barak, Dov
   Israel Institute for Biological Research, Department of Organic Chemistry
   P.O.B. 19, 70450 Ness-Ziona, ISRAEL

Beeri, Rachel
   Department of Biological Chemistry, The Life Science Institute
   The Hebrew University, 91904 Jerusalem, ISRAEL

Berman, Harvey A.
   Department of Biochemical Pharmacology, School of Pharmacy
   State University of New York at Buffalo,
   313 Hochstetter Hall, Buffalo, NY 14260, USA

Brimijoin, Stephen
   Department of Pharmacology, Mayo Clinic, 200 First Street S.W.
   Rochester, Minnesota 55905, USA

Brodbeck, Urs
   Institute of Biochemistry and Molecular Biology, University of Bern
   Buhlstrasse 28, P.O.B. 98, CH-3012 Bern, SWITZERLAND

Colin, Malcolm
   School of Biological Sciences, Queen Mary and Westfield College
   Mile End Road, London E1 4NS, ENGLAND

Cygler, Miroslaw
   Biotechnology Research Institute, National Research Council of Canada.
   6100 Royalmount Ave., Montreal, Quebec H4P 2R2, CANADA

Doctor, Bhupendra P.
   Division of Biochemistry, Department of the Army
   Walter Reed Army Institute of Research, WRAMC
   Washington D.C. 2037-5100, USA

Enz, Albert
   Preclinical Research, Sandoz Pharma Ltd., Bld. 386 762
   CH-4002 Basel, SWITZERLAND

Fournier, Didier
   INRA - Laboratoire de Biologie des Invertébrés
   BP. 2078, 06606 Antibes Cedex, FRANCE

Greenfield, Susan A.
University Department of Pharmacology, Mansfield Road
Oxford OX1 3QT, ENGLAND

Harel, Michal
Department of Structural Biology, Weizmann Institute of Science
76100 Rehovot, ISRAEL

Hirth, Christian
LICMP-DIEP, Centre d'Etude de Saclay, Bâtiment 152, 91191 Gif-sur-Yvette
Cedex, FRANCE

Hjalmarsson, Karin
National Defence Research Establishment, Department of NBC Defence
S-90182 Umea, SWEDEN

Hucho, Ferdinand
Freie Universitat Berlin, Institut fur Biochemie, Thielallee 63
1000 Berlin 33, GERMANY

Inestrosa, Nibaldo
Department of Cell and Molecular Biology, Catholic University of Chile
Casilla 114-D, Santiago, CHILE

Layer, Paul G.
Max-Planck-Institut fur Entwicklungsbiologie, Spemannstrasse 35/IV
D-7400 Tubingen 1, GERMANY

Lockridge, Oksana
University of Nebraska Medical Center, Eppley Institute, 600 S. 42nd. Street
Omaha, NE 68198-6805, USA

Masson, Patrick
Centre de Recherches du Service de Santé des Armées
Unite de Biochimie, BP. 87, 38702 La Tronche Cedex, FRANCE

Massoulié, Jean
Laboratoire de Neurobiologie, Ecole Normale Supérieure
CRNS-URA 295, 46 rue d'Ulm, 75230 Paris Cedex 05, FRANCE

Primo-Parmo, Sergio L.
Departments of Pharmacology and Anesthesiology, University of Michigan
1500 East Medical Center Drive, R 4038 Kresge Medical Research II
Ann Arbor, MI 48109-0572, USA

Quinn, Daniel M.
Department of Chemistry, The University of Iowa, Iowa City
Iowa 52242, USA

Rosenberry, Terrone L.
Department of Pharmacology, Case Western Reserve University
School of Medicine, 10900 Euclid Avenue, Cleveland, OH 44106-4965, USA

Rotundo, Richard L.
University of Miami, School of Medicine, Department of Cell Biology and
Anatomy, R-124, P.O.B. 016960, Miami, Florida 33101, USA

Shafferman, Avigdor
Israel Institute for Biological Research, Department of Biochemistry
P.O.B. 19, 70450 Ness-Ziona, ISRAEL

Silman, Israel
Department of Neurobiology, Weizmann Institute of Science
76100 Rehovot, ISRAEL

Sketelj, Janez
Institute of Pathophysiology, School of Medicine
61105 Ljubljana, SLOVENIA

Soreq, Hermona
Department of Biological Chemistry, The Life Science Institute
The Hebrew University, 91904 Jerusalem, ISRAEL

Sussman, Joel L.
Department of Structural Biology, Weizmann Institute of Science
76100 Rehovot, ISRAEL

Taylor, Palmer
Department of Pharmacology, School of Medicine
University of California, San Diego
9500 Gilman Drive, La Jolla, CA 92093-0636, USA

Velan, Baruch
Israel Institute for Biological Research, Department of Biochemistry
P.O.B. 19, 70450 Ness-Ziona, ISRAEL

Weinstock, Marta
Department of Pharmacology, School of Medicine, The Hebrew University
91904 Jerusalem, ISRAEL

Williamson, Martin S.
AFRC Institute of Arable Crops Research, Rothamsted Experimental Station
Harpenden, Herts, AL5 2JQ, ENGLAND

Zakut, Haim
The Edith Wolfson Medical Center, Department of Obstetrics & Gynecology
P.O.B. 5, 58100 Holon, ISRAEL

Zhorov, Boris S.
Pavlov Institute of Physiology of the Russian Academy of Sciences
Nab. Makarova 6, St. Petersburg B-034, RUSSIA

Christian Hirth 1944-1992

Christian Hirth succumbed to a sudden heart attack on May 28, 1992; he was only 48 years old at the time. He was a professor of organic chemistry at Université Louis Pasteur in Strasbourg, where he was director of a CNRS research unit.

Christian obtained his Ph.D. thesis at Strasbourg in the laboratory of Jean-François Biellmann in 1972 and was a postdoctoral fellow under Georges Cohen at the Pasteur Institute in Paris, where he extended his interests to biology and made many fruitful and long-lasting associations. On his return to Strasbourg in 1975, he set up, together with his friend and colleague, Maurice Goeldner, a research team specializing in bioorganic chemistry, in the laboratory of Professor Guy Ourisson.

It was in this framework that he carried out his most important research, which involved the design and utilization of photosuicide compounds as topographic probes of biological macromolecules, among which the acetylcholine receptor and acetylcholinesterase were particularly noteworthy subjects.

His achievements led to the establishment of the CNRS research unit which he directed. He had just recently started to organize a new research team in the Protein Engineering Laboratory of the CEA at Saclay.

Christian's science was both rigorous and original. At the time of his death, he was making a major contribution to our understanding of the structure of cholinergic recognition sites. He was an invited speaker at the 36th Oholo Conference in Eilat which, sadly, was the last meeting he attended.

Christian was liked and respected by his collaborators and by those who encountered him at scientific meetings. He was a warm and loyal friend and colleague, with a great sense of humor, often sharp but never unkind. His death, in the midst of a brilliant scientific carrier, is a great loss to all of us, and we will miss him very much.

PREFACE

Acetylcholinesterase plays a key role in cholinergic transmission. By rapid hydrolysis of the neurotransmitter acetylcholine the enzyme terminates the chemical impluse, thereby allowing rapid repetitive responses. Inhibitors of cholinesterases have important applications in agriculture - pest control, and in medicine - treatment of various disorders such as myasthenia gravis, glaucoma and management of Alzheimer's disease. These are some of the reasons for the special attraction of scientists from multiple disciplines such as neurobiology, pharmacology, chemistry, biochemistry, molecular biology and biotechnology to cholinesterase research.

We have been fortunate to organize the 36th OHOLO conference on *Multidisciplinary Approaches to Cholinesterase Functions* at a time when this field of research is bursting with new concepts and ideas, brought about by the recent structural resolution of the catalytic subunit of acetylcholinesterase and by application of methodologies related to molecular genetics and protein engineering. This volume summarizes most of the important advances resulting from these developments as reflected in the OHOLO meeting.

The first article in this volume provides an overview of the large body of investigations accumulated over a century and a perspective on past and future studies on cholinesterases. This is followed by articles dedicated to biochemical and molecular genetic studies that reveal the origin and extent of the intriguing multiplicity of the molecular forms of cholinesterases. Papers describing the regulated expression of acetylcholinesterase in hematopoietic, muscle and nervous systems, as well as in engineered mammalian and bacterial high level expression systems, shed light on various postranscriptional and postranslational modifications of cholinesterases. The section dealing with polymorphism and structure includes papers describing the three dimensional structures of the *Torpedo* acetylchoninesterase and different enzymes sharing common folding patterns, despite the absence of sequence identity. The report on the crystal structure of two complexes of *Torpedo* acetylcholinesterase with ligands of clinical importance is yet another step in unravelling the unique catalytic functions of acetylcholinesterase. These reports together with articles describing site directed labeling and analysis of site directed mutants of cholinesterases, have identified amino acids constituting the catalytic triad and residues related to the peripheral and catalytic anionic subsites. Further insight into functional domains of cholinesterases and the role of particular amino acids, is provided by kinetic studies with various inhibitors and the analysis of natural mutants of human acetylcholinesterase and butyrylcholinesterase, as well as of *Drosophila* acetylcholinesterase insecticide resistant strains. Some articles are involved in the classical challenges of the physiological and developmental functions of cholinesterases, as well as in the possibility that these molecules fulfill functions other than inactivation of acetylcholine at the synapses.

A few papers address the clinical implications of acetylcholinesterase as a pretreatment drug for organophosphate toxicity and the potential use of different acetylcholinesterase inhibitors in the treatment of Alzheimer's disease. Finally, the contribution from a number of researchers from different labs provides a set of recommendations for a unifying nomenclature for the cholinesterase genes, their transcription and translation products, the numeration of amino acids and secondary structural motifs of acetylcholinesterases and related proteins.

We are obliged to all the contributors to this volume and to our colleagues: Palmer Taylor, Jean Massoulie, Hermona Soreq, Gabriel Amitai, Yaacov Ashani and Israel Silman, who were instrumental in formulating the scientific scope of the meeting. We hope that the volume contains a wealth of information that will generate new directions in cholinesterase research.

Avigdor Shafferman
Baruch Velan

June 1992
Israel Institute For Biological Research
Ness-Ziona, Israel

ACKNOWLEDGEMENTS

The organizing committee of the 1992 OHOLO conference gratefully acknowledges the generous support of the following organizations (Alphabetical order):

American Cyanamid Company, Pearl River, NY, U.S.A.

Amgen Center, Thousand Oaks, CA, U.S.A.

International Society for Neurochemistry, Chapel Hill, NC, U.S.A.

Joseph Meyerhoff Fund Inc., Baltimore, MD, U.S.A.

Merck Research Laboratories, Rahway, NJ, U.S.A.

Ministry of Science & Technology, Jerusalem, Israel

Ministry of Tourism, Jerusalem, Israel

U.S. Army Research and Developments, European Branch, London, U.K.

U.S Office of Naval Research, European Office, London, U.K.

ACKNOWLEDGMENTS

CONTENTS

IMPACT OF RECOMBINANT DNA TECHNOLOGY AND PROTEIN STRUCTURE DETERMINATION ON PAST AND FUTURE STUDIES ON ACETYLCHOLINESTERASE

Palmer Taylor

Department of Pharmacology
University of California, San Diego
La Jolla, CA 92093-0636

It is a pleasure and challenge to open the scientific sessions of the 36th Oholo Conference. The pleasure comes from the opportunity to participate in a conference organized at a most opportune time in the development of the field of cholinesterases. Avigdor Shafferman, Baruch Velan and their colleagues who have served as local organizers have planned this conference with unwavering enthusiasm in what has been a politically difficult period for Israel. My challenge stems from attempting to assemble from a large body of investigations a perspective on past and future studies on the cholinesterases. Studies on cholinesterases have been propelled by advances in recombinant DNA technology, protein chemistry and crystallography over the past six years, and our field of endeavor now fully encompasses the span from gene to protein, both in terms of structure and function. The work of Joel Sussman, Israel Silman and their colleagues has provided us with structural resolution of the catalytic subunits of acetylcholinesterase (AChE) extending to 2.8Å resolution (Sussman et al., 1991). This development now enables us to relate aspects of catalytic function and inhibitor binding to structural domains and to particular amino acid residues. With structural resolution of the cholinesterases at an atomic level, virtual reality emerges within the field.

At the other end of the spectrum, the genes encoding mammalian AChE and butyrylcholinesterase have been characterized (Li et al., 1991; Arpagaus et al., 1990) and localized to the chromosomal positions 7q22 and 3q26 respectively in man (Getman et al., 1992; Gaughan et al., 1991; Allderdice et al., 1991) and the AChE gene localizes to distal region of chromosome 5 in mouse, a region syntenic with human 7q (Rachinsky et al., 1992). While the structural resolution of chromosome localization is orders of magnitude less than that yielded by the crystal structure of AChE, recombinant DNA technology and nucleic acid sequencing enable one to determine rapidly essential elements in gene structure. Moreover, the power of genetics can be applied to the cholinesterases as can be seen from the correspondence of AChE and the QT-blood group antigen (Zelinski et al., 1991; Bartels and Lockridge, 1992). Virtually all of what we discuss over the next 4 days will either relate to the boundaries of the gene and protein themselves or extend to intermediary processes of gene transcription, mRNA stabilization, mRNA translation, peptide processing and secretion of the newly biosynthesized enzyme.

WHERE WE'VE BEEN

Cholinesterase research predates its very identification as an enzyme and, in fact, extends back to the mid-19th century. Various investigators were intrigued with the pharmacologic actions of the calabar bean in the early 1860's where both animal and self-administration were used to study its pharmacology. Argyll-Robinson's accounts of his responses to self-administration of extracts into one of his eyes using the other eye as a corresponding control are particularly enlightening (Argyll-Robinson, 1863). The first organophosphate inhibitor of cholinesterases was synthesized in the same period by Clermont.

.It was not until 1914 and the appearance Sir Henry Dale's classic manuscript on the nicotine- and muscarinic-like actions of the esters of choline that the concept of physostigmine inhibiting an enzyme responsible for the degradation of a natural neurotransmitter in the cholinergic nervous system emerged (Dale, 1914). Nevertheless, the name cholinesterase was coined in a paper by the organic medicinal chemist, Stedman some 18 years later (Stedman et al., 1932).

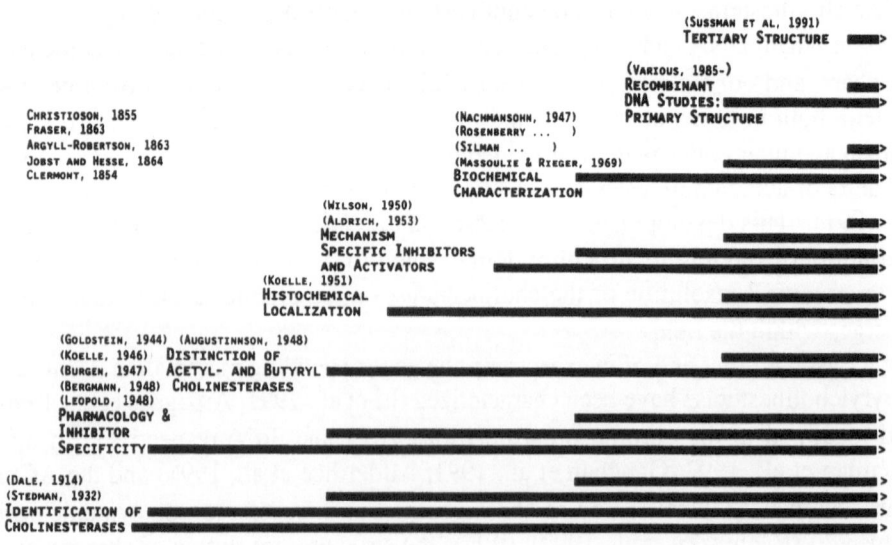

Fig. 1. Developments in the Field of Cholinesterases. An attempt is made to show escalating developments in the field with some of the critical studies and investigators that initiated new avenues of endeavor. New approaches also contributed substantially to ongoing research currents; major synergistic contributions are shown by the staircase additions. The early studies in the mid-19th century were concerned only with the pharmacologic effects of inhibitors without reference to the enzyme and are not included in the staircase.

A review of cholinesterase research since Dale's time reveals several important trends, some of which are depicted in Fig. 1. First, momentum in the field has been and continues to be facilitated by two intrinsic attributes of the enzyme. These are its high catalytic capacity and its susceptibility to a plethora of natural and man-made inhibitors.

The high substrate turnover enabled measurements in heterogeneous systems by crude Warburg manometric methods in the early years, facilitated later histological detection of activity (Koelle, 1951) and permitted the identification of enzyme isoforms on the basis of hydrodynamic properties (Massoulie' and Reiger, 1969). The susceptibility of cholinesterases to inhibitors allowed one to identify the acetyl and butyryl subtypes and to study mechanisms of catalysis through the use of hemisubstrates and reactivators. Furthermore, the inhibitors of the cholinesterase revealed applications to therapeutics and toxicology which underwrote investigations of specificity of ligand interactions.

Research in the field is earmarked by clusters of innovation and, as illustrated in Fig. 1, new developments invariably reinforce ongoing research in established areas. Hence, one creates a staircase effect in the overall endeavor. Lastly, progress in the field is very much dependent on new technological developments and applications.

Following the second world war and the concerns about the insidious use of organophosphates, a major focus of research was directed to the specificity and pharmacological actions of inhibitors of pharmacologic and toxicologic import. This period found a substantial fraction of investigators in the pharmacology community working on the cholinesterases and contributions took on a true international flavor. Shortly thereafter, important developments occurred in histochemical techniques and in the distinction of the acetyl- from the butyrylcholinesterases. The contributions of Irwin Wilson in the 1950's on site direction of enzyme inhibitors and reactivators and on delineating individual steps in the reaction mechanism are particularly noteworthy for they preceded investigations of others on site-directed labeling and affinity-based methods to modify and purify enzymes (Wilson, 1959). The understanding of mechanisms of action of inhibitors and their wide structural diversity now provide a rich pharmacology for site-specific mutagenesis research, an area whose potential will become evident at this conference. The biochemical applications in purifying and characterizing AChE in Wilson's and David Nachmansohn's laboratories at Columbia (cf: Nachmansohn and Wilson, 1951) provided the substrate and training for Israel Silman and Terrone Rosenberry, two preeminent contributors to the biochemical and kinetic characterization of the enzyme for the past two decades. Jean Massoulié's and Francois' Rieger's finding in 1969 of a dimensionally asymmetric, tail-containing form of AChE spawned a new dimension of the study of polymorphism of the cholinesterases (Massoulie' and Reiger, 1969). It is noteworthy that the collagen-containing structure of AChE stood unique among membrane proteins and it was only recently that other cell surface proteins have been found to be assembled in a similar manner.

Recombinant DNA technology was a natural direction for cholinesterase research in the 1980's. The initial cloning of the AChE gene indicated that it shared global sequence homology only with thyroglobulin (Schumacher et al., 1986), but over the past 6 years, proteins homologous to cholinesterase have enlarged to an extensive, but functionally eclectic, family (Fig. 2), not all of which function as hydrolases. The family now includes the tactins which are important for heterologous cell contacts. This structural similarity may provide an indication for non-catalytic function of the cholinesterases in developmental processes. The recent report of a high resolution crystal structure of Torpedo AChE adds the critical third dimension to structure (Sussman et al., 1991). A comparison of three dimensional structures also indicates that certain carboxypeptidases and dehalogenases share a common folding pattern in their backbone despite the absence of appreciable sequence identity. Both the recombinant DNA and crystallographic studies depended critically on protein chemistry to provide known sequence for molecular cloning and for the crystalline enzyme itself.

Explorations with the cholinergic nervous system have continued to yield many of the seminal discoveries in pharmacology and neurobiology; these include identification of the first neurotransmitter, the initial distinction of receptor subtypes, demonstration of quantal release of neurotransmitter, the first molecular characterization and reconstitution of a neurotransmitter receptor, and biophysical measurements of ligand-gated channels through patch clamping and noise analysis. To this, we can now add the crystal structure of AChE for it appears to be the first synapse-localized protein integral to neurotransmission that has a known three dimensional structure.

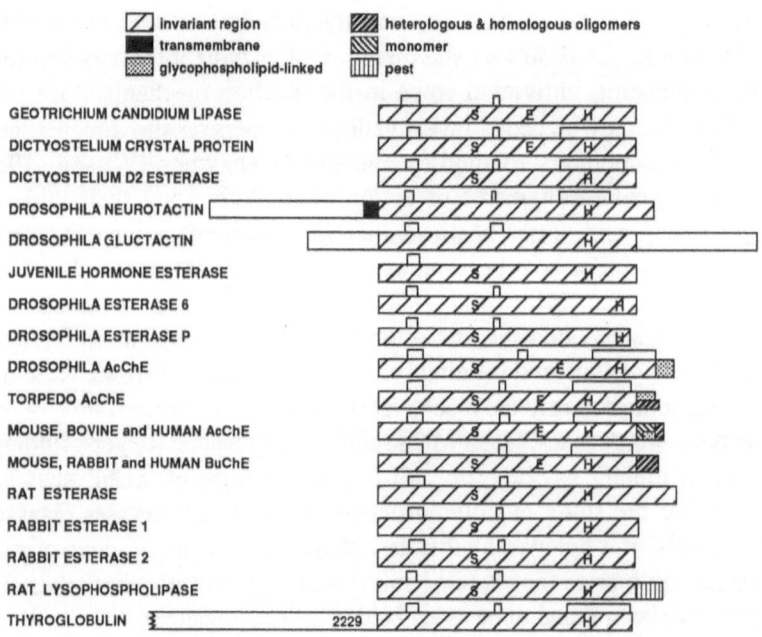

Fig. 2. Relationships of Proteins with Sequence Homology to the Cholinesterases. Proteins containing serines, histidines and glutamates in homologous positions to Ser 200, His 440, and Glu 327 in Torpedo are shown by S, H and E, respectively. Placement of cysteines to form loops A, B, and C (MacPhee-Quigley et al., 1986) is shown by the linkages on top of the structure.

FROM GENE TO PROTEIN

The recent characterization of the entire genes encoding Torpedo, mouse and human AChE's (Maulet et al., 1990; Li et al., 1991; Li et al., 1992a) provide a structural basis for the study of the regulation of expression of the gene encoding AChE. Already several unique facets of AChE expression are apparent. First, AChE expression is precisely regulated in the hematopoietic system, the nervous system and in muscle. In the hematopoietic system only cells of certain lineages express the enzyme, whereas in skeletal muscle overall expression appears to be coordinated with that of the nicotinic acetylcholine receptor (nAChR) in muscle cell differentiation and synaptogenesis (Fig. 3). While details of expression of the nicotinic receptor and AChE differ in the muscle cell differentiation program, similar increases in overall gene expression occur upon stem cell differentiation to myoblasts, myoblast to myotube formation and with innervation. It, therefore, will be of interest to compare transcriptional rates, mRNA stability and translation efficiencies associated with such events for these two proteins. Studies of run-on transcription rates for AChE and nAChR, promoter-reporter gene expression and degradation of mRNA in the presence of protein synthesis inhibitors indicate that an enhanced level of mRNA associated with myoblast fusion to myotubes is due to stabilization of AChE mRNA, while transcriptional activation is responsible for increasing the mRNA encoding nAChR subunits in the identical fusion step (Fuentes and Taylor, 1992). Thus, it appears that the AChR and AChE use distinct points of regulation in muscle cell differentiation.

Structure of the AChE Gene in Relation to Gene Expression

Another noteworthy feature of AChE gene expression is that a single AChE gene shows coordinate regulation with cholinergic receptors which are encoded by a great multiplicity of genes. Moreover, receptors use this gene multiplicity to achieve specificity in expression during development. At present, five subtypes of muscarinic receptors have been cloned and at least six distinct α-subunits and four β-subunits for nicotinic receptor exist in neurons. Embryonic or extra-junctional nicotinic receptors in skeletal muscle contain four distinct subunits $\alpha\beta\gamma\delta$. Upon maturation and formation of muscle endplates, the γ-subunit is replaced by an ϵ-subunit. To coordinate expression in the synapse with the multigenic receptor systems, AChE has developed distinct transcriptional, mRNA stabilization, processing and translational regulatory events. A diagram of the structure of the mammalian AChE gene summarizes the points of regulation of gene expression (Figure 4).

The 5'-end of the AChE gene is found to have two transcriptional start sites. The major site is the most 3' of the two and appears to be the only one used in the hematopoietic system and muscle (Li et al., 1992a) while the second site is located further in the 5' direction. To date its use has only been found in brain.

Transcription from the second cap site (1α) results in splicing to a point 5' of the first cap site. To date only the major promoter has been characterized; it features tandem Egr-1 consensus sequences in both mouse and human (Li et al., 1992a;

Getman et al., 1992b). Their function is being characterized by deletion analysis and site specific mutagenesis (Li et al., 1992a), but it appears that this region is responsible for tissue specific expression. Also, alternative splice acceptor sites are

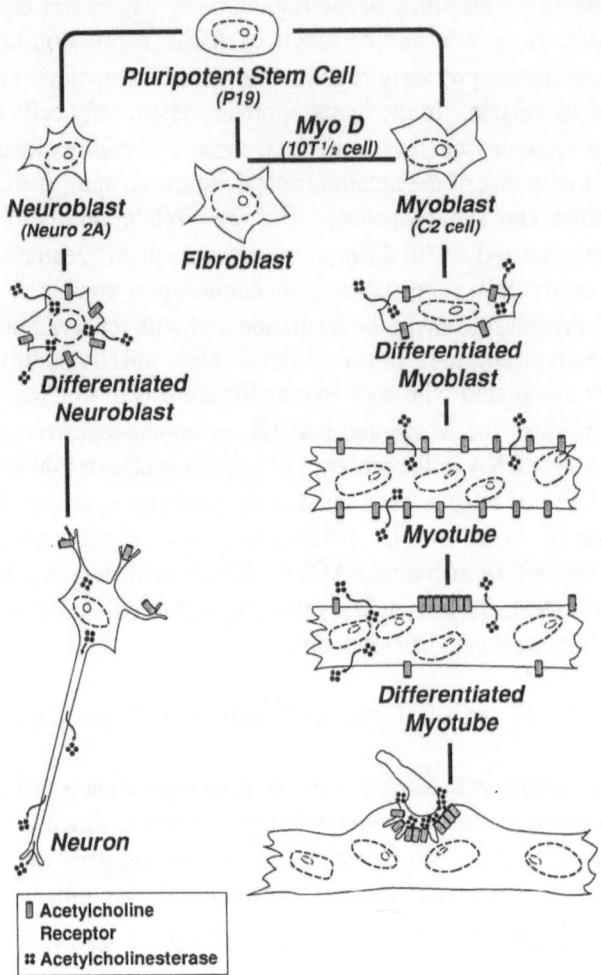

Fig. 3. Schematic Representation of Stem Cell Differentiation to Form Neuronal and Muscle Lineages. Embryonal stem cells or embryoid bodies are able to be differentiated into cells that form fibroblastic, myoid (muscle) or neuronal lineages. In turn, the neuronal and muscle program leads to differentiated neurons and myotubes. Innervation of muscle results in clustering of both AChE and nicotinic receptors in synaptic junctional areas and then the disappearance of these proteins in extrajunctional areas. Overall, coordinate expression is seen for AChE and the nicotinic receptor. Although intact tissues, primary culture and cell lines show differences in developmental factors controlling gene expression, the simplified cellular systems are essential starting points for understanding gene expression in relation to development. For example, C-2 myoblasts upon changing growth conditions will fuse to myotubes with concomitant increases and clustering of synaptic proteins; 10T1/2 cells, upon treatment with myoD, will differentiate along a muscle program and P-19 embryonal stem cells will differentiate into neurons or muscle depending on treatment with exogenous agents. Hence, cell culture becomes useful for studying individual phases of differentiation. The scheme also depicts the coordinated expression of the nicotinic acetylcholine receptor and AChE.

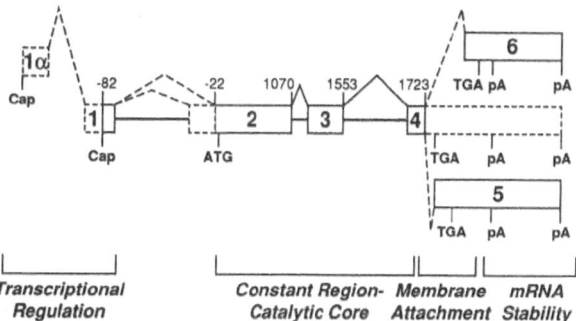

Fig. 4. Schematic View of the Mouse and Human Acetylcholinesterase Genes Showing Alternative Splicing Positions (----) and Invariant Splicing (——) Sites for the start of transcription (cap), start of translation (ATG), end of translation (TGA) and polyadenylation (pA) are marked. Distances approximate exon and intron distances in both the mouse and human genes.

found in the 5'-untranslated region which may influence the stability of mRNA species. The promoter region is devoid of TATA or CAAT boxes and contains GC-rich sequence. As found for other proteins expressed in muscle, elements which regulate transcription are likely to be found not only in the 5'-flanking region but also in downstream introns.

This alternatively spliced 5'-region of the gene then leads into an invariant region (exons 2, 3 and 4) which encodes the catalytic core of the molecule. All of the essential catalytic residues (Ser 200, His 440 and Glu 327) are encoded by these exons as are all of the cysteines that form intrasubunit disulfide bonds.

Proceeding in the 3' direction, we again find splicing to alternative exons before reaching the translation termination codons. In the mammalian AChE gene three alternatives exist. Splicing to the most proximal acceptor site gives rise to a nucleotide sequence encoding a peptide whose sequence predicts that it will be processed by cleavage of its hydrophobic carboxyl-terminus and addition of a glycophospholipid. This splice alternative gives rise to a mRNA encoding the glycophospholipid form of the enzyme which is abundant in nerve and muscle cells of poikiotherms but in mammals is largely found in the hematopoietic system. That this sequence encoded in the open reading frame of the alternatively spliced exon is both necessary and sufficient for yielding the glycophospholipid-linked form of the enzyme can be shown by linking this portion of the genomic sequence to the invariant exons through loop-out mutagenesis. Expression of the intron deleted sequence in COS cells gives rise to a species of AChE which is retained on the outer surface of the cell, incorporates ethanolamine and is released by addition of phosphatidylinositol specific phospholipase C (Gibney et al., 1990; Li et al., 1992b). By contrast, splicing from the invariant exon 4 to exon 6 gives rise to a hydrophilic species which is secreted into the medium and retained within the cell (fig. 5).

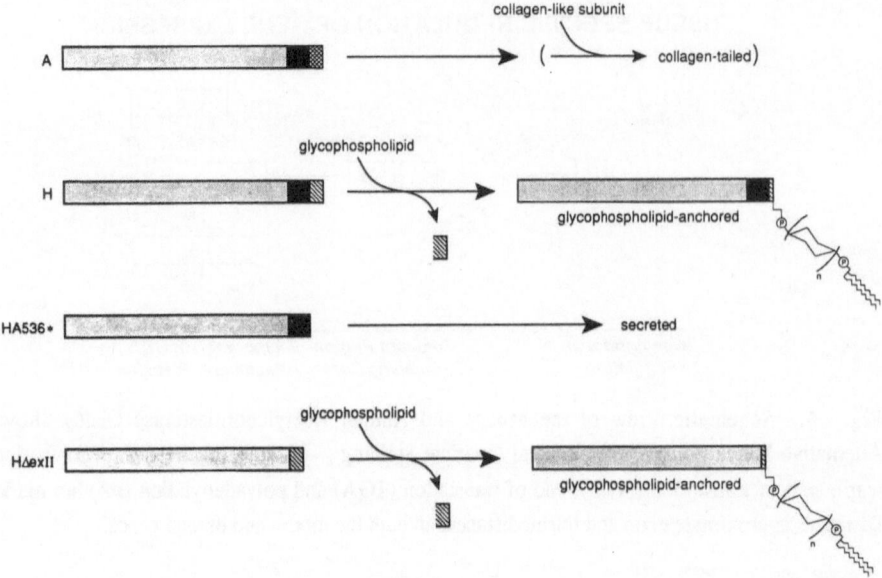

Fig. 5. Schematic Representation of Fates of Wild-Type and Carboxyl-Terminal Mutant AChE Proteins from Torpedo. Depicted are wild-type and mutant AChE proteins specified in the vectors SV-A, SV-H, SV-HA536*, and SV-HΔexII are indicated. A and H proteins differ only in their most distal carboxyl-terminal regions after residue 535. The carboxyl-terminal domain of A is hydrophilic while that of H is hydrophobic. The HA536* protein shares the same 535-amino acid sequence but lacks either carboxyl-terminal domain. The HΔexII protein has the sequence encoded by exons for residues 1-479 and for residues 536-565 in fusion as a result of the in-frame deletion of an intervening exon from its DNA. The results of transfection experiments in COS cells, when taken together, indicate that the A protein that is expressed remains intracellular and is secreted. As expressed in its native context, however, the AChE subunit assembles into tetramers in association with a collagen-like tail and is secreted. The expressed H enzyme, like its native counterpart, is processed with the attachment of glycophospholipid (denoted by ethanolamine, a polymer of six-membered rings [monosaccharides and inositol] and the diacylglycerol) to a newly defined carboxyl-terminal residue with the cleavage of the hydrophobic terminal peptide. It becomes localized on the cell surface. Truncated A536* is secreted to a significant degree, whereas the HΔexII fusion protein is processed with glycophospholipid and also remains on the cell surface. Thus, the carboxyl-terminal domains encoded by alternative exons each dictate a specific fate other than rapid secretion into the extracellular space and the sequence denoted by the hatched line is both necessary and sufficient for glycophospholipid anchoring (from Gibney and Taylor, 1990).

The third mRNA arising from alternative splicing and detected by RNase protection is a direct continuation of exon 4 to its first in-frame stop codon. Analysis of the "read-through" sequence shows that in mouse the stop codon will dictate a peptide of 30 residues extending from the region encoded by exon 4. This peptide is hydrophilic and the extended exon 4 encoded form of the enzyme expresses a hydrophilic AChE species. In human the length of this extended peptide is far longer. However, in human erythroleukemia cells the mRNA transcribed by the read-through sequence cannot be detected. Moreover, transfection of a plasmid containing the human AChE genomic sequence allowing for the exon 4-acceptor splicing alternatives into murine erythroleukemia cells gives rise to only the sequences found in the human erythroleukemia cells (Li et al., 1992b). Hence, splicing preference seems to be controlled by the gene sequence rather than splicing factors or other cryptic factors present within the human or mouse erythroleukemia cell.

Further 3' in the gene we note two polyadenylation signals separated by 230 bp. An AT-rich region, which may influence mRNA stability, intervenes between the two polyadenylation signals. Use of the two polyadenylation signals is tissue specific wherein the first signal is predominantly used in brain, whereas the second is favored in muscle. The same polyadenylation signals are used regardless of the splice acceptor from exon 4. The absence of down-stream exons also contributes to the compact nature of the mammalian AChE gene.

In contrast to the mammalian butyrylcholinesterase which is encoded in over 60 kb of the genome, the mammalian AChE gene, even with all of its alternative exons, is found localized within 6 - 7 kb of genomic sequence. The compact nature of the mammalian AChE gene is striking; not only are intron distances short but common donor sites splice to nearby alternative acceptor positions and common polyadenylation signals are used at the 3' end of the gene. Thus, a structural framework for examining tissue and developmentally specific gene expression of AChE is now available. To understand the complexities associated with myogenesis, neurogenesis, innervation and denervation (fig. 3), expression of synapse-specific proteins in each developmental step requires study.

Post-Translational Modifications of Acetylcholinesterases

While alternative mRNA processing provides the initial commitment to molecular diversity of AChE, further refinements occur at the post-translational level. I have already alluded to the post-translational processing that occurs for glycophospholipid-linked species. Carbohydrate processing occurring post-translationally is also not identical for the various species of AChE (Doctor et al., 1983; Bon et al., 1987). Assembly of subunits gives rise to further diversity. The heteromeric species consist of catalytic subunits disulfide bonded to either a lipid-linked subunit or to a triple-helix of collagen-containing subunits (fig. 6). They all appear to arise from splicing from the invariant exon (exon 4 to exon 6); exon 6 contains a cysteine four residues from the carboxyl-terminus which is used for linkage to other catalytic subunits or to the two types of structural subunits.

The homomeric forms exist as monomers, dimers and tetramers and can have as many as three alternative splice positions. Splicing to exon 6 yields a hydrophilic species which by virtue of the sulfhydryl can exist as monomers, dimers and tetramers. Splicing to exon 5 yields a hydrophobic carboxyl terminus which provides a signal for its cleavage and addition of a glycophospholipid. This species, like the lipid-linked heteromeric species, is tethered to the outer surface of the membrane. A third species is encoded by a continuation of exon 4 without splicing. This continuation sequence is hydrophilic and lacks a sulfhydryl group to form oligomers.

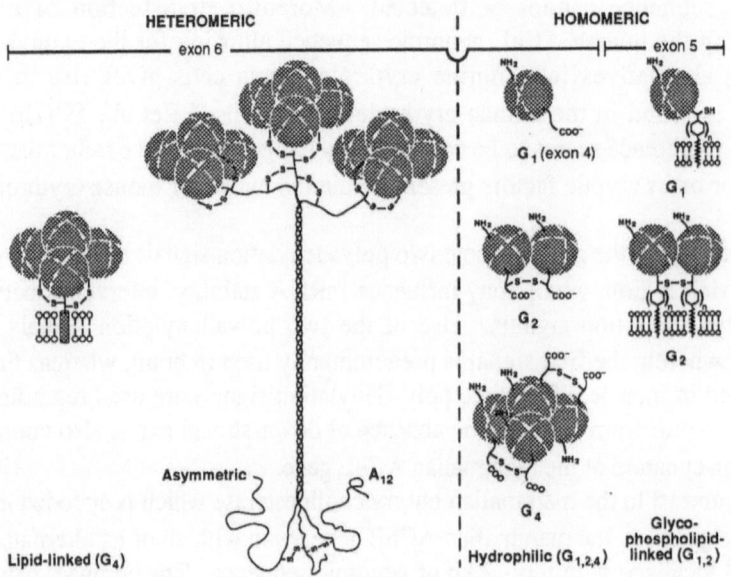

Fig. 6. Molecular Species of Mammalian Acetylcholinesterase. The species are divided into two classes: The heteromeric class consists of catalytic subunits disulfide bonded to either a lipid-linked subunit or a triple helix of collagen-containing subunits. The homomeric class exists in various oligomeric forms and can be subdivided into soluble (hydrophilic) or glycophospholipid-linked species. Also shown are the three alternative exons which encode the different species.

CHOLINESTERASE STRUCTURES AT MOLECULAR AND ATOMIC RESOLUTION

A variety of interacting ligands, many of which have a well-defined pharmacology, has provided for several decades an opportunity to examine aspects of the reactivity and ligand specificity of the cholinesterases. Complementary endeavors whereby site-specific modifications of AChE are examined in relation to substrate and inhibitory ligand specificity are also underway (Gibney et al., 1990b; Velan et al., 1991; Neville et al., 1990). In addition to expanding classical studies of enzyme mechanisms and structure-activity relationships by mutating individual residues in the enzyme, it is worthwhile to reexamine many of the previous studies in relation to the crystal structure of AChE. Interpretations of such studies were limited to a molecular or submolecular level, but now have a framework at an atomic level of resolution to refine or extend the conclusions. The most informative of these studies typically relied on spectroscopy. Moreover, the general approaches remain as important complements to crystallography for they often yield information on solution dynamics of the enzyme.

Active Center Conformation and Configuration of Bound Residues

In 1973 Shinitsky, Dudai and Silman provided evidence by absorption and fluorescence spectroscopy for a charge-transfer complex between bound acridinium and pyridinium ligands and a tryptophan in AChE. Ongoing studies of the crystal structure clearly show Trp 84 to be the prime candidate for this interaction. The crystal structure of the ligand-enzyme complex should yield a detailed picture of the nature of ring stacking.

In the late 1960's, Belleau and colleagues (cf: Belleau, 1971) showed that *bis*-quaternary and monoquaternary ligands modified the reactivity of AChE towards inactivation by methanesulfonylation in an unusual manner. Rather than simply protecting the enzyme, under certain conditions enhanced reactivity was observed. Hence, these ligands do not merely obscure the active site serine but rather facilitate interactions with small ligands. The availability of purified enzyme in the 1970's enabled Taylor and Jacobs (1974) to follow-up on this observation. By using *bis*-quaternary analogues that quench AChE fluorescence by Förster energy transfer, prior phosphorylation or sulfonylation of the active center serine was found not to affect the affinity of *bis*-quaternary ligands for AChE unless either (a) the acylating agent was large or (b) a large steric volume existed around the quaternary group. The overlap distance for steric occlusion was equivalent to a four carbon atom chain which should approximate the distance between the serine and the quaternary moiety of acetylcholine. Hence, these studies also point to certain ligands not fully precluding access of small acylating agents to the active center serine. Such ligands presumably bind in an orientation in which one end of the *bis*-quaternary is localized to the active center position of the choline head group and the other end is pointed out of the active center gorge so as not to obscure the active center serine (an exo-orientation). Tryptophan fluorescence quenching also showed that each subunit behaved independently and there was little or no intersubunit tryptophan quenching by ligand that would be expected if the binding site resided close to an axis or center of symmetry. This is also borne out in the crystallographic studies which reveal the active centers on opposite sides of the dimer and not near an axis of symmetry (Sussman et al., 1991).

Barnett and Rosenberry (1977) showed that N-methylacridinium binding to the enzyme will enhance the rate of AChE catalyzed hydrolysis of neutral but not cationic esters, again pointing to the simultaneous access of two compounds to the active center and the inability of this tricyclic cation to obscure the active center serine. This finding also suggests conformational flexibility within the active center.

Chignell and colleagues have shown that spin-labeled *bis*-quaternary ligands bind to AChE in a manner in which both quaternary ends are highly immobilized. The lack of spin-dipole coupling in the bound state also indicates that these ligands bind in an extended conformation. Unusually high affinity is found for *bis*-quaternary inhibitors such as ambenonium or certain benzoquinonium analogues (Taylor & Jacobs, 1974; Taylor and Lappi, 1975). In an extended conformation the centers of the two aromatic residues linked to the ammonium moieties are separated by 25Å. Hence, these ligands might be expected to interact with the Trp 84 and Phe 330 region at the base of the gorge and with other as yet undefined aromatic residues at the lip of the gorge.

Berman and Taylor (1978) showed by phosphorylation of the active center serine of AchE with pyrenebutylmethylphosphonofluoridate that the spectrum of pyrene shifts to wavelengths greater than can be replicated by any organic solvent in which pyrene absorption is measured. While the wavelength maxima of the pyrene spectra correlate best with solvent polarizability, the unusually large shift is indicative of an environment of immobilized aromatic side chains. The aromatic patch within the gorge provides the obvious surface for pyrene association.

Studies by Wilson and Quan (1958) show the importance of a *m*-hydroxyl group in the binding of phenyltrimethylammonium analogues to the active center of AChE and that orientation for presumed hydrogen bond formation is critical. Crystallographic studies combined with site-specific mutagenesis of the residues should now enable one to dissect out the molecular details of this hydrogen bonding interaction.

Studies of Massoulie' and colleagues (1975) have demonstrated that edrophonium stabilizes AChE conformation against denaturation. The presence of a large number of water molecules in the gorge and their displacement by edrophonium with concomitant formation of new van der Waals interactions, hydrogen bonds and Coulombic interactions provide a molecular explanation for ligand stabilization of the structure of the enzyme.

The Peripheral Site

Studies by J.P. Changeux (1966) and those of several others that followed demonstrated from steady-state catalysis that reversible inhibition of AChE could not be explained simply on the basis of a single site but rather certain ligands affect catalysis through binding to an allosteric site. The ligand, propidium, shows enhanced fluorescence upon binding to AChE and is selective for the peripheral site (Taylor and Lappi, 1974). Propidium's spectroscopic properties allow one to compare its inhibition kinetics with direct analyses of binding. Fluorescence energy transfer enables one to measure the distance between the propidium acceptor dipole and a donor dipole achieved by reacting fluorescent phosphorylating agents with the active center serine. These studies show that propidium is some 20Å away from the active center. Presumably this would place it near the lip of the gorge where its two positive charges may gate access of ligands to the gorge. A site 20Å distant is also consistent with the locations of distinct labeled peptides protected by propidium and an active center selective ligands (Weise et al., 1990). Propidium, by virtue of an allosteric action, affects active center conformation as can be shown by differences in the spectra of reporter ligands (Epstein et al., 1979). Similarly, modification of the active center affects the quantum yield of bound propidium (Fibroulet et al., 1991 and unpublished). Finally, substrates which produce substrate inhibition dissociate propidium at concentrations which match those producing substrate inhibition. Hence, this site becomes the candidate for the binding of a second substrate molecule that causes substrate inhibition (Radic et al, 1991). Examination of structure of the propidium complexes should further elucidate structural details of peripheral site inhibition.

WHERE WE ARE GOING

The studies on gene structure of AChE provide but one example of how recombinant DNA studies will be applied to uncovering details on the regulation of cholinesterase expression. Figure 7 also illustrates several other areas of cholinesterase research likely to be developed in the final decade of this century. Both studies on gene expression and on the physical structure of the enzyme are likely to proceed rapidly over the next few years. What is remarkable is that virtually all of these areas should begin to wane by the end of the century. Studies of gene expression, if they can be pursued as structural studies extending to a third dimension, could carry us for longer term, but even this research must be predicated on currently unavailable technologies which will yield better structural resolution. Future anticipated areas of endeavor and many that one cannot yet foresee will be absolutely dependent on new technological developments. Our great strides in understanding cholinesterase have relied on such developments in the past.

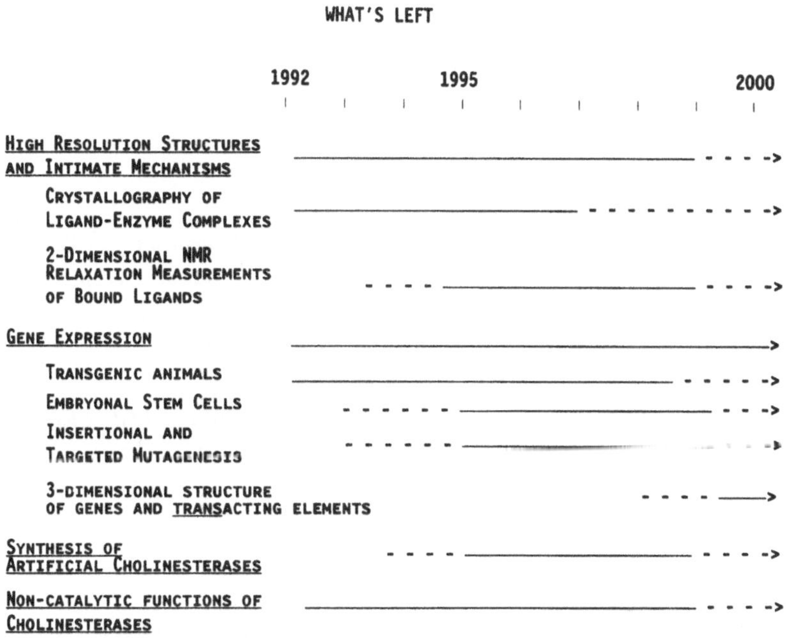

WHAT'S LEFT

Fig. 7. Projected Areas of Future Work on the Cholinesterases. The dotted lines denote anticipated periods of start-up and periods where the field has matured and the level of activity is declining.

Acknowledgements

Research work from our laboratory discussed in this review is supported by the USPHS (GM24437,GM18360), USARDC (DAMD 17-9-C-1058) and the Muscular Dystrophy Association. I wish to thank Shelley Camp, Ying Li, Maria Elena Fuentes, Damon Getman, Zoran Radic', Gretchen Gibney and Tara Rachinsky who have been major contributors to the studies on gene structure alluded to in this presentation.

REFERENCES

Allderdice, P.W., Gardner, M.A.R., Calutiea, D., Lockridge, O., LaDu, B.N. and McAlpine, P.J., 1991, *Genomics* 11, 452-454.

Argyll-Robinson, D., 1863, *Edinb. Med. J.* 8, 815-820.

Arpagaus, M., Kolt, M. Vatsu, K.P., Bartels, C.F., La Du, B.N. and Lockridge, O., 1990, *Biochemistry* 29, 124-131.

Barnett, P. and Rosenberry, T.L., 1977, *J. Biol. Chem.* 252, 7200-7206.

Bartels, C. and Lockridge, O., 1992 This Volume.

Belleau, B., 1971, *Adv. Chem. Ser.* 108, 141-165.

Bon, S., Me'Flah, K., Musset, F., Grassi, J. and Massoulie', J., 1987, *J. Neurochem.* 49, 1720-1731.

Changeux, J.P., 1966, *Mol. Pharmacol.* 2, 369-392.

Dale, H.H., 1914, *J. Pharmacol. Exp. Therap.* 6, 147-190.

Doctor, B.P., Camp, S., Gentry, M.K., Taylor, S.S. and Taylor, P. 1983, *Proc. Natl. Acad. Sci. (USA)* 18, 5767-5771

Epstein, D.J., Berman, H.A. and Taylor, P., 1979, *Biochemistry* 18, 4749-4754.

Fibroulet, A., Rieger, F., Amitai, G. and Taylor, P., 1990,, *Biochemistry* 29, 914-920.

Froede, H.C. and Wilson, I.B., 1971, In: The Enzymes (Boyer, P.D., ed.) 3rd Ed. Vol 5, pp. 87-114, Academic Press, New York.

Fuentes, M.E. and Taylor, P., 1992, submitted for publication.

Ganghan, G., Park, H., Priddle, J., Craig, I. and Craig, S., 1991, *Genomics* 11, 455-458.

Getman, D., Eubanks, J., Evans, G. and Taylor, P., 1992, *Am. J. Human Genetics*, in press.

Getman, D., Camp, S. and Taylor, P., 1992b, in preparation.

Gibney, G. and Taylor, P., 1990, *J. Biol. Chem.* 265, 12576-12583.

Koelle, G.B., 1951, *J. Pharmacol. Exp. Thera p.* 103, 153-171.

Li, Y., Camp, S., Getman, D., Rachinsky, T.L. and Taylor, P., 1991, *J. Biol. Chem.* 260, 23082-23091.

Li, Y., Camp, S., Rachinsky, T.L. and Taylor, P., 1992a, submitted for publication.

Li. Y., Camp, S. and Taylor, P., 1992b, submitted for publication.

MacPhee-Quigley, K., Vedvick, T.S., Taylor, P. and Taylor, S.S., 1986, *J. Biol. Chem.* 261, 13565-13570.

Massoulié, J., Bon, S., Rieger, F. and Vigny, M., 1975, *Croat. Chem. Acta.* 47, 163-179.

Massoulié, J. and Rieger, F., 1969, *Eur. J. Biochem.* 14, 430-439.

Maulet, Y., Camp, S., Gibney, G., Rachinsky, T.L., Ekström, T.J. and Taylor, P., 1990, *Neuron* 4, 289-301.

Nachmansohn, D. and Wilson, I.B., 1951, *Adv. Enzymol.* 12, 259-339.

Neville, L.F., Gnatt, A., Padan, R., Seidman, S. and Soreq, H., 1990, *J. Biol. Chem.* 265, 20735-20738.

Rachinsky, T.L., Crenshaw, E.B. and Taylor, P., 1992, *Genomics*, in press.

Radic, Z., Reiner, E. and Taylor, P., 1991, *Mol. Pharmacol.* 39, 98-104.

Schumacher, M., Camp, S., Maulet, Y., Newton, M., MacPhee-Quigley, K., Taylor, S.S., Friedmann, T. and Taylor, P., 1986, *Nature* 319, 407-407.

Shinitsky, M., Dudai, Y. and Silman, I., 1973, *FEBS Lett.* 30, 125-128.

Stedman, E., Stedman, E. and Easson, L.H., 1932, *Biochem. J.* 26, 2056,2066.

Sussman, J.L., Harel, M., Frolow, F., Oefner, C., Goldman, A., Toker, L. and Silman, I., 1991, *Science* 253, 872-879.

Taylor, P. and Jacobs, N.M., 1974, *Mol. Pharmacol.* 10, 93-108.

Taylor, P. and Lappi, S., 1975, *Biochemistry* 14, 1989-1997.

Velan, B., Grosfeld, H., Kronman, C., Leitner, M., Gozes, Y., Flashner, Y., Marcus, D., Cohen, S. and Shafferman, A., 1992, *J. Biol. Chem.* 266, 23977-23984.

Wee, V.T., Sinha, B.K.. Taylor, P. and Chignell, C.F., 1976, *Mol. Pharmacol.* 12, 667-674.

Weise, C., Kreienkamp, H.J., Raba, R., Pedak, A., Aavikssar, A. and Hucho,   F.,1990 *EMBO J.* 9, 3885-3888.

Wilson, I.B., 1959, *Fed. Proc.* 18, 752-758.

Wilson, I.B. and Quan, C., 1958, *Arch. Biochem. Biophys.* 235, 2312-2315.

Zelinski, T., White, L., Coghan, G. and Philipps, S., 1991, *Genomics* 11, 165-167

BIOSYNTHESIS OF THE MOLECULAR FORMS OF ACETYLCHOLINESTERASE

Jean Massoulié[1], Suzanne Bon[1], Alain Anselmet[1], Jean-Marc Chatel[1], Françoise Coussen[1], Nathalie Duval[2], Eric Krejci[1], Claire Legay[1] and François Vallette[3]

[1]Laboratoire de Neurobiologie, URA CNRS 295, Ecole Normale
   Supérieure, 46 rue d'Ulm, 75005 Paris - France
[2]Laboratoire des Venins, Institut Pasteur, 25 rue du Dr. Roux
   75015 Paris - France
[3]Laboratoire de Neurobiologie Physico chimique, URA
   CNRS 1112, rue Pierre et Marie Curie, 75005 Paris - France

POLYMORPHISM OF CHOLINESTERASES: A SINGLE GENE GENERATES MULTIPLE FORMS, DIFFERING IN THEIR QUATERNARY STRUCTURE AND MODE OF ANCHORING

Vertebrates possess a variety of forms of acetylcholinesterase (AChE), as schematically illustrated in Figure 1. These forms present the same catalytic activity, but differ in their quaternary structure and in their interactions with membranes or with the extra-cellular matrix (basal lamina) (for reviews, see Massoulié and Bon, 1982; Massoulié and Toutant, 1988; Massoulié et al., in press). The catalytic subunits are generated from a single gene, they undergo various post-translational modifications and in some cases associate with structural subunits. Alternative splicing, together with the use of several transcription origins and polyadenylation signals, generate multiple mRNAs (Sikorav et al., 1987, 1988; Schumacher et al., 1988; Maulet et al., 1990). The coding sequence, however, seems to be modified only by alternative splicing of its 3' region. In *Torpedo*, the major part of this coding sequence, common to all mRNAs, is included in a large exon I (1678 nucleotides, from the initiation codon of *T. marmorata* AChE) and a small exon II (167 nucleotides) (Maulet et al., 1990). Exon II may be spliced to either exon III$_H$ or exon III$_T$, generating the two kinds of catalytic subunits, H and T. These subunits constitute, respectively, the two main molecular forms

Multidisciplinary Approaches to Cholinesterase Functions, Edited by
A. Shafferman and B. Velan, Plenum Press, New York, 1992

17

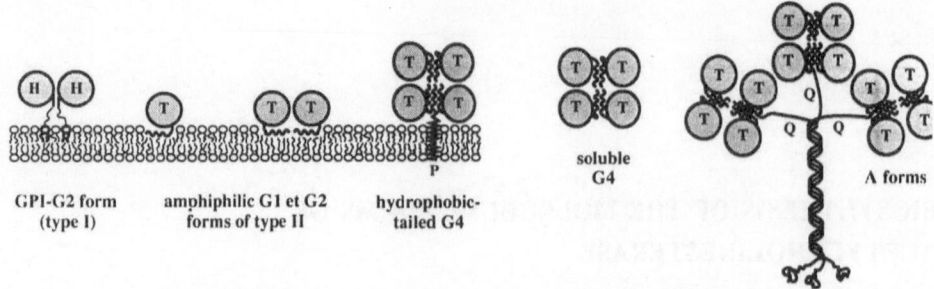

Figure 1. Catalytic subunits and molecular forms of acetylcholinesterase

occurring in the electric organ: the glycolipid (GPI)-anchored dimers and the collagen-tailed asymmetric forms. We do not designate the alternative exons and the corresponding catalytic subunits by the same letters as the globular (G) and asymmetric (A) molecular forms, because the distinction based on the general quaternary structure of the molecules does not coincide with the nature of their catalytic subunits: while collagen-tailed forms A forms contain only T subunits, different types of globular G forms may include either H or T subunits. In addition, a third type of catalytic subunit may be generated by an mRNA structure in which exon II is continued by the adjacent genomic sequence. Such a read-through structure has been identified in a *Torpedo* cDNA (Sikorav et al., 1988) and in a mouse cDNA (Rachinsky et al., 1990). The existence of the resulting catalytic subunit (R), has not been demonstrated yet.

In *Torpedo* asymmetric forms, a cysteine residue located at the fourth position upstream of the C-terminus, forms disulfide bridges between dimers of T subunits, (MacPhee-Quigley et al., 1986) or between a T subunit and one of the structural subunits of the collagen tail. An homologous cysteine residue forms inter-subunit linkages in the soluble tetrameric forms of human serum BuChE (Lockridge et al., 1987), and is responsible for the disulfide linkage between T subunits and the hydrophobic structural subunit, P, in the hydrophobic-tailed membrane-bound tetramers of mammalian brain AChE (Roberts et al., 1991).

In addition to the collagen-tailed forms and GPI-anchored dimers (amphiphilic dimers of type I), we have shown that some *Torpedo* tissues contain amphiphilic AChE dimers which we classified as type II, because they clearly differ in their amphiphilic properties (solubilization, aggregation in the absence of detergent) from type I dimers (Bon et al., 1988a, b). The abundance of amphiphilic dimers of type II was too low to allow a complete biochemical characterization, and identification of their catalytic subunits (H, T, or other) and the structure of their hydrophobic domain remains uncertain. Such forms may, however, be important because amphiphilic dimers and monomers of type II represent a major component of AChE activity in a number of tissues of higher vertebrates, e.g. chicken and rabbit muscles, murine neuroblastoma cells (Bon et al., 1991).

PRODUCTION OF INACTIVE ACHE IN TRANSFECTED CELLS AND *IN VIVO*

We transfected COS cells with H and T catalytic subunits of *Torpedo* AChE (Duval et al., 1992a). The resulting immunoreactive subunits formed a doublet of 70-72 kD in the case of the T subunit, and two doublets of 66-68 and 80-82 kD in the case of the H subunit. These various proteins were glycosylated, as indicated by their sensitivity to N-glycanase, but remained as doublets after deglycosylation. The existence of doublets probably reflects incomplete processing, possibly incomplete cleavage of the N-terminal signal peptide, a feature common to both subunits, and of the C-terminal glycophosphatidyl (GPI) addition signal peptide, which is specific for the H subunit.

Untransfected COS cells contain a very low endogenous cholinesterase activity, consisting exclusively of amphiphilic G_1 and G_2 forms. When the transfected COS cells were maintained at 37°C, we did not observe any catalytic activity corresponding to *Torpedo* AChE. They did, however, produce a significant *Torpedo* AChE activity, more than an order of magnitude above the endogenous level, when they were transferred to a lower temperature, e.g. 27°C. Even under these conditions, however, active AChE represented only a very minor part of the immunoreactive AChE protein, as shown by comparison of the staining intensity obtained in Western blots, revealed with antibodies directed against *Torpedo* AChE, with equivalent activities in extracts from electric organ and transfected cells.

The presence of inactive, probably misfolded, AChE protein in transfected cells clearly results from the fact that correct maturation is not efficiently achieved in a foreign cellular environment. This situation may not be entirely artefactual, however, because we also observed variations in the ratio of immunoreactive protein, quantified by two-site ELISA, and AChE activity in different fractions from chicken tissues *in vivo* (Chatel et al., in preparation). Centrifugation in sucrose gradients showed that inactive forms displayed exactly the same sedimentation coefficients as active forms, and were more abundant as monomers, G_1, and dimers, G_2. These inactive forms did not bind to an affinity column of N-methylacridinium, in agreement with the lack of a functional active site. In addition, they were not recognized by wheat germ agglutinin and were sensitive to endoglycosidase H, suggesting that they were retained in the endoplasmic reticulum. The major pool of inactive AChE observed *in vivo* did not seem to correspond to the rapidly degraded pool which was demonstrated in muscle cultures by Rotundo et al. (1989), because it was not metabolically labeled within 4 hours. It thus appears to constitute a set of stable molecules, similar in their quaternary associations to the active ones, and resident in the endoplasmic reticulum.

During development of the brain, in chicken and quail, G_4 becomes largely predominant over the smaller molecular forms, reaching about 80% of the total activity before hatching. In quail, we found that the contribution of inactive AChE protein increases markedly between E8 and E16; at this stage, it is higher than in the adult (Anselmet et al., in preparation). The abundance of inactive AChE varies in the different brain areas, in the following order: optic lobe > brain stem > neuroretina > cerebellum. It remains to be determined whether inactive AChE protein has any physiological significance.

MOLECULAR FORMS OF *TORPEDO* ACHE PRODUCED IN COS CELLS

We analyzed the active molecular forms produced by transfected COS cells expressing *Torpedo* H and T subunits at 27°C, and rat T subunits at 37°C (Duval et al., 1992a).

In the case of the H subunits, we obtained GPI-anchored dimers, which were largely exposed at the cell surface, indicating that COS cells correctly process the *Torpedo* GPI-addition signal. The cells also released non-amphiphilic dimers into the culture medium, probably representing lytic derivatives of the membrane-bound GPI-G_2 form.

In the case of the T subunit, the situation was more complex, because the cells contained several molecular forms, including non-amphiphilic tetramers, together with amphiphilic dimers and monomers. We characterize amphiphilic forms by their capacity to bind detergent micelles, as indicated by displacement of their sedimentation coefficient (Triton X-100 vs Brij-96), or charge-shift electrophoresis (Triton X-100 ± deoxycholate). The amphiphilic G_1 and G_2 forms resemble the amphiphilic forms of class II (Bon et al., 1989). It is interesting that about half of AChE activity was accessible to the external medium, suggesting that amphiphilic forms of type II may be anchored at the cell surface.

The cells also released AChE into the culture medium, consisting mostly of non amphiphilic G_4 and amphiphilic G_2 forms. Thus, amphiphilic forms, in spite of their affinity for detergents, may appear in a soluble form, as already observed in *Torpedo* plasma (Bon et al., 1988a) and in murine neuroblastoma cultures (Lazar et al., 1984).

We obtained the same types of molecular forms with the rat T subunits. It may be noted, however, that G_2 was predominant over G_1 in the case of *Torpedo*, whereas G_1 was predominant in the case of rat. A similar difference is also observed in the tissues of these species, and thus probably reflects a different tendency of these subunits to form oligomers.

Our transfection experiments show that amphiphilic forms of type II may be generated from T subunits (Figure 1). This is consistent with the fact that only T subunits seem to be expressed in the mammalian central nervous system (Rachinsky et al., 1990) in which certain areas contain a high level of amphiphilic G_1 form (Legay et al., submitted).

ROLE OF THE C-TERMINAL T PEPTIDE; EXPRESSION OF A TRUNCATED TΔ SUBUNIT

The C-terminal peptides encoded by the alternatively spliced III_H and III_T exons are entirely different, and most of the H_C peptide is removed in the mature GPI-anchored dimers. It appeared likely, therefore, that these C-terminal peptides are not necessary for the catalytic activity of AChE.

We constructed a truncated TΔ subunit, which retained only the first 4 amino acids of the 40 amino acid-T_C peptide (Duval et al., 1992a). When expressed in COS cells, this truncated TΔ subunit generated only non-amphiphilic monomers. This demonstrates that the T$_C$ peptide contains the hydrophobic domain of amphiphilic forms of type II, and that it is involved in the formation of oligomers, in agreement with the fact that the C-terminal cysteine is responsible for inter-subunit disulfide bonds. Indeed, Velan *et al.* (1991) recently showed that only monomers of human AChE were obtained after replacement of this cysteine by an alanine residue.

THE COLLAGENIC TAIL SUBUNIT, Q

We determined partial peptide sequences of the collagenic subunit, isolated from purified asymmetric *Torpedo* AChE (Krejci et al., 1991). This allowed us to obtain a corresponding probe by amplification (PCR), using two oligonucleotide primers, and to obtain cDNA clones in a *Torpedo* electric organ library. Several clones seem to present variations, possibly due to alternative splicing. The deduced primary structure of a tail subunit consists of several domains: a signal peptide, an N-terminal non collagenic domain (40 amino acids), a collagenic domain, and a C-terminal non collagenic domain. The length of the collagenic domain (187 amino acids) corresponds to a collagenic triple helix of about 50 Å, in agreement with the length of the tail observed in electron micrographs of isolated molecules. This collagenic domain is flanked by two pairs of cysteines, which may establish disulfide bonds between each pair of strands and thus lock the triple helical structure. In addition, two adjacent cysteines are located in the middle of the N-terminal domain, and 10 cysteines in the second half of the C-terminal domain. The C-terminal domain is retained in the mature protein, as indicated by the presence of a peptide which contains its C-terminus.

The Q subunits combine with catalytic T subunits when both proteins are co-expressed in transfected COS cells, producing asymmetric forms which display a characteristic sensitivity to collagenase. Thus, the synthesis of asymmetric forms, which is restricted, *in vivo*, to differentiated muscle and neural cells simply requires the availability of both catalytic and structural subunits. This suggests that the expression of the structural subunit may be the limiting factor in the assembly of asymmetric forms, e.g. during muscle differentiation.

INTERACTION BETWEEN COLLAGENIC AND CATALYTIC SUBUNITS

We co-expressed *Torpedo* catalytic H, T and TΔ catalytic subunits with the collagenic Q subunit, in COS cells: only the complete T subunit was found to assemble with Q, producing asymmetric forms (Duval et al., 1992b). This shows that the T$_C$ peptide is necessary for heterologous association with the structural subunits, as well as for homologous associations between T subunits.

We also co-expressed the rat T subunit with the *Torpedo* Q subunit, and obtained hybrid asymmetric forms (Legay et al., submitted). Therefore, the complementarity between the catalytic and collagenic subunits is well conserved among vertebrates. The C-terminal

T$_C$ peptide presents a higher level of identity between *Torpedo* and mammals than the common catalytic domain (75 % vs 52 %).

Both N and C-terminal domains of the Q subunit contain cysteine residues, and could, therefore, be linked to the catalytic T subunits. In order to determine which of these domains is responsible for this association, we raised an antiserum against a fusion protein containing the C-terminal domain (Q$_C$) (Duval et al., 1992b). This antiserum bound intact asymmetric forms, from electric organs, and from transfected COS cells, but did not bind modified molecules obtained after limited collagenase digestion. These molecules sediment faster than the intact forms, indicating that their friction coefficient is reduced by removal of the distal part of the collagenic tail. This experiment demonstrates that the C-terminal domain is located at the distal end of the tail. The N-terminal domain, Q$_N$, thus appeared to be that part of the tail polypeptide involved in attachment to the T subunit tetramers.

To verify this hypothesis, we constructed a chimeric protein in which this domain, Q$_N$, was fused to the C-terminal peptide of the H subunit, which was thought to contain the GPI-addition signal (Gibney and Taylor, 1990). When this chimeric protein, Q$_N$/H (Figure 2), was co-expressed with the catalytic T subunit, we observed the formation of amphiphilic tetramers, and in smaller proportions dimers and monomers, which were sensitive to a glycophosphatidylinositol specific phospholipase C (PI-PLC), and therefore GPI-anchored. A large fraction of the major GPI-G$_4$ form appeared exposed at the cell surface. By creating this novel type of membrane-anchored AChE form, which formally resembles the hydrophobic-tailed tetramers, we thus confirmed the capacity of the Q$_N$ domain to assemble with T subunits.

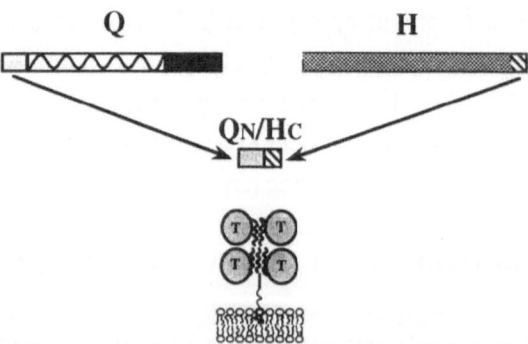

Figure 2. Construction of GPI-anchored AChE tetramers

CONCLUSIONS

Our transfection experiments, together with previous biochemical analyses, showed that H subunits produce GPI-anchored dimers, while T subunits produce all other known molecular forms: amphiphilic forms of type II, soluble non-amphiphilic tetramers, as well as heteroligomeric collagen-tailed and hydrophobic-tailed forms. The possible existence of additional subunits thus remains a puzzle.

We have shown that the C-terminal T_C peptide is not necessary for the catalytic activity, but is responsible for the hydrophobic interactions of the amphiphilic forms of type II, and for the formation of homo- and heteropolymers. The assembly of collagen-tailed forms only requires the simultaneous synthesis of both catalytic T subunit and structural Q subunits. The exclusive presence of these forms in differentiated muscle and nerve cells, *in vivo*, thus appears to be controlled by the expression of the collagenic subunit, Q. Similarly, it is likely that the production of the hydrophobic-tailed G_4 form depends on the expression of the 20 kD hydrophobic structural subunit (Gennari et al., 1987; Inestrosa et al., 1987). It is interesting that the levels of the A forms and of the G_4 form are controled in distinct manners by innervation and activity in muscles, suggesting that the production of the various AChE forms is largely dependent on the expression of their associated structural subunits. The similarity of the quaternary structure of the collagen-tailed and hydrophobic-tailed forms further suggests that the binding domains of the collagenic and hydrophobic structural subunits may be homologous or even genetically related. It will be extremely interesting to analyse the structure of these AChE-associated subunits and the regulation of their expression in further detail.

REFERENCES

Anselmet, A., Fauquet, M., Chatel, J.-M., Maulet, Y., Massoulié, J. and Vallette, F.-M. Evolution of acetylcholinesterase expression in the developing central nervous system of the quail. In preparation.

Bon, S., Toutant, J.P., Méflah, K. and Massoulié, J. (1988a) Amphiphilic and nonamphiphilic forms of *Torpedo* cholinesterases: I. Solubility and aggregation properties. *J. Neurochem.* 51:776-785.

Bon, S., Toutant, J.P., Méflah, K. and Massoulié, J. (1988b) Amphiphilic and nonamphiphilic forms of *Torpedo* cholinesterases: II. Existence of electrophoretic variants and of phosphatidylinositol phospholipase C-sensitive and -insensitive forms. *J. Neurochem.* 51:786-794.

Bon, S., Rosenberry, T.L. and Massoulié, J. (1991) Amphiphilic, glyco-phosphatidylinositol-specific phospholipase C (PI-PLC)-insensitive monomers and dimers of acetylcholinesterase. *Cell. Mol. Neurobiol.* 11:157-172.

Duval, N., Massoulié, J. and Bon, S. (1992a) H and T subunits of acetylcholinesterase from *Torpedo*, expressed in COS cells, generate all types of globular forms. *J. Cell Biol.*, in press.

Duval, N., Krejci, E., Grassi, J., Coussen, F., Massoulié, J. and Bon, S. (1992b) Molecular architecture of acetylcholinesterase collagen-tailed forms; construction of a glycolipid-tailed tetramer. *EMBO J.*, in press.

Chatel, J.-M., Frobert, Y., Massoulié, J. and Vallette, F.-M. Existence of an inactive pool of acetylcholinesterase in chicken. In preparation.

Gennari, K., Brunner, J. and Brodbeck, U. (1987) Tetrameric detergent-soluble acetylcholinesterase from human caudate nucleus: subunit composition and number of active sites. *J. Neurochem.* 49:12-18.

Gibney, G. and Taylor, P. (1990) Biosynthesis of *Torpedo* acetylcholinesterase in mammalian cells. Functional expression and mutagenesis of the glycophospholipid-anchored form. *J. Biol. Chem.* 265:12576-12583.

Inestrosa, N.C., Roberts, W.L., Marshall, T. and Rosenberry, T.L. (1987) Acetylcholinesterase from bovine caudate nucleus is attached to membranes by a novel subunit distinct from those of acetylcholinesterase in other tissues. *J. Biol. Chem.* 262:4441-4444.

Krejci, E., Coussen, F., Duval, N., Chatel, J.M., Legay, C., Puype, M., Vandekerckhove, J., Cartaud, J., Bon, S. and Massoulié, J. (1991a) Primary structure of a collagenic tail subunit of *Torpedo* acetylcholinesterase: co-expression with catalytic subunit induces the production of collagen-tailed forms in transfected cells. *EMBO J.* 10:1285-1293.

Lazar, M., Salmeron, E., Vigny, M. and Massoulié, J. (1984) Heavy isotope labeling study of the metabolism of monomeric and tetrameric acetylcholinesterase forms in the murine neuronal-like T28 hybrid cell line. *J. Biol. Chem.* 259:3703-3713.

Lockridge, O., Eckerson, H.W. and La Du, B.N. (1979) Interchain disulfide bonds and subunit organization in human serum cholinesterase. *J. Biol. Chem.* 254:8324-8330.

MacPhee-Quigley, K., Vedvick, T.S., Taylor, P. and Taylor, S. Profile of the disulfide bonds in acetylcholinesterase. J. Biol. Chem. 261, 13565-13570.

Massoulié, J. and Bon, S. (1982) The molecular forms of cholinesterase in vertebrates. *Annu. Rev. Neurosci.* 5:57-106.

Massoulié, J. and Toutant, J.P. (1988) Vertebrate cholinesterases: structure and types of interactions. *Handb. Exp. Pharmacol.* 86:167-224.

Massoulié, J., Pezzementi, L., Bon, S., Krejci, E. and Vallette, F.-M. (1992) Molecular and cellular biology of cholinesterases. *Prog. Neurobiol.,* in press.

Maulet, Y., Camp, S., Gibney, G., Rachinsky, T., Ekstrom, T.J. and Taylor, P. (1990) Single gene encodes glycophospholipid-anchored and asymmetric acetylcholinesterase forms: alternative coding exons contain inverted repeat sequences. *Neuron* 4:289-301.

Rachinsky, T.L., Camp, S., Li, Y., Ekstrom, T.J., Newton, M. and Taylor, P. (1990) Molecular cloning of mouse acetylcholinesterase: tissue distribution of alternatively spliced mRNA species. *Neuron* 5:317-327.

Roberts, W.L., Doctor, B.P.., Foster, J.D. and Rosenberry, T.L. (1991) Bovine brain acetylcholinesterase primary sequence involved in intersubunit disulfide linkages. *J. Biol. Chem.* 266:7481-7487.

Rotundo, R.L., Thomas, K., Porter-Jordan, K., Benson, R.J.J., Fernandez-Valle, C. and Fine, R.E. (1989) Intracellular transport, sorting, and turnover of acetylcholinesterase. Evidence for an endoglycosidase H-sensitive form in Golgi apparatus, sarcoplasmic reticulum, and clathrin-coated vesicles and its rapid degradation by a non-lysosomal mechanism. *J. Biol. Chem.* 264:3146-3152.

Schumacher, M., Maulet, Y. Camp, S. and Taylor, P. (1988) Multiple messenger RNA species give rise to the structural diversity in acetylcholinesterase. *J. Biol. Chem.* 263:18979-18987.

Sikorav, J.L., Duval, N., Anselmet, A., Bon, S., Krejci, E., Legay, C., Osterlund, M., Reimund, B. and Massoulié, J. (1988) Complex alternative splicing of acetylcholinesterase transcripts in *Torpedo* electric organ; primary structure of the precursor of the glycolipid-anchored dimeric form. *EMBO J.* 7:2983-2993.

Velan, B. Grosfeld, H., Kronman, C., Leitner, M., Gozes, Y., Lazar, A., Flashner, Y., Marcus, D., Cohen, S., and Shafferman, A. (1991) *J. Biol. Chem.* 266:23977-23984.

BINDING OF ASYMMETRIC (A_{12}) ACETYLCHOLINESTERASE TO C_2 MUSCLE CELLS AND TO CHO MUTANTS DEFECTIVE IN GLYCOSAMINOGLYCAN SYNTHESIS

Nibaldo C. Inestrosa,[1] Herman Gordon,[2]
Jeffrey D. Esko[3] and Zach W. Hall[4]

[1]Department of Cell and Molecular Biology
Catholic University of Chile, Santiago, CHILE
[2]Department of Anatomy, University of Arizona
Tucson, AZ 85721, USA
[3]Department of Biochemistry, University of Alabama at Birmingham
Birmingham, AL 35294, USA
[4]Department of Physiology, University of California
San Francisco, CA 94143, USA

INTRODUCTION

The principal biological role of acetylcholinesterase (AChE) is termination of impulse transmission at cholinergic synapses by rapid hydrolysis of the neurotransmitter acetylcholine (Katz, 1966). AChE is a polymprphic enzyme which may be classified in globular and asymmetric forms (Bon et al., 1979). Of these, the tailed asymmetric A_{12} form is found in high concentrations at endplates (Hall, 1973). The occurrence of a collagenous domain in the tailed enzyme suggests that it interacts with the extracellular matrix (Lwebuga-Mukasa et al., 1976; Inestrosa et al., 1982). It has been postulated that this interaction occurs through attachment to heparin-like glycosaminoglycans (GAGs) present in proteoglycans (PGs) (Brandan et al., 1985). In the present study we have directly evaluated the binding of purified A_{12} AChE to cultured wild-type and mutant mouse C_2 skeletal muscle cells. Also, to investigate the potential role of heparan sulfate proteoglycans (HSPGs) as A_{12} AChE cell surface receptors, we studied the interactions of A_{12} AChE with chinese hamster ovary (CHO) cells that synthesize varying amounts of cell surface heparan sulfate (HS) and other GAGs (Esko, 1991).

EXPERIMENTAL PROCEDURES

Materials: The acridinium resin for AChE purification was kindly provided by Dr. T.L. Rosenberry, Case Western Res. Univ., Cleveland, OH. Heparin, HS and heparinase were obtained from Sigma Chem Co. Heparitinase was obtained from ICN Biochem, Cleveland, OH. Collagenase (form III) was obtained from Advances Biofactures Corp.

Multidisciplinary Approaches to Cholinesterase Functions, Edited by
A. Shafferman and B. Velan, Plenum Press, New York, 1992

A$_{12}$ AChE purification: AChE was purified from the electric organ of the eel Electrophorus electricus as described by Rosenberry and Richardson (1977).

Cells lines: (a) The mouse C$_2$ muscle cells were maintained as exponentially growing myoblasts in a medium consisting of Dulbecco´s modified Eagle´s medium (DME) with 1 g/l glucose, suplemented with 20% fetal calf serum and 0.5% chick embryo extract. After confluency the growth medium was removed and replaced with 0.2 ml/cm^2 of DME plus 2.5% horse serum. After 15 h, myoblasts fuse into myotubes that increased in size and branching complexity during the next days (Inestrosa et al., 1983). (b) GAG variants of the C$_2$ muscle cell line were obtained as described by Gordon and Hall (1989). Only the clone S 27 mainly deficient in cell surface HSPGs but also chondroitin sulfate (CS) was studied. (c) Parental wild-type CHO cells (CHO-KI) and mutant CHO cells (pgsD-677 and pgsE-606) were routinely grown in Ham´s F 12 medium supplemented with 10% bovine fetal calf serum and L-glutamine. Mutant pgsD-677 does not synthesize HS and overexpresses CS by a factor of 3 so that the total amount of sulfated GAG is comparable in pgsD-677 and wild-type cells (Esko et al., 1988). Mutant pgsE-606 is an HS-N-sulfotransferase-deficient cell line that expresses normal levels of HS and CS, but the HS is undersulfated by a factor of 2-3 (Bame and Esko, 1989).

Binding of AChE to C$_2$ myotubes: For binding experiments, the purified A$_{12}$ AChE was initially diluted with 100 mM sodium phosphate (pH 7.0) and 100 mM NaCl, then in DME containing 150 mM NaCl and 2 mg/ml of bovine serum albumin (BSA). The cells were normally plated into 24-mm wells of Linbro trays at a density of 10^5 cells per well and cultivated as described above. In a typical binding assay, after the medium was removed, the myotubes or the CHO cells were washed three times with 2 ml of DME plus 2 mg/ml BSA, and then incubated with 1 ml of diluted AChE (in DME, 150 mM NaCl and 2 mg/ml BSA) under humidified 90% air, 10% CO$_2$ atmosphere at 37°C. At the end of the incubation, the medium was removed, and the myotubes and the CHO cells were rinsed three times with 2 ml of DME and 5 ml of phosphate saline buffer, and then incubated with 1 ml of 10 mM sodium phosphate (pH 7.4), 2 mg/ml BSA and 1 M NaCl, in order to remove the AChE bound to the cell surface. Aliquots of this saline extract were tested for AChE activity. In some experiments, at the end of the incubation, the myotubes and the CHO cells peeled off the wells due to the extensive washes. It was therefore necessary to harvest the cells in centrifuge tubes before to make the final extraction in 1 M NaCl. Empty wells blanks were run for all the experiments to monitor the efficacy of the rinsing procedures. Because the A$_{12}$ AChE is found associated with the C$_2$ muscle surface, a blank in which C$_2$ myotubes were not incubated with AChE, but were washed with 1 M NaCl was also included.

AChE assay: AChE activity was measured by the method of Ellman et al. (1961)., with 0.75 mM acetylthiocholine as the substrate.

RESULTS

Binding of Purified A$_{12}$ AChE to C$_2$ Myotubes

The time course of A12 AChE binding to C2 mouse myotubes is presented in Fig. 1. Binding reached a maximum by 30 min at 37°C, and the amount of bound AChE remained constant for at least 2 h. The time course of AChE binding was essentially the same when the incubation was carried out at 4°C. In all subsequent equilibrium-binding experiments, incubation was conducted at 37°C for a period of 1 h.

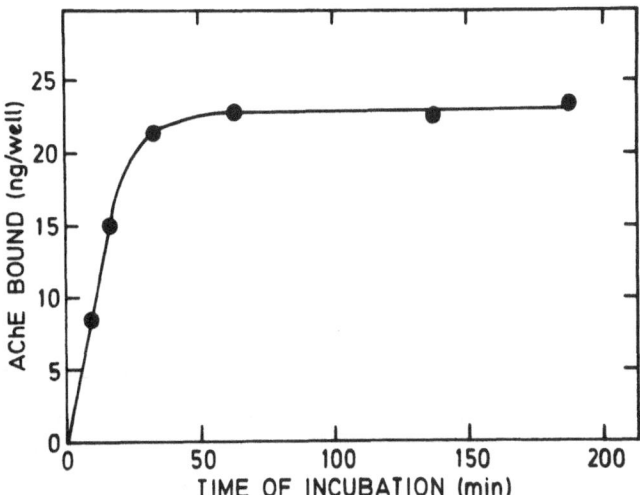

Figure 1. Time course of binding of AChE to C2 myotubes Two ug of Electrophorus AChE was added to each well in an incubation volume of 0.5 ml. The experiments were carried out at 37°C, similar results were obtained when the experiments were carried out at room temperature.

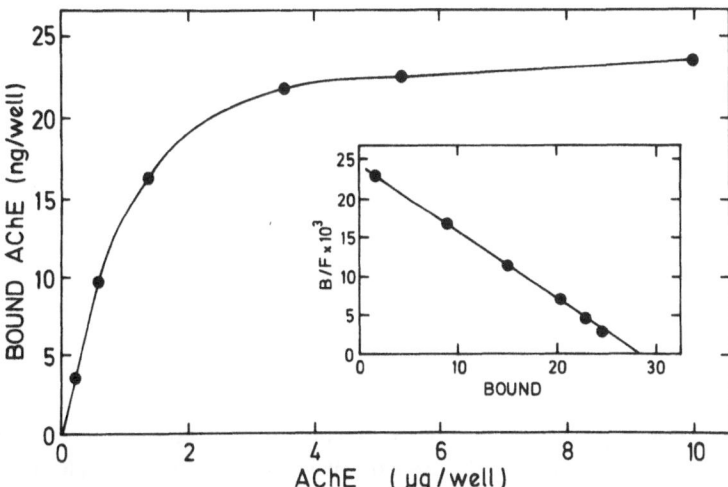

Figure 2. Saturable binding of AChE to C_2 mouse myotubes. A Scatchard analysis (inset) of the saturation curve data is shown. Each point is the average of at least three determinations. Replicate experiments gave similar results.

When increasing amounts of A_{12} AChE were incubated with the C_2 myotubes, the quantity of asymmetric AChE bound to the cells reached a plateau, which indicates saturation (Fig. 2). The mean maximal amount of AChE bound to cells was 23 ± 2.7 ng/well (n=6). This value corresponds to a maximum of 0.6% of the total incubated AChE. Examination of the binding data in the form of Scatchard plots (Fig. 2 inset) indicated the presence of a single class of binding sites. The calculated association constant and maximum binding capacity for this experiment were $K = 0.65 \times 10^7$ M^{-1} and Bmax= 27 ng/well, respectively. The above calculations assume a molecular weight of 1,150,000 for the A_{12} AChE (Rieger et al., 1976). The total endogenous surface AChE activity present in C_2 myotubes (Inestrosa et al., 1982) represent less than 3% of the amount of Electrophorus AChE bound to the cells in the first points of the saturation curve. Parallel experiments with DFP-treated myotubes, did not change the saturation curve, nor the association constant or the maximal binding capacity, suggesting that there are an excess of AChE-binding sites on the muscle cell surface.

Effect of Collagenase Treatment of A_{12}AChE on the Binding to C_2 Muscle Cells

In order to test whether the collagen tail is required for the binding of the AChE to the C_2 myotubes, AChE was treated with highly purified collagenase and subsequently tested for their binding capacity.

Table 1. Effect of collagenase treatment on the binding of asymmetric AChE to muscle cells.

Treatment	Binding (%)
Control	100
Collagenase + EDTA	92
Collagenase + EDTA + Heparin	22
Collagenase	8
NaCl (1 M)	2

Twenty five ug of AChE were incubated with 800 units of purified collagenase in the presence of 0.1 M Tris-HCl (pH 7.4) plus 10 mM $CaCl_2$ in a total volume of 150 ul for 1 h at 37° C.. In control experiments the AChE was incubated in the presence of the collagenase plus 10 mM EDTA, or collagenase + 10 mM EDTA + 5 mg/ml heparin. Results represent the mean values in one of at least two independent experiments.

As Table 1 indicates, the treatment with collagenase almost totally abolished (90%) the interaction of the asymmetric AChE with C_2 myotubes. These data suggest that the collagenous tail is required for the interaction between collagen-tailed AChE and the muscle cell surface.

Effect of Heparin and Heparinase on the Binding of AChE to C_2 Myotubes

Increasing amounts of heparin in the binding assay decreased the percent of AChE bound to the cell surface (Fig. 3A). At 0.5 mg/ml the value observed was only 40% of the initial. These experiments suggests that the binding of A_{12} AChE to C_2 myotubes is partially mediated by heparin-like macromolecules.

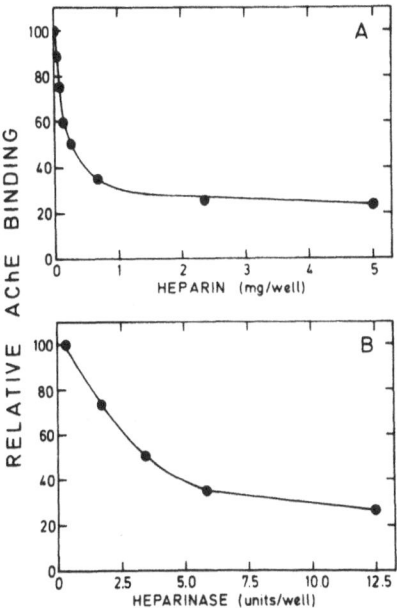

Figure 3. Effect of heparin and heparinase treatment on AChE binding to C_2 myotubes. **A**: AChE was bound to muscle cells in an incubation mixture containing increasing amounts of heparin. **B**: After C_2 myotubes where rinsed, they were incubated with *Flavobacte heparinum* heparinase and 2 mg/ml BSA, at 37°C for 1 h, then the cells where washed and the binding assay was carried out as described previously. Values are duplicates of two determinations.

To determine whether A_{12} AChE interacts specifically with HSPGs on the muscle cell surface, we incubated the C_2 myotubes with heparinase, an enzyme which specifically degrades heparin and HS (Jackson et al., 1991). As shown in Fig. 3B, the enzymatic treatment caused an important inhibition on the binding of A_{12} AChE to C_2 mouse myotubes.

Binding of A_{12} AChE to Wild-Type and PG-Deficient (S_{27} Clone) C_2 Muscle Cells

To investigate the possible role of PGs on the anchorage of A_{12} AChE to the muscle surface, a mutant of C_2 muscle cells, that lack mainly HSPGs but also CS (Gordon and Hall, 1989) was examined for their ability to bind A_{12} AChE. As Fig. 4B shows, the S_{27} variant bound only 60% of the amount of AChE bound by the wild-type C_2 cells. A rather similar value was obtained when wild-type cells were pre-treated with heparitinase (Fig. 4A), an enzyme that specifically degrades HS residues (Jackson et al., 1991).

When HS was used in the binding assays in wild-type cells almost 50% of binding sites for AChE were blocked by this GAG (Fig. 4B). In the case of the S_{27} clone, the proportion of total binding inhibitable by GAG was smaller, but significant, suggesting that a residual amount of HSPGs were still present in the S_{27} variant. Taken as a whole, the above results indicate that HSPGs are an important factor in the association of A_{12} AChE to the muscle cell surface. However it is also clear, from the results with wild-type and mutant C_2 cells, that some of the AChE bound to the cells apparently interacts with another cell surface component as suggested by the fact that heparin removes most of the AChE bound (Fig. 4A and B). This possibility is supported by previous evidence showing that heparin was able to solubilize a dermatan sulfate proteoglycan (DSPGs) from rat muscle (Brandan and Inestrosa, 1987).

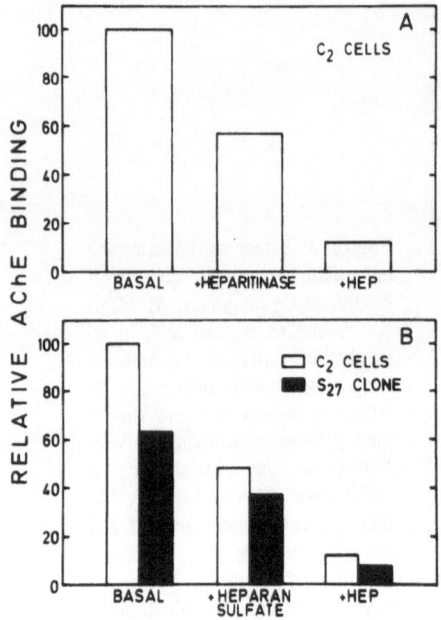

Figure 4. Binding of AChE to Wild-type and Mutant (S_{27} clone) C_2 Muscle Cells. Effect of GAG Competition or Removal. Binding of A_{12} AChE was carried out as described in Fig. 2, in the absence or presence of 0.2 units/ml heparitinase, 1 mg/ml heparin (HEP) and 0.5 mg/ml HS. All the binding data are normalized to wild-type C_2 muscle cells in the absence of exogenous GAG or after substraction of nonspecific binding. Results indicate the means of three independent experiments. A: Effect of heparitinase on C_2 cells, B: Effect of HS and HEP on C_2 and S_{27} muscle cells.

Binding of A_{12} AChE to Wild-Type and CHO Cell Mutants

CHO cells mutants with specific alterations in proteoglycan synthesis has provided an important tool for exploring the biological activity of proteoglycans in living cells. This strategy has many advantages over conventional approaches that involve enzymatic removal of GAG chains or that require purified proteoglycans or GAGs in competitive assays (Esko, 1991). Wild-type CHO cells (CHO-KI) as well as CHO cell mutant defective at different sites in GAG biosynthesis (Esko et al., 1988) were used as a model system for analyzing binding sites for A_{12} AChE. Wild-type CHO-KI cells were capable of binding AChE in a manner comparable to that found for C_2 muscle cells (Fig. 5A).

The total amount of AChE bound to CHO-KI cells corresponds to 31% of that observed in C_2 cells. As in the case of muscle cells, pre-treatment of CHO-KI cells with heparitinase determined a 50% decreases in the binding of A_{12} AChE. Mutant CHO-677 cells, which has undetectable levels of HSPGs but overexpresses the CS did bind only 40% of the amount of A_{12} AChE bound by wild-type CHO-KI cells (Fig. 5B). That is similar to the amount of AChE bound to CHO-KI cells after the treatment with heparitinase (Fig. 5A). Finally, the binding of A_{12} AChE was reduced by more than 75% in CHO clone 606, a mutant expressing undersulfated HSPGs. This finding is consistent with a previous report suggesting that the degree of sulfatation of heparin can significantly alter its ability to interact with A_{12} AChE (von Bernhardi and Inestrosa, 1990). Also it is possible that an undersulfated HSPG lead to alterations in the assembly of cell surfaces components relevant to the anchorage of A_{12} AChE.

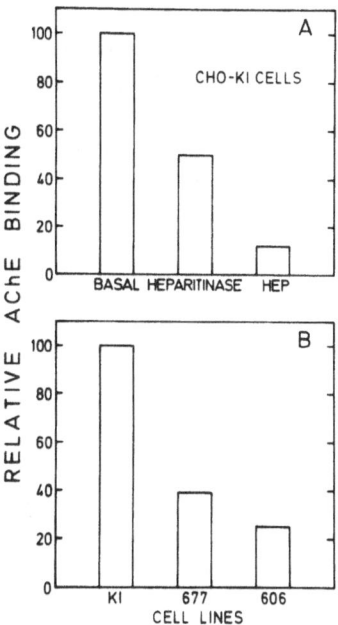

Figure 5. Binding of A_{12} AChE to wild-type and HS-deficient cells. Cells were incubated for 1 h at 37°C with a saturating concentration of A_{12} AChE. **A:** Shows the effect of 0.2 units/ml heparitinase and 1 mg/ml heparin (HEP) on CHO-KI parental, wild-type CHO cells. **B:** Indicate the relative binding of wild-type CHO cells (KI) and the clones 677 and 606. Results represent the mean values in one of at least three independent experiments.

DISCUSSION

These studies provide definitive evidence that cell surface PGs mediate the binding of A_{12} AChE to both C_2 muscle cells and CHO cells. Several of our findings lead to the conclusion that collagen-tailed AChE binds primarily to polysaccharides chains of HSPGs on the cell surface:

(a) Enzymatic removal of HS chains partially inhibits the AChE binding to wild-type C_2 and CHO cells.

(b) Competitive assays in presence of HS partially block the AChE binding.

(c) HS-deficient C_2 cells (S_{27} clone) had reduced levels of binding.

(d) Mutant CHO cells with either undetectable levels of HSPGs (677) or undersulfated HS (606) present clear reduced binding capacity of AChE.

Our binding studies also suggest that surface components other than the HS residues are involved in the binding of asymmetric AChE to the cell surface. In fact, despite that heparitinase treatment removed most of the HS residues from the cell surface, both C2 and CHO-KI cells were still able to bind a substantial amount of asymmetric AChE. The same behavior was observed in HS-deficient cells in C2 (S27 clone) and CHO-677 mutants. On the other hand, it is likely that other PGs or surface components are involved in the binding of AChE. Consistent with this possibility is the fact that heparin was able to eliminate almost all the heparitinase-insensitive binding of AChE. Previous studies have shown that heparin releases DSPG together with asymmetric AChE from rat muscle (Brandan and Inestrosa, 1987). Also DS was able to solubilize 30% of the AChE activity released by heparin (von Bernhardi and Inestrosa, 1990), therefore it is entirely possible that DSPGs or others are also involved in the binding of AChE to the cell surface.

Significant progress has been made in understanding the function of PGs, many of the advances derive from the characterization of animal cell mutants altered in GAGs synthesis (Esko, 1991). In this context, our studies with the CHO-677 cells that lack HS and overexpress CS (Esko et al., 1988), indicates: first, that HS is the primary cell surface mediator of AChE binding (60%) and, second that the extra CSPG present in the mutant may be of sufficient density on the cell surface to mediate AChE binding (40%). The results with the CHO clone 606, the mutant expressing undersulfated HSPGs with no apparent modification in CSPGs (Bame and Esko, 1989) also demonstrate how important is the presence of HS residues in the binding of AChE. However, because only small amount of AChE was bound to such cells (20%), it is possible that the decreased charge density of the polysaccharide may directly affect the interaction of either CS/DSPGs with the asymmetric AChE. In any case, further studies are necessary to stablish which other cell surface PGs or macromolecules, besides HSPGs are involved in the additional binding of AChE to the cell surface.

ACKNOWLEDGEMENTS

We thank Eliseo O. Campos and Enrique Brandan for their help with the manuscript. Supported by CONICYT-Fundacion Andes to NCI and, MDA and NIH to ZWH.

REFERENCES

Bame, K.J., and Esko, J.D., 1989, Undersulfated heparan in a chinese hamster ovary cell mutant defective in heparan sulfate N-sulfotransferase. J. Biol. Chem. 264: 8059.

Bon, S., Vigny, M., and Massoulie, J., 1979, Asymmetric and globular forms of acetylcholinesterase in mammals and birds Proc. Natl. Acad. Sci. USA 76: 2546.

Brandan, E., and Inestrosa, N.C., 1987, Co- solubilization of asymmetric acetylcholinesterase and dermatan sulfate proteoglycan from the extracellular matrix of the rat skeletal muscle FEBS Lett. 213: 159.

Brandan, E., Maldonado, M., Garrido, J., and Inestrosa, N.C., 1985, Anchorage of collagen-tailed acetylcholinesterase to the extracellular matrix is mediated by heparan sulfate proteoglycans J. ` Cell Biol. 101: 985.

Ellman, G.L., Courtney, K.D., Andres, V.Jr., and Featherstone R.M., 1961, A new and rapid colorimetric determination of acetylcholinesterase activity. Biochem. Pharmacol 7: 88.

Esko, J.D., 1991, Genetic analysis of proteoglycan structure, function and metabolism Current Opinion in Cell Biol. 3: 805.

Esko, J.D., Rostand, K.S., and Weinke, J.L., 1988, Tumor formation dependent on proteoglycan biosynthesis Science 41: 1092.

Gordon, H., and Hall, Z.W., 1989, Glycosaminoglycanariants in the C_2 muscle cell line. Dev. Biol. 135:1.

Hall, Z.W., 1973, Multiple forms of acetylcholinesterase and their distribution in endplate and non-endplate of rat diaphragm muscle. J. Neurobiol. 4: 343.

Inestrosa, N.C., Silberstein, L., and Hall Z.W., 1982, Association of the synaptic form of acetylcholinesterase with extracellular matrix in cultured mouse muscle cells. Cell 29: 71.

Inestrosa, N.C., Miller, J.B., Silberstein, L., ZiskindConhaim, L., and Hall, Z. W., 1983, Development and regulation of 16 S acetylcholinesterase and acetylcholine receptors in a mouse muscle cell line. Exp. Cell Res. 147:93.

Jackson, R.L., Busch, S.J., and Cardin, A.D., 1991, Glycosaminoglycans: molecular proteins interactions, and role in physiological processes. Physiol. Rev. 71: 481.

Katz, B., 1966, "Nerve, Muscle, and Synapse", McGraw-Hill, New York.

Lwebuga-Mukasa, J.S., Lappi, S., and Taylor, P., 1976, Molecular forms of acetylcholinesterase from Torpedo californica. Their relationship to synaptic membranes. Biochem. 15:1425.

Rosenberry, T.L., and Richardson, J.M., 1977, Structure of 18 S and 14 S acetylcholinesterase. Identification of collagen-like subunits that are linked by disulfide bonds to catalytic subunits. Biochem. 16: 3550.

von Bernhardi, R., and Inestrosa N.C., 1990, Dermatan sulfate and de-sulfated heparin solubilized collagen- tailed acetylcolinesterase from the rat neuromuscular junction. Brain Res. 529: 91

SUBUNIT ASSEMBLY AND GLYCOSYLATION OF

MAMMALIAN BRAIN ACETYLCHOLINSTERASE

Urs Brodbeck and Jian Liao

Institute of Biochemistry and Molecular Biology
University of Bern
Bühlstrasse 28
3012 Bern, Switzerland

INTRODUCTION

Mammalian acetylcholinesterase (AChE) exists in brain as globular tetrameric form (G_4 form) of which approximately 80% are amphiphilic membrane bound and about 20% are hydrophilic non-membrane bound enzyme. (for review see Massoulié and Toutant, 1988). The amphiphilic membrane bound G_4 form consists of two pairs of catalytic subunits. In one of the two pairs, the catalytic subunits are thought to be linked together by one disulfide bond which is located near the C-terminus. In the other pair, the catalytic subunits are each linked through the same C-terminal disulfide bonds to a structural subunit of approximately 20 kDa in molecular mass which carries the hydrophobic membrane anchor (Gennari et al., 1987; Inestrosa et al., 1987; Roberts et al., 1991; Heider and Brodbeck, 1992). On the other hand, AChE from mammalian erythrocytes consists of two disulfide linked catalytic subunits (G_2 form) which are membrane bound through a glycosyl phosphatidylinositol (GPI) moiety covalently attached to the C-terminus of each subunit (for review see Silman and Futerman, 1987).

Reports on the primary structure of AChE revealed a high homology between mammalian brain and erythrocyte forms (Bon et al., 1986; Chhajlani et al., 1989; Soreq and Prody, 1989; Soreq et al., 1990; Doctor et al., 1990; Rachinsky et al., 1990; Heider et al., 1991) but relatively little is known about the glycosylation of the two enzyme forms. Available cDNA sequences only provided clues about the potential N-glycosylation sites but it is not clear whether all of these sites are used and whether the two forms of mammalian AChE differ in glycosylation or not. Tissue specific differences in glycosylation of ChE were shown by differences in lectin interaction (Rotundo, 1984; Méflah et al., 1984) and by different reactivities towards monoclonal antibodies (mAb) (Musset et al., 1987; Bon et al., 1987). G_2 and G_4 AChE further differ in the extent of N-glycosylation. G_2 AChE from human and bovine red cell membranes is more heavily N-glycosylated than human and bovine G_4 form. Bovine AChE (both G_2 and G_4 forms) are more heavily N-glycosylated than the corresponding human forms (Liao et al., 1991). N-acetylgalactosamine could neither be detected in human and bovine brain nor in bovine erythrocyte AChE indicating that both enzyme forms contain no O-linked carbohydrates (Liao et al., 1992). Certain mAbs specifically reacted with N-linked carbohydrates of G_4 AChE but not with the G_2 form indicating that G_2 and G_4 AChE undergo different post-synthetic modifications leading to different subsets of N-linked carbohydrates (Liao et al., 1992).

Relatively little is known about the structure of the hydrophobic anchor of mammalian brain G_4 AChE. Its hydrophobic domain can be labelled with the photoactivatable

Multidisciplinary Approaches to Cholinesterase Functions, Edited by
A. Shafferman and B. Velan, Plenum Press, New York, 1992

33

hydrophobic reagent ^{125}I-trifluorophenyl-diazirine (^{125}I-TID), an affinity reagent which partitions into the hydrophobic phase of membranes or into detergent micelles. With this label, a non-catalytic structural subunit of about 19 to 23 kDa molecular mass was identified in AChE from human brain (Gennari et al., 1987) and in bovine brain (Inestrosa et al., 1987). Since this entity is sensitive to proteolysis, it is assumed to consist of a hydrophobic peptide (Fuentes and Inestrosa, 1988, Fuentes et al., 1988). An additional hydrophobic constituent was suggested to be associated with the anchor (Inestrosa et al., 1987; Stieger and Brodbeck, 1988; Heider et al., 1991) but AChE from bovine brain was shown not to contain the typical constituents of the GPI-anchor, i.e. inositol, ethanolamine, or glucosamine (Inestrosa et al., 1987). Furthermore, little is known about the antigenicity of this anchor domain.

Antibodies have long been recognized as exquisitely sensitive probes of the structure of protein molecules. With the advent of the monoclonal antibody technology, it was possible to obtain a variety of homogeneous antibodies each directed against a given epitope and endowed with a unique equilibrium constant for the antigen. Many mAbs had been produced against AChE from different sources in the past several years which provided us a new approach of gaining insight into the structure of AChE molecules (for rewiew see Rakonczay and Brimijoin, 1988). In the present study, monoclonal antibodies were specifically raised against different domains of AChE and used in the study of the quaternary structure of mammalian AChE. New insights were gained on the assembly of the subunits and their attachment to the hydrophobic anchor domain.

METHODS

Antibodies

mAbs were obtained from the Hybridoma Laboratory of the Statens Seruminstitut, Copenhagen, Denmark. The two non-inhibitory mAbs (132-3 and 132-5) which were used in the present study, were raised against detergent-soluble tetrameric form of AChE from human brain.

Enzyme Purification

The different forms of mammalian AChE were purified essentially as described by Sørensen et al. (1982) and Brodbeck et al. (1983). Two consecutive steps of affinity chromatography provided pure preparations as judged by SDS-PAGE and immunochemical methods. BtChE from human serum was kindly donated by Dr. Oksana Lockridge.

SDS-Polyacrylamide Gel Electrophoresis (SDS-PAGE)

SDS-PAGE was performed essentially according to the method described by Laemmli (1970) using 5-15% polyacrylamide gradient gels. Protein was stained with Coomassie Brilliant Blue R-250.

Electroblotting and Western Blot on Nitrocellulose Membranes

After electrophoresis, proteins from SDS-PAGE were electrophoretically transferred to nitrocellulose membranes in a semi dry blot system (Ancos, Denmark) as described by Ploug et al. (1989). After the transfer, nitrocellulose membranes were rinsed 3 times with 50 mM Tris-HCl buffer, pH 7.4, containing 0.14 M NaCl. Then, 20 mM Tris-HCl buffer, pH 7.4, containing 0.14 M NaCl and 3% BSA, was added, and blocking was carried out at room temperature for 2 hours.

RESULTS AND DISCUSSION

As shown previously (Heider and Brodbeck, 1992), tetrameric detergent soluble bovine caudate nucleus AChE could be reduced and alkylated in conditions which retain at least 95% of initial activity. This treatment alone did neither result in monomerization of

AChE nor did it create a hydrophilic enzyme. However, in presence of SDS, the enzyme became monomerized. Incubation of AChE with trypsin in the presence of the reversible inhibitor edrophonium rendered the enzyme hydrophilic and lead to catalytically active monomers. SDS-PAGE of this preparation in non-reducing conditions revealed a small decrease in the subunit molecular mass only. N-terminal sequencing of the enzyme, before and after trypsin treatment, yielded identical N-termini showing that the enzyme was monomerized subsequent to C-terminal tryptic cleavage. From our results, we concluded that the most C-terminal cysteine residue is involved in intersubunit disulfide bonding as well as in the attachment of AChE to the membrane anchor. Our results are in agreement with those of Roberts et al. (1991) who independently showed that the most C-terminal cys residue is involved in both inter-subunit disulfide bonding and anchor attachment. In human AChE, substitution of the same cys residue by ala resulted in impairment of interchain disulfide bridge formation and led to secretion of monomeric enzyme (Velan et al., 1991). As further shown by Heider and Brodbeck (1992), G_4 brain AChE is not only held toghether by inter-subunit disulfide bonding but also by a hydrophobic contact area in the C-terminal region of the primary structure which strengthens the forces holding the subunits as well as the membrane anchor together.

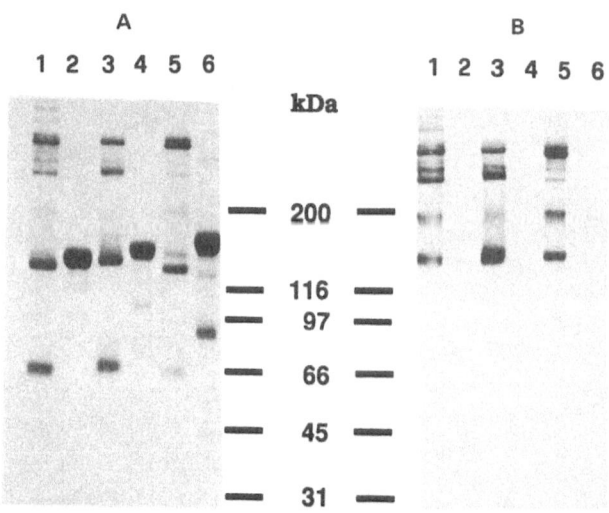

Figure 1. SDS-page and Western blot of AChE from different mammals. Electrophoresis was carried out in on-reducing conditions on a 5-15% gradient polyacrylamide gel. (A) Proteins stained with Coomassie brilliant blue R-250; (B) proteins revealed by western blotting with mAb 132-5. Lane 1, detergent-soluble form of AChE from human brain; lane 2, detergent-soluble form of AChE from human erythrocytes; lane 3, detergent-soluble form of AChE from bovine brain; lane 4, detergent-soluble form of AChE from bovine erythrocytes; lane 5, detergent-soluble form of AChE from monkey brain; lane 6, BtChE from human serum. The gels were calibrated with phosphorylase b (97 kDa), bovine serum albumin (66 kDa), ovalbumin (45 kDa), and carbonic anhydrase (31 kDa).

The subunit assembly was further assessed with two different monoclonal antibodies. One of the mAb 132-3 (IgG_1) recognized a highly conserved epitope in all forms of mammalian AChE of different origin and thus, could be used to detect all forms of AChE on Western blots with high sensitivity. The other mAb, i.e. mAb 132-5 (IgG_1), specifically recognized the G_4 brain enzyme (figure 1) of different mammalian origin but not with amphiphilic G_2 AChE from eryhtrocytes. Figure 1 further shows that mAb 132-5 apparently did not react with subunits of G_4 AChE from brain devoid of the hydrophobic anchor.

Since mAb 132-5 detected subunits only to which the hydrophobic anchor was attached while mAb 132-3 targeted the catalytic subunits regardless of the absence or presence of the hydrophobic anchor, the two mAbs could be used to further refine the picture on the subunit assembly of G4 AChE from mammalian brain. As shown in figure 2, freshly prepared G4 AChE from monkey brain revealed on a Western blot after SDS-PAGE in non-reducing conditions substantial amounts of monomer which appeared at 65 kDa and was devoid of the anchor. Similar patterns have been observed with G4 AChE from bovine and human brain showing that in freshly prepared enzyme, not all of the subunits are disulfide bonded. Since sucrose density gradient centrifugation did not reveal a corresponding amount of monomeric enzyme, it must be assumed that those subunits which appeared as monomer on SDS-PAGE in non-reducing conditions, were assembled in the tetramer by hydrophobic forces only and not by additional disulfide bonding.

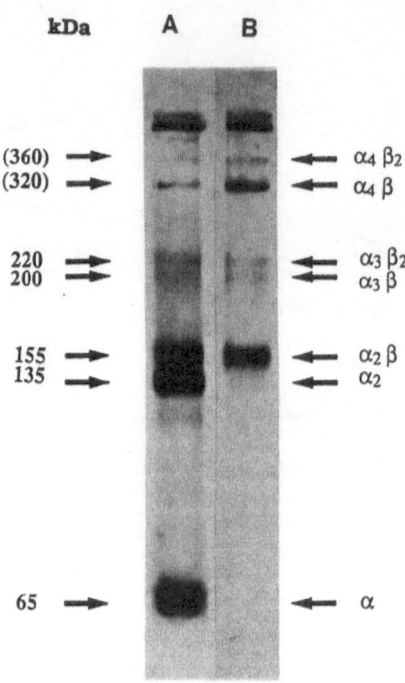

Figure 2. Western blot of amphiphilic G4 AChE from monkey brain revealed by mAb 132-3 (lane A) and mAb 132-5 (lane B). SDS-PAGE was carried out in non-reducing conditions on a 5-15% polyacrylamide gradient gel followed by transfer of the protein to PVDF membrane. The molecular weights were calculated from the position of the marker proteins shown in figure 1. α: catalytic subunit, β: hydrophobic anchor.

Figure 2 further depicts the existence of light and heavy dimers appearing at 135 and 155 kDa, respectively, in agreement with previously published patterns for the human (Gennari et al., 1987) and bovine brain enzyme (Inestrosa et al., 1987). Interestingly, another doublet of bands is seen in the molecular mass range of 200 and 220 kDa which most likely correspond to heavy and light trimers. Several slower moving bands further up in the gel are also clearly seen. Their molecular masses, however, are subject to uncertainty as their position in the gel is beyond the calibration range of the marker proteins. The results obtaind with mAb 132-5 are interesting as this antibody stained bands corresponding to the heavy dimer, both bands composed of trimers and aggregates of higher molecular mass. From the molecular masses of the two trimers and the finding that mAb 132-5 only reacted with subunits containing the hydrophobic anchor, it follows that the band appearing at 200 kDa is most likely composed of three catalytic subunits and one anchor ($\alpha_3\beta$), and that appearing at 220 kDa, contains three catalytic subunits and two anchors ($\alpha_3\beta_2$). Likewise, the bands appearing in the estimated molecular weight range of 320 and 360 kDa are probably tetramers containing one and two hydrophobic anchors ($\alpha_4\beta$ and $\alpha_4\beta_2$, respectively). The topmost two bands contain aggregates of even higher molecular masses. The existence of small amounts of such forms even in presence of Triton X-100, is evident in results from sucrose density gradient centrifugation (see e.g. figures 2a and 3a in Heider and Brodbeck, 1992). Such forms might be assembled in a way similar to the asymmetric A_8 and A_{12} forms of AChE. Alternatively, they could have been formed by a tail to tail association of the hydrophobic anchor similar to the dumbbell-shaped forms described in AChE from *Electrophorus electricus* (Dudai et al., 1973, Bon et al., 1978). In any case, the analogous assembly of the amphiphilic G4 brain AChE and of the collagen tailed asymmetric forms implies that the hydrophobic anchor of the brain form might be structurally related to the collagen tail. Since the two structural subunits serve the same purpose, they could be of common origin and might have arisen by either gene duplication or by different splicing of the same gene.

ACKNOWLEDGMENTS

We are grateful to Drs. C. Koch, V. Mortensen, and B. Nørgaard-Pedersen for their collaboration in raising the mAbs deployed in this study.

REFERENCES

Bon, S., Cartaud, J., and Massoulié, J., 1978, Dumbbell-shaped associations of tailed *Electrophorus* acetylcholinesterase molecules, *Mol. Biol. Rep.* 4:61.

Bon, S., Chang, J.-Y., and Strosberg, A.D., 1986, Identical N-terminal peptide sequences of asymmetric forms and of amphiphilic low-salt-soluble and detergent-soluble dimers of *Torpedo* acetylcholinesterase: comparison with bovine acetylcholinesterase, *FEBS Lett.* 209:206.

Bon, S., Méflah, K., Musset, F., Grassi, J., and Massoulié, J., 1987, An immunoglobulin M monoclonal antibody recognizing a subset of acetylcholinesterase molecules from electric organs of *Electrophorus* and *Torpedo*, belongs to the HNK-1 anti-carbohydrate family, *J. Neurochem.* 49:1720.

Brodbeck, U., Gentinetta, R., and Ott, P., 1983, Purification by affinity chromatography of red cell membrane acetylcholinesterase, *in*: "Membrane Proteins. A Laboratory Manual," A. Azzi, U. Brodbeck, and P. Zahler, eds, Springer-Verlag, Berlin.

Chhajlani, V., Derr, D., Earles, B., Schmell, E., and August, T., 1989, Purification and partial amino acid sequence analysis of human erythrocyte acetylcholinesterase, *FEBS Lett.* 247:279.

Doctor, B.P., Chapman, T.C., Christner, C.E., Deal, C.D., De la Hoz, D.M., Gentry, M.K., and Wolf, A.D., 1990, Complete amino acid sequence of fetal bovine serum acetylcholinesterase and its comparison in various regions with other cholinesterases, *FEBS Lett.* 266:123.

Dudai, Y., Herzberg. M., and Silman, I., 1973, Molecular structures of acetylcholinesterase from electric organ tissue of the electric eel, *Proc. Natl. Acad. Sci. USA* 70:2473.

Fuentes, M.E., and Inestrosa, N.C., 1988, Characterization of a tetrameric G4 form of acetylcholinesterase from bovine brain: a comparison with the dimeric G2 form of the electric organ, *Mol. Cell. Biochem.* 81:53.

Fuentes, M.E., Rosenberry, T.L., and Inestrosa, N.C., 1988, A 13 kDa fragment is responsible for the hydrophobic aggregation of brain G4 acetylcholinesterase, *Biochem. J.* 256:1047.

Gennari, K., Brunner, J., and Brodbeck, U., 1987, Tetrameric detergent-soluble acetylcholinesterase from human caudate nucleus: subunit composition and number of active sites, *J. Neurochem.* 49:460.

Heider, H., and Brodbeck, U., 1992, Monomerization of tetrameric bovine caudate nucleus acetylcholinesterase, *Biochem. J.* 281:279.

Heider, H., Litynski, P., Stieger, S., and Brodbeck, U., 1991, Comparative studies on the primary structure of acetylcholinesterases from bovine caudate nucleus and bovine erythrocytes, *Cell. Mol. Neurobiol.* 11:105.

Inestrosa, N.C., Roberts, W.L., Marshall, T.L., and Rosenberry, T.L., 1987, Acetylcholinesterase from bovine caudate nucleus is attached to membranes by a novel subunit distinct from those of acetylcholinesterases in other tissues, *J. Biol. Chem.* 262:4441.

Laemmli, U.K., 1970, Cleavage of structural proteins during the assembly of the head of bacteriophage T4, *Nature* 227:680.

Liao, J., Heider, H., Sun, M.-C., and Brodbeck, U., 1992, Different glycosylation in acetylcholinesterases from mammalian brain and erythrocytes, *J. Neurochem.* 58:1230.

Liao, J., Heider, H., Sun, M.-C., Stieger, S., and Brodbeck, U., 1991, The monoclonal antibody 2G8 is carbohydrate-specific and distinguishes between different forms of vertebrate cholinesterases, *Eur. J. Biochem.* 198:59.

Massoulié, J., and Toutant, J.-P., 1988, Vertebrate cholinesterases: structure and types of interaction, *in*: "Handbook of Experimental Pharmacology," Vol. 86, V.P. Whittaker, ed., Springer-Verlag, Berlin.

Méflah, K., Bernard, S., and Massoulié, J., 1984, Interactions with lectins indicate differences in the carbohydrate composition of the membrane-bound enzymes acetylcholinesterase and 5'nucleotidase in different cell types, *Biochimie* 66:59.

Musset, F., Frobert, Y., Grassi, J., Vigny, M., Boulla, G., Bon, S., and Massoulié, J., 1987, Monoclonal antibodies against acetylcholinesterase from electric organs of *Electrophorus* and *Torpedo*, *Biochimie* 69:147.

Ploug, M., Jensen, A.L., and Barkholt, V., 1989, Determination of amino acid compositions and NH_2-terminal sequences of peptides electroblotted onto PVDF membranes from tricine-sodium dodecyl sulfat-polyacrylamide gel electrophoresis: application to peptide mapping of human complement component C3, *Anal. Biochem.* 181:33.

Rachinsky, T.L., Camp, S., Li, Y., Ekström, T.J., Newton, M., and Tylor, P., 1990, Molecular cloning of mouse acetylcholinesterase: tissue distribution of alternatively spliced mRNA species, *Neuron* 5:317.

Rakonczay, Z., and Brimijoin, S., 1988, Monoclonal antibodies to human brain acetylcholinesterase: properties and applications, *Cell. Mol. Neurobiol.* 8:85.

Roberts, W.L., Doctor, B.P., Foster J.D., and Rosenberry, T.L., 1991, Bovine brain acetylcholinesterase primary sequence involved in intersubunit disulfide linkages, *J. Biol. Chem.* 266:7481.

Rotundo, R.L., 1984, Asymmetric acetylcholinesterase is assembled in the Golgi apparatus, *Proc. Natl. Acad. Sci. USA* 81:479.

Silman, I., and Futerman, A.H., 1987, Modes of attachment of acetylcholinesterase to the surface membrane, *Eur. J. Biochem.* 170:11.

Sorensen, K., Gentinetta, R., and Brodbeck, U., 1982, An amphiphile dependent form of human brain caudate nucleus acetylcholinesterase: purification and properties, *J. Neurochem.* 39:1050.

Soreq, H., and Prody, C.A., 1989, Sequence similarities between human acetylcholinesterase and related proteins: putative implications for therapy of anticholinesterase intoxication, *in*: "Progress in Clinical and Biological Research," Vol. 289, R. Rein and A. Golombek, eds, Alan R. Liss, New York.

Soreq, H., Ben-Aziz, R., Prody, C.A., Seidman, S., Gnatt, A., Neville, L., Lieman-Hurwitz, J., Lev-Lehman, E., Ginzberg, D., Lapidot-Lifson, A., and Zakut, H., 1990, Molecular cloning and construction of the coding region for human acetylcholinesterase reveals a G+G-rich attenuating structure, *Proc. Natl. Acad. Sci. USA* 87:9688.

Stieger, S., and Brodbeck, U., 1988, Assay and purification of PI-specific phospholipase C from *Bacillus cereus* using commercially available phospholipase C, *in*: "Post-translational Modification of Proteins by Lipids", U. Brodbeck and C. Bordier, eds, Springer-Verlag, Berlin.

Velan, B., Grosfeld, H., Kronman, C., Leitner, M., Gozes, Y., Lazar, A., Flashner, Y., Marcus, D., Cohen, S., and Shafferman, A., 1991, The effect of elimination of intersubunit disulfide bonds on the activity, assembly, and secretion of recombinant human acetylcholinesterase - expression of acetylcholinesterase Cys-580->Ala Mutant, *J. Biol. Chem.* 266:23977.

MOLECULAR ORGANIZATION OF RECOMBINANT HUMAN ACETYLCHOLINESTERASE

Baruch Velan[1]., Chanoch Kronman[1], Moshe Leitner[1], Haim Grosfeld[1], Yehuda Flashner[1], Dino Marcus[2], Arie Lazar[2], Anat Kerem[3], Shoshana Bar-Nun[3], Sara Cohen[1] and Avigdor Shafferman[1]

[1]Department of Biochemistry
[2]Department of Biotechnology
 Israel Inst. Biol. Res. Ness-Ziona, 70450, Israel
[3]Department of Biochemistry
 Tel-Aviv University, Tel-Aviv, Israel

INTRODUCTION

Acetylcholinesterase (abbreviated AChE) occurs in multiple molecular forms in different tissues of vertebrates and invertebrates (reviewed in Massoulie and Bon, 1982; Silman and Futerman, 1987; Chatonnet and Lockridge, 1989). This heterogeneity is generated through tissue-specific associations of various catalytic and structural subunits. Characterized catalytic subunits are divided into two major types, the T-type and the H-type, both derived from a single gene by alternative splicing (Schumacher et al., 1988; Sikarov et al.,1988). Structural subunits include the collagen-like structure (Krejci et al., 1991) that allows attachment to the basal lamina and the 20kD lipid-linked hydrophobic subunit (Inestrosa et al., 1987) which is associated with the mammalian brain enzyme.

Studies on the assembly and post-translation processing of AChE catalytic subunits have been hindered by the complexity of the components involved. Cells can assemble the various types of subunits into intracellular, surface bound and secreted molecular forms in an array of complex configurations. To allow dissection of catalytic subunit synthesis and processing, we have established a recombinant expression system in which a single species of HuAChE subunit is expressed. Human 293 kidney cells, transfected with a very efficient expression vector (Kronman et al., 1992) carrying the cDNA sequences of the T-type AChE polypeptide (Soreq et al., 1990), were found to secrete the soluble, oligomeric form of the human enzyme (Velan et al., 1991a).

Multidisciplinary Approaches to Cholinesterase Functions, Edited by
A. Shafferman and B. Velan, Plenum Press, New York, 1992

39

This system is used here to study post-translation events in the processing of HuAChE and their effect on the structure, activity and secretion of the enzyme. This analysis reveals heterogeneity in N-glycosylation level and in signal-processing pattern of the catalytic subunit as well as in the extent of homo-oligomerization. In addition, mutagenesis studies in which signals for assembly or glycosylation were eliminated indicate that impairment in oligomerization does not affect secretion whereas side chain oligosaccharides are required for efficient export. Enzymatic activity, on the other hand, depends neither on oligomerization nor on N-glycosylation.

METHODS

Construction of expression vectors for rHuAChE : Plasmids construction, isolation of DNA fragments, cloning and bacterial transformation were performed essentially as described in Current Protocols in Molecular Biology (Ausubel *et al.*, 1987). Multipartite expression vectors carrying *ache* cDNA (Soreq *et al.*, 1990) bacterial *neo* r (Southern and Berg, 1982) and mutant *dhfr* (Simonsen and Levinsen, 1983) were generated as described previously (Velan *et al.*, 1991a; Kronman *et al.*, 1992). Mutagenesis was performed by DNA cassette replacement (Shafferman *et al.*, 1987) into a series of HuAChE sequence variants which conserve the wild type coding specificity but carry new unique restriction sites (Shafferman *et al.*, 1992a).

Transfection of 293 cells : CsCl purified plasmid preparations were used to transfect 293 cells using the calcium phosphate method (Wigler *et al.*, 1978). Transient transfections were carried on as described previously (Velan *et al.*, 1991a). Cells were transferred 24 hours after transfection to medium containing 10% cholinesterase-depleted fetal bovine serum (Shafferman *et al.*, 1992a) . For stable transfections cells were cultured in presence of 0.8mg/ml G418 for 3 weeks. Stable pools were generated by trypsinization (Velan *et al.*, 1991b) and stable clones were obtained by limiting dilution (Kronman *et al.*, 1992). The high producer colonies were expanded to establish cell clones.

Production of rHuAChE : Bench scale production of HuAChE was performed in monolayer cultures in a multitray unit (Nunc) of 6,000 cm^2 surface area. 3×10^8 cells of the AChE producer clone (clone R11) were seeded in 2 liter Iscove's modified Dulbecco's medium supplemented with 10% ChE-depleted fetal bovine serum. Medium was harvested every 48 hours and replaced with fresh medium. Four consecutive harvests allowed collection of 45mg rHuAChE. Cell culture supernatants were concentrated and dialyzed by ultrafiltration, AChE was absorbed to procainamide-sepharose (Ralston *et al.*, 1985) and eluted with decamethonium (0.15M). Further purification, achieved by a second procainamide chromatography, yielded an enzyme with a specific activity of ca. 6,000 U/mg at a 50-60% recovery.

Analysis of rHuAChE : AChE activity in cell growth medium was assayed (Ellman *et al.*, 1961) at 27°C and monitored by a Thermomax microplate reader. AChE protein mass was determined by a specific ELISA (Shafferman *et al.*, 1992a). Sucrose gradients, SDS PAGE and immunoblot analysis were performed as described previously (Velan *et al.*, 1991a). N-Glycanase (0.62 Units; Genzyme) treatment was performed on 30µg denatured AChE. Amino terminus of HuAChE was determined by subjection to 12 cycles of Edman degradation.

Pulse chase : Metabolic labeling of methionine-starved recombinant 293 cells (30mm dish confluent cultures) was achieved by incubation for 1 h with ^{35}S-methionine (100 µCi/ml; 100Ci/mmol; Amersham) followed by chase in methionine containing DMEM (Gibco) . At various chase periods, medium was collected and cells were lysed in phosphate-buffered saline containing 0.5% Nonidet P-40, 0.2M iodoacetamide, 1mM PMSF and 100 U/ml aprotinin. rHuAChE was immunoprecipitated from cleared lysates and from medium with rabbit anti-rHuAChE antibodies and protein-A bacterial adsorbent (Biomakor). Following non-reducing SDS PAGE and fluorography, bands intensity was determined by densitometry. Endoglycanase H treatment (0.3 IU/sample) was performed on AChE aliquots eluted from the adsorbent.

RESULTS AND DISCUSSION

High Level Expression of Recombinant HuAChE in Stably Transfected 293 Cells

Stable colonies of transfected 293 cells were generated by using plasmid vectors in which *ache* expression was controlled by the cytomegalovirus immediate early promoter-enhancer while the *neo* selector gene was driven by the low efficiency promoter of H_2L^d (Kronman *et al.*, 1992). About 700 colonies were screened for AChE production of which 13 were further propagated to establish stable clones. AChE secretion in the selected clones reached a level of 50-150 mU AChE per 10^6 cells per 24 hrs, which amounts to 8-25 pg AChE per cell. It should be mentioned that this high value (ca. 10% of total protein produced per cell) was reached without resorting to gene amplification.

Three clones, designated R7,R8 and R11 were evaluated for adaptation to various anchorage-dependent large scale propagation systems. The most effective systems, relies on propagation of R11 cells in multitray units or cultisphere microcarrier beads. The latter system results in productivity of 5mg/L/day (Lazar *et al.*, in preparation). Affinity chromatography was used to purify milligram amounts of rHuAChE from bench-scale cultures. Homogeneous enzyme preparations displaying the expected specific activity (6,000 units/mg-protein; Velan *et al.*, 1991a) were used for characterization of the rHuAChE. All catalytic features examined, including substrate specificity, selectivity to inhibitors, catalytic rate constants and response to increased substrate concentration suggest that the recombinant enzyme is an authentic AChE (Kronman *et al.*, 1992).

Heterogeneity in N-Glycosylation, N-Terminus Processing and Oligomerization of rHuAcE

Analysis of the purified enzyme by SDS-PAGE suggests that the rHuAChE is an oligomer of catalytic subunits which display heterogeneity in size. The 140kD band revealed at non-reducing condition resolves upon addition of ß-mercapthoethanol into two bands of 70kd and 67kd (Fig 1A), both recognized by antibodies directed against a peptide derived from the HuAChE catalytic subunit (Velan *et al.*, 1991b). The difference in size of the rHuAChE subunits appears to result from heterogeneity in glycosylation levels. Removal of N-linked carbohydrate side chains by digestion with N-Glycanase converts the two fuzzy rHuAChE bands into a single sharp band of 62kD which is recognized by the specific antibody (Fig 1A; lanes 3,6). Heterogeneity in glycosylation could account also for the heterogeneous band of primate brain AChE polypeptide (Brodbeck and Liao, 1992) and may be related to inefficient utilization of one of the glycosylation signals on the subunit molecule.

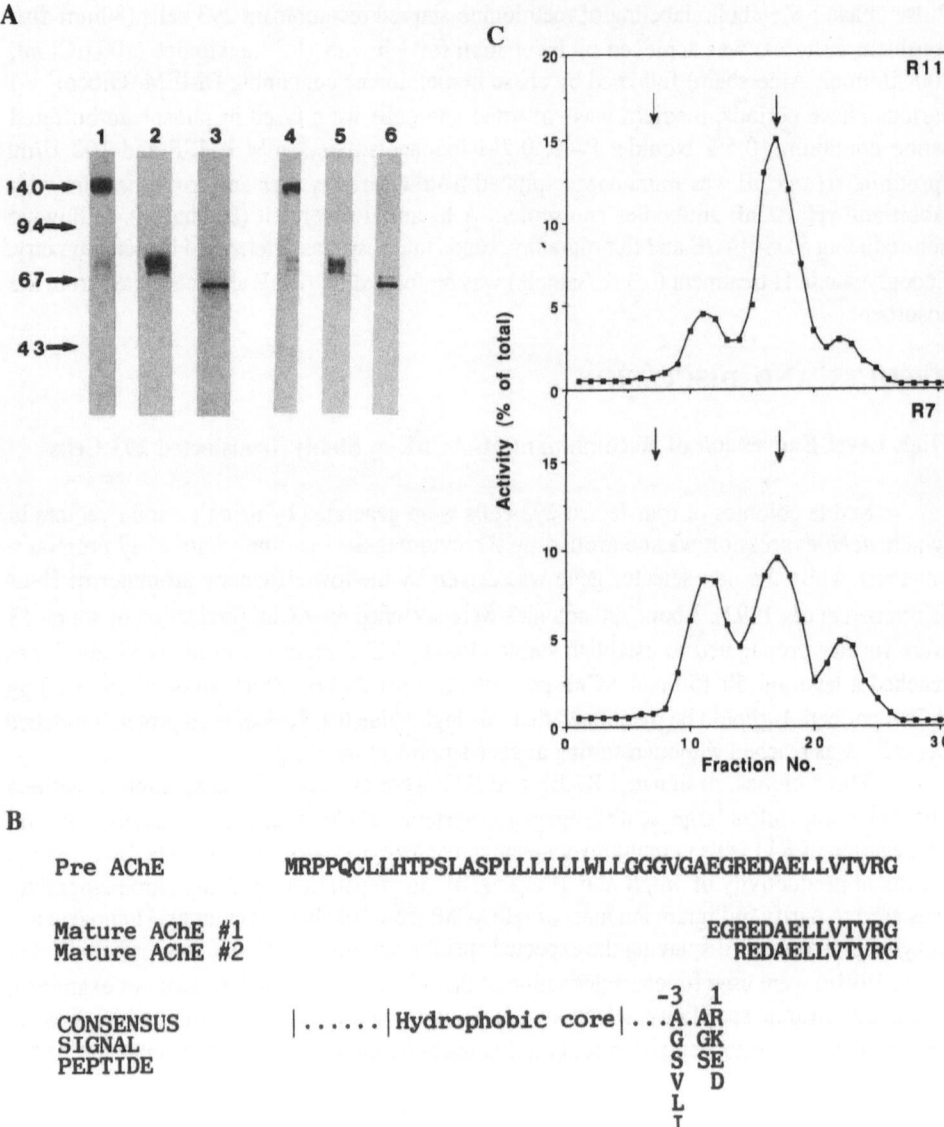

Figure 1. Molecular heterogeneity of rHuAChE: A). SDS-PAGE and immunoblot; B). N-terminus sequence; C). Sucrose gradient.

A). Lanes 1,2 and 3 represent Coomassie blue staining and lanes 4,5 and 6 represent immunoblots stained with anti-HuAChE antibodies. Lanes 1,4: purified non-reduced rHuAChE , lanes 2,5: purified reduced rHuAChE, lanes 3,6: reduced HuAChE treated with N-Glycanase. B). Pre-HuAChE N-terminal sequence is compared to that of the two mature forms. Signal peptidase consensus cleavage site is derived from Perlman and Halvorson (1983). C). Sucrose gradient profiles of HuAChE from the two stable lines R7 and R11 are compared. Alkaline phosphatase (right arrow; 6S) and β-galactosidase (left arrow; 16S) serve as sedimentation markers.

Analysis of the N-terminus sequence of mature HuAChE by Edman degradation reveals two populations of molecules (Fig. 1B) differing in length by two amino acid residues. Comparison of the two HuAChE sequences to the consensus sequence of signal-peptidase cleavage sites suggests that both molecules could have been generated by legitimate signal processing events. Indeed, the same alternative processing of the HuAChE polypeptide was also identified in enzyme isolated from human erythrocyte AChE (Haas and Rosenberry, 1985). Interestingly, the stoichiometry in the erythrocyte enzyme (0.66 Glu/ 0.34 Arg) is similar to that of the two AChE populations revealed in the recombinant system (Kronman *et al.*, 1992). A possible regulatory role for coexisting alternative processing in AChE has yet to be elucidated. It should be noted, however, that in the bovine enzyme a similar pattern of processing would be prevented by the presence of a proline (Doctor *et al.*, 1990) instead of the arginine at amino-acid position 3 in the mature polypeptide (Fig 1B).

Another aspect of polymorphism in the recombinant enzyme is related to extent of oligomerization (Velan *et al.*, 1991a). Molecular forms, whose sedimentation coefficients correspond to monomers, homodimers and homotetramers were detected in all tested preparations , yet the relative ratios between these species varied significantly (Fig. 1C). The relative amount of G4 appears to be inversely related to production rate, whereas the relative amount of monomers is a function of the culture age and is probably related to proteolytic disassembly of dimers (Velan *et al.*, 1991b).

rHuAChE Subunits are Efficiently Assembled and Secreted from 293 Cells

Assembly of rHuAChE into oligomers was monitored by metabolic labeling of a recombinant 293 cells (line R11) by [^{35}S]-methionine followed by chase, AChE immunoprecipitation and non-reducing SDS-PAGE. Newly synthesized monomers were found to assemble in the cell into oligomers, and the oligomeric forms were secreted to the medium (Fig. 2A). The small fraction of monomeric forms detected in the medium may result from degradation of the extracellular material, since addition of the protease inhibitor aprotinin leads to reduction in their amount (data not shown).

Monitoring the secretory pathway of rHuAChE from 293 cells by endoglycanase H analysis (Fig 2A), lectin binding experiments and use of transport blockers (Kerem *et al.*, in preparation) suggest that subunit dimerization occurs in the endoplasmic reticulum (ER). Oligomers are then shuttled through the various cisternae of the Golgi apparatus, while they undergo terminal glycosylation and are subsequently secreted as mature oligomers. Dimerization and export from the ER appear to be coupled and are relatively slow events, whereas transport from the trans-Golgi is relatively rapid.

The overall secretory process of rHuAChE in 293 cells appears to be efficient and is characterized by the high recovery (70-80%) of the intracellularly synthesized rHuAChE in the medium (Fig 2B). This high recovery is significantly different from the 20% recovery observed in avian muscle cells (Rotundo, 1988). The difference between the recombinant system and the native system may be the reflection of a specific intracellular degradation mechanism which has evolved in excitatory cells (Rotundo *et al.*, 1989).

Elimination of Dimerization Signal Does Not Affect Secretion of rHuAChE

Coupling between AChE subunit oligomerization and secretion was tested by characterizing a HuAChE mutant in which cysteine 580 was replaced by alanine (C580A). Of the seven conserved cysteine residues in cholinesterase catalytic subunits (see sequence compilation in Gentry and Doctor, 1991) six were implicated in intrasubunit disulfide bonds (MacPhee-Quingly *et al.*, 1986; Lockridge *et al.*, 1987). The cysteine at the carboxy terminus of the polypeptide (Cys580 in HuAChE) was implicated in intersubunit bond formation by chemical modification studies (Lockridge *et al.*, 1979; Roberts *et al.*, 1991) and more recently by direct mutagenesis studies (Velan *et al.*, 1991b). Efficient expression of the rHuAChE C580A mutant monomers in the embryonal 293 cells as well as in the neuroblastoma cell line SK-N-SH indicates that AChE secretion is not restricted to assembled molecular forms.

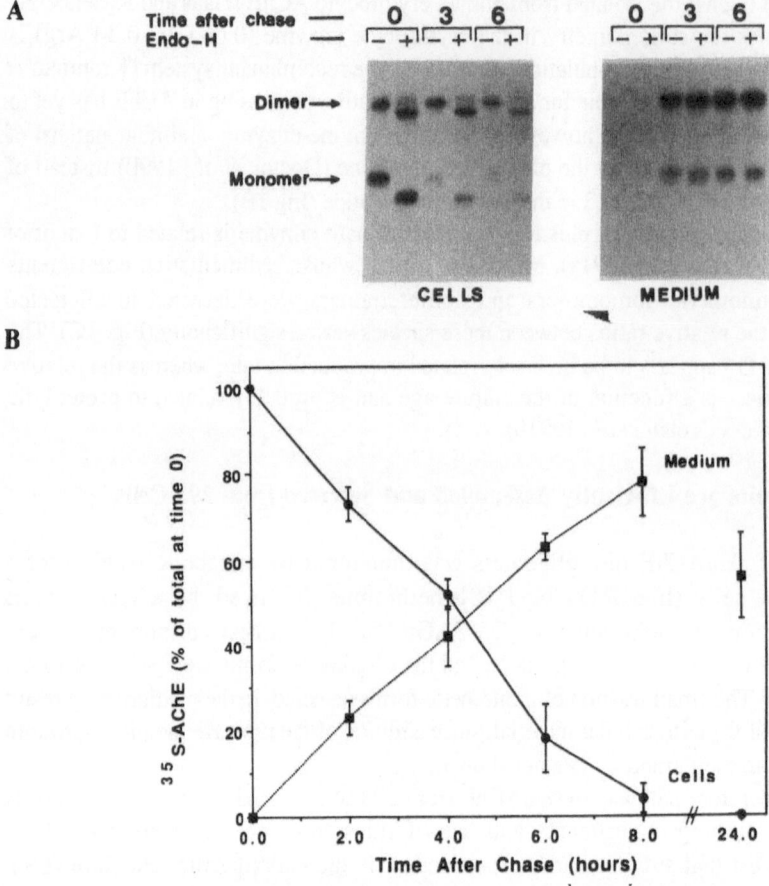

Figure 2. Assembly and secretion of rHuAChE from 293 cells.

A). rHuAChE immunoprecipitated from cell lysates or medium of metabolically labeled R11 cells was treated with (+) or without (-) endoglycanase H (Endo-H) and resolved by non-reducing SDS-PAGE. B). Recovery of HuAChE polypeptides in cell medium: rHuAChE was immunoprecpitated during chase from cells (circles) and medium (squares) and quantified by densitometry analysis. Each value represents an average of three experiments (±S.D).

Secretion of the C580A AChE monomers from transfected 293 cells was monitored by pulse chase and by differential lectin binding and compared to that of wild type. No change in secretion rates of mutant monomers was revealed by the pulse chase experiment (Fig. 3). Most of the intracellular monomers were exported from the cell within six hours and the recovery level appears to be similar to that of wild type recombinant oligomers. The rate determining steps in monomer processing are, as in wild type, those preceding terminal glycosylation. This is suggested by the large pool of intracellular, endoglycanase H resistant AChE molecules in both wild type and C580A AChE (Figs 2,3).

Moreover, no difference between the secretory profiles of wild type oligomers and mutant monomers was revealed by lectin blot analysis of intracellular and secreted enzyme (Kerem *et al.*, in preparation). Concanavalin A, which binds to terminal-mannose detected intracellular AChE in both mutant and wild-type, but did not stain extracellular material. Wheat Germ Agglutinin, which binds to terminal-GlcNAc, revealed in the both systems a minute intracellular pool but stained efficiently the extracellular material. The Ricinus agglutinin , which binds to terminal-galactose, stained exclusively the extracellular mutant and wild type HuAChE.

In summary, the monomeric mutant AChE subunit which lacks Cys580 can be shuttled through the entire secretory pathway at the same efficiency and at the same rate as the wild type multimeric form. In this respect HuAChE is different from many other multisubunit ecto-proteins in which interference with assembly leads to rapid ER degradation of the unassembled forms (For review see Hurtley and Helenius, 1989).

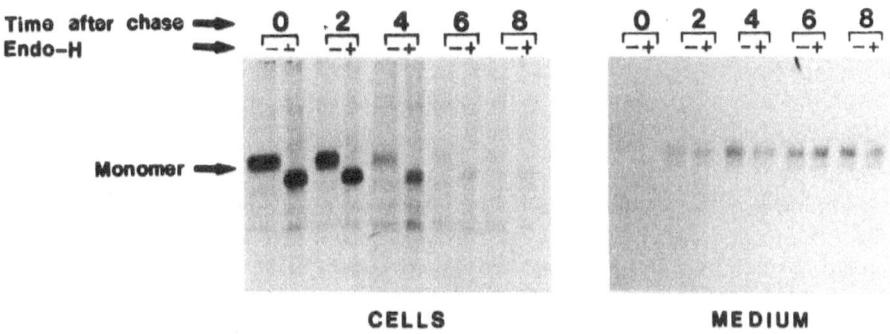

Figure 3. Secretion of C580A mutant rHuAChE from transfected 293 cells.

A 293 clone expressing mutant monomers (D9) was used in the pulse chase experiments similar to those described in Fig 2.

It appears that the secretory pathway does not exert its "quality control" function (Rose and Doms, 1988; Fra *et al.*, 1991) to abort AChE mutant monomer secretion. It should be noted that in secretory IgM and probably in other oligomeric proteins the domains involved in assembly are also those responsible for ER-retention when assembly is perturbed (Davis *et al.*, 1988; Sitia *et al.*, 1990; Gething and Sambrook, 1992). Whether the effective secretion of mutants monomeric AChE attests to the physiological secretion of wild type globular soluble AChE monomers remains to be elucidated. Nevertheless, the recent observation (Li *et al.*, 1991) that mammalian AChE mRNA can be spliced to generate an enzyme which lacks C-terminal cysteine could indicate that unassembled, non-tethered AChE catalytic subunits may have an *in-vivo* function.

N-Glycosylation is Required for Efficient Secretion of rHuAChE but Not for Catalytic Activity

The HuAChE catalytic polypeptide carries three putative signals for N-glycosylation (Soreq *et al.*, 1990) involving the asparagine·residues at positions 265, 350 and 464. The 3-dimensional model of HuAChE (Barak *et al.*, 1992) based on the TcAChE X-ray structure (Sussman *et al.*, 1991) locates the three glycosylation sites on the surface and suggests that all three are functional. DNA cassette replacement mutagenesis (Shafferman *et al.*, 1992a) was used to substitute each one of these asparagines thereby generating the AChE mutants N265Q, N350Q and N464Q. These single mutants were then used to engineer the double mutants [N265Q;N350Q], [N265Q;N464Q], [N350Q;N464Q] and the triple mutant

[N265Q;N350Q;N464Q], devoid of any N-glycosylation sites. All forms were integrated into multipartite expression vectors and were expressed in 293 cells.

Analysis of the AChE activity and AChE polypeptide mass by CAT normalized ELISA assays (Shafferman *et al.*, 1992a,b) was used to assess the involvement of glycosylation in AChE secretion and AChE activity. Blocking of a single site, two sites and all three sites resulted in secretion yields of 20-30%, 2-3% and 0.5% respectively. Interestingly, the decrease in secretion is independent of the site of mutation, and appears to relate to the overall glycosylation level (data not shown). In contrast to the significant effect on polypeptide secretion, blocking of N-glycosylation sites had no major effect on AChE catalytic activity. Specific activities calculated from the enzymatic activities of the various glycosylation mutants and the respective ELISA values (Shafferman *et al.*, 1992a) were found to be comparable to the wild type value of ca. 6 mU/ng. This observation conforms well with the fact that none of the surface-exposed, N-linked olygosaccharides are mapped in the vicinity of the active-site gorge rim.

ACKNOWLEDGEMENT

We thank Gila Friedman, Tamar Sery Nechama Zeliger and Yoel Papier for their excellent technical assistance. This work was supported by the U.S. Army Research and Development Command, under contract No. DAMD17 -89-C 9117

REFERENCES

Ausubel, M.F., Brent, R., Kingston,R.E., Moore, D.D., Smith, J.A., Seidman, J.G., and Struhl, K. (eds). 1987. *"Current Protocols In Molecular Biology"* Wiley Interscience. New-York.

Barak, D., Ariel, N.,Velan, B.,and Shafferman, A. 1992. This volume.

Brodbeck, U. and Liao, J. 1992. This volume.

Chatonnet, A. and Lockridge, O. 1989. *Biochem. J.* 260:625-634.

Davis, A.C., Roux, K.H. and Shulman, M.J. 1988. *Eur.J.Immunol.* 18:1001-1008.

Doctor, B.P., Chapman, T.C., Christner, C.E., Deal, C.C., De La Hoz, M.K., Gentry, R.K., Orget, R.A., Rush, R.S., Smyth, K.K. and Wolfe, A.D.1990. *FEBS Let.* 266:123-127.

Ellman, G.L., Courtney, K.D., Andres, V. and Featherstone, R.M. 1961. *Biochem. Pharmacol.* 7:88-95.

Fra, A.M., Alberini, C., Bet, P., Finazzi, D., Valetti, C. and Sitia, R. 1991. *Ann. Biol. Clin.* 49:283-286.

Gentry, M.K. and Doctor, B.P. 1991, in *Cholinesterases: Structure, Function, Mechanism, Genetics, and Cell Biology*. (Eds Massoulie, J., Bacou, F., Barnard, E.A. Doctor,B.P. and Quinn, D.M.) Am. Chem. Soc. Washington pp. 394-398.

Gething, M.-J. and Sambrook, J. 1992. *Nature* 355:33-45.

Haas, R. and Rosenberry T.L. 1985. *Analyt. Biochem.* 148:154-162.

Hurtley, S.M. and Helenius, A. 1989. *Annu. Rev. Cell Biol.* 5:227-307.

Inestrosa, N., Roberts, W.L., Marshall, T.L. and Rosenberry, T.L. 1987. *J. Biol. Chem.* 262:4441-4444.

Krejci, E., Coussen, F., Duval, N., Chatel, J-M., Legay, C., Puype, M., Vandekerckhove, J., Cartaud, J., Bon, S. and Massoulie, J. 1991. *EMBO J.* 10:1285-1293.

Kronman, C.,Velan, B., Gozes, Y., Leitner, M., Flashner, Y., Lazar, A., Marcus, D., Sery, T., Papier, A., Grosfeld, H., Cohen, S. and Shafferman, A. 1992. *Gene.* in press.

Li, Y., Camp, S., Rachinsky, T.L., Getman, D. and Taylor, P. 1991. *J. Biol. Chem.* 266:23083-23090.

Lockridge,O., Eckerson, H.W. and La Du, B.N. 1979. *J. Biol. Chem.* 254:8324-8330.

Lockridge, O., Adkins, S. and La Du, B.N. 1987. J. *Biol. Chem.* 262:12945-12952.

MacPhee-Quigley, K., Vedvick, T.S., Taylor, P. and Taylor, S.S. 1986. *J. Biol. Chem.* 261:13565-13570.

Massoulie, J. and Bon, S. 1982. *Annu. Rev. Neurosci.* 5:57-106.

Perlman, D., and Halvorson, H.O. 1983. *J. Mol. Biol.* 167:391-409.

Ralston, J S., Rush, R.S., Doctor, B.P. and Wolfe, D. 1985. *J. Biol.Chem* 260: 4312-4318.

Roberts, W.L., Doctor, B.P., Foster, J.D. and Rosenberry T.L. (1991) *J. Biol. Chem.* 266:7481-7487.

Rose, J.K. and Doms R.W. 1988. *Annu. Rev. Cell. Biol.* 4:257-288.

Rotundo, L.R. 1988. *J. Biol. Chem.* 263:19398-19406.

Rotundo, R.L., Thomas, K., Porter-Jordan, K., Benson, R.J.J., Fernandez Valle, C. and Fine R.E. 1989. *J. Biol. Chem.* 264:3146-3152.

Schumacher, M., Maulet, Y., Camp, S. and Taylor, P. 1988. *J. Biol. Chem.* 263: 18979-18987.

Shafferman, A., Velan, B., Cohen. S., Leitner, S., and Grosfeld, H.1987. *J. Biol. Chem.* 262:6227-6237.

Shafferman, A. Kronman, C.,Flashner, Y., Leitner, M.,Grosfeld, H.,Ordentlich, A., Gozes, Y., Cohen,S.,Ariel, N., Barak, D., Harel, M., Silman, I., Sussman J.L., and Velan, B. 1992a. *J. Biol. Chem.* in press.

Shafferman, A., Velan, B., Ordentlich, A., Kronman, C., Grosfeld, H., Leitner, M., Flashner, Y., Cohen, S., Barak, D. and Ariel N. 1992b. Submitted.

Sikorav, J.L., Duval, N., Anselmet, A., Bon, S., Krejci, E., Legay, E., Osterlund, M., Reimund, B. and Massoulie, J. 1988. *EMBO J.* 7:2983-2993.

Silman, I. and Futerman, A.H. 1987. *Eur. J. Biochem.* 170;11-22.

Simonsen. C.C. and Levinson, A.D. 1983. *Proc. Natl. Acad. Sci.* USA 80, 2495-2499.

Sitia, R., Neuberger,M., Alberini, C., Bet, P., Fra, A.M., Valetti, C., Williams, G. and Milstein C. 1990. *Cell* 60:781-790.

Soreq, H., Ben-Aziz, R., Prody, C.A., Gnatt, A., Neville, A., Lieman Hurwitz, J., Lev- Lehman,E., Ginzberg, D., Seidman, S., Lapidot-Lifson , Y. and Zakut, H. 1990. *Proc. Natl. Acad. Sci.* USA 87:9688-9692.

Southern, P.J. and Berg, P. 1982. *J. Mol. Appl. Genet.* 1:327-341.

Sussman, J.L., Harel, M., Frolow, F., Oefner, C., Goldman, A, Toker, L. and Silman, I. 1991. *Science* 253: 872-879.

Velan, B., Kronman, C., Grosfeld, H., Leitner, M., Gozes, Y., Flashner, Y., Sery, T., Cohen, S., Benaziz, R., Seidman, S., Shafferman, A. and Soreq, H. 1991a. *Cell. Mol. Neurobiol.* 11:143-156.

Velan, B., Grosfeld, H., Kronman, C.,Leitner, M., Gozes, Y., Lazar, A., Flashner, Y., Marcus. D., Cohen, S. and Shafferman, A. 1991b.*J. Biol Chem.* 266:23977-23984.

Wigler, M., Silverstein, S., Lee, L.S., Pellicer, A., Cheng, Y. C. and Axel, R. 1977. *Cell* 11:223-232.

EXPRESSION AND REFOLDING OF FUNCTIONAL HUMAN

BUTYRYLCHOLINESTERASE FROM *E. COLI*

Patrick Masson [1], Steve Adkins [2], Philippe Pham-Trong [1], and
Oksana Lockridge [3]

[1] Centre de Recherches du Service de Santé des Armées
Unité de Biochimie, BP 87, 38702 La Tronche Cedex, France
[2] University of Michigan Medical School
Toxicology Department, Ann Arbor, MI, USA
[3] University of Nebraska Medical Center
Eppley Institute, Omaha, NE 68198-6805, USA

INTRODUCTION

Human butyrylcholinesterase (BuChE ; EC.3.1.1.8) is directed by a single gene [1-3]. The BCHE gene is polymorphic (cf. other papers in this book) and some of its alleles produce plasma BuChE allozymes with abnormal kinetic properties. Though no physiological function has yet been assigned to this enzyme, it is of pharmacological importance since plasma BuChE is involved in the hydrolysis of various ester-containing drugs [4]. Moreover this enzyme has proven to be an efficient scavenger to protect against organophosphate poisons [5].

For the purpose of studying structure-function relationships of BuChE, the human sugar-free enzyme was produced by recombinant DNA techniques. Human cDNA encoding a 28 aminoacid signal peptide and the 574 aminoacids of the mature BuChE was cloned into a plasmid vector and expressed in *E.coli*.

In this paper we describe the conditions to obtain functional BuChE and some properties of the active refolded enzyme.

MATERIALS AND METHODS

Bacterial Expression Vector

A full-length BuChE cDNA including the signal peptide was obtained by ligating the 0.9 kb PvuII/ BamHI fragment from cDNA clone Z13[6] to the 2 kb BamHI/EcoRI fragment of cDNA clone Z19. This 2.9 kb DNA fragment was ligated into the SmaI and EcoRI sites of the polylinker region of pGEM 3 Zf- (Promega, Madison, WI, USA) used as plasmid expression vector. The 29 aminoacids coding for the SP 6 promoter, the lac Z peptide, and the polylinker are on the 5' side of the cDNA. Expression of BuChE cDNA was under the control of the lac operon promoter.

Multidisciplinary Approaches to Cholinesterase Functions, Edited by
A. Shafferman and B. Velan, Plenum Press, New York, 1992

Expression of BCHE

Competent *E.coli* strain HB101 was transfected and transformants were selected by ampicillin resistance.

Transformed cells were grown at 30°C in 100 ml LB medium containing 0.1mg/ml ampicillin. Isopropyl-D-thiogalacto-pyranoside(1mM final) was added during log phase growth as inducer of the lac promoter to make BuChE fusion protein. Incubation was continued for 2-6 h.

Cells were pelleted by centrifugation. Periplasmic proteins were extracted from the cell pellet with 10 ml of 50 mM Tris/HCl pH 8.0 containing 1mM EDTA, 25% sucrose and 1mg/ml lysozyme. Inclusion bodies and membrane proteins were separated from cytoplasmic proteins by centrifugation [7].

Detection of BuChE Protein in Cell Fractions

The different cell lysate fractions (10 ml) were analyzed by non denaturing acrylamide gel electrophoresis [8], and SDS electrophoresis. The presence of BuChE protein was tested by Western blotting on nitrocellulose membrane followed by immunodetection. Four different mouse nonoclonal antibodies directed toward human plasma BuChE were used. Bound mAb were visualized with peroxidase-labelled sheep antibodies to mouse IgG. Non denaturing gels were also stained for BuChE activity using the method of Karnovsky and Roots [9] with butyrylthiocholine iodide as substrate.

Denaturation and Renaturation of BuChE

Cell lysate fraction proteins were denatured by 8 M urea in 50 mM Tris/HCl pH 8.5 containing 1 mM EDTA and 10 mM DTT for 30 min at 20°C.

To promote correct disulfide bond formation and to minimize protein aggregation process, renaturation was achieved in 5 ml by diluting 10-fold the samples in precooled 50 mM Tris/HCl pH 8.5 containing 10 mM EDTA, 2mM reduced glutathione, and 1mM oxidized glutathione [10]. Refolding reaction was conducted at 4, 10, and 25°C for 24-96 h. For each reaction, at the end of the incubation period, samples were concentrated to 100 µl on Centrisart 1 (Sartorius A.G., Gottingen, FRG).

Analysis of Refolded Enzyme

Refolded enzyme was analyzed on non denaturing slab gels stained for BuChE activity. Isoelectric point was determined by electrofocusing on Pharmacia Phast gels pH 3.5-9. Apparent molecular mass of active enzyme was estimated by polyacrylamide gradient gel electrophoresis and by disc-electrophoresis in multiple gels of different acrylamide concentrations. The enzyme was assayed in 0.067 M Na/K phosphate pH7.4 at 25°C with benzoylcholine (50 µM) as substrate.

RESULTS AND DISCUSSION

Transfected bacterial cells expressed BuChE protein as judged by reactions on Western blots with the 4 monoclonal antibodies against human BuChE. However, the recombinant BuChE was inactive. Yet, the misfolded rBuChE was capable of ligand binding : it bound to procainamide-gel and could be eluted by 10 mM decamethonium in 20 mM phosphate, pH 7.0.

Most of the immunoreative proteins were found in the inclusion body fraction. On SDS gels in the presence of β-mercaptoethanol, two major immunoreactive bands of 65 and 70 kDa were detected (Fig.1). Their sizes are consistent with size of the fusion protein and size of processed enzyme. Indeed, the expected size of the fusion protein subunit is 71 kDa,and the size of the sugar-free rBuChE in which the 29 aminoacids of the vector and the 28 aminoacids of the signal peptide have been cleaved off is 65 kDa. Minor fast-migrating

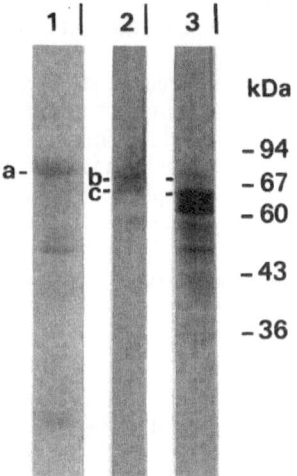

Fig.1
Immunodetection of rBuChE on Western blot after SDS- electrophoresis in the presence of β-mercaptoethanol.
Lane 1, human plasma BuChE (a, monomer : 85 kDa)
Lane 2, fresh inclusion body proteins (20 µl), b : 70 kDa ; c : 65 kDa ;
Lane 3, inclusion body proteins stored for 21 days at 4°C (50 µl).

bands were also detected ; they are presumably proteolytic degradation products because their intensity increases upon storage.

Since bacteria are known to be unable to form correct disulfide bonds, inactivity of the rBuChE may be due to "scrambled" protein structure. Urea unfolding and refolding of the protein in proper oxido/reducing conditions led to functional enzyme. Non denaturing gel analysis of the refolded enzyme after 24 h incubation in the refolding buffer showed that the amount of active refolded enzyme was maximun at 4°C and decreased with increasing temperature. Such an unusual temperature-dependent reaction is in accordance with the expected behavior for complex reactions like protein folding [11]. Intensity of active bands decreased as the incubation period increased ; this may be attributed to enzyme degradation by corefolded bacterial proteases. Four active bands were observed : the major band is the tetramer, minor forms are monomer, dimers and trimer (Fig.2). Active aggregates were also detected. The apparent size of the tetramer is 310 kDa. This overestimation of size suggests that rBuChE deviates from spherical shape. The total activity found in the inclusion body fraction was 8 µmol benzoylcholine hydrolyzed / h / 100ml culture

The pI of the refolded active enzyme forms were found to be 4.04 , 5.20 and 6.30, respectively. The pI of the major form was 6.30 ; other forms could be degraded enzyme.

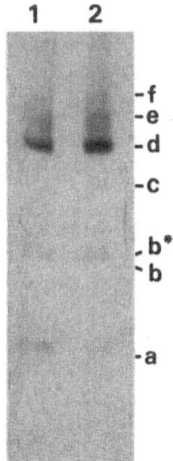

Fig.2
Polyacrylamide gradient gel electrophoresis (4-30%) of refolded rBuChE stained for activity
Lane 1, periplasmic fraction.(15 µl)
Lane 2,inclusion body fraction.(15µl)
a, monomer ; b, b*,dimers ;
c,trimer ; d, tetramer ; e,f,aggregates.

The theoretical pI calculated from aminoacid composition [12] using the PREDICT 89 computer program is 6.4 ; the pI of the fully glycosylated human plasma enzyme is 4.02 ± 0.04.

It is not yet known whether minor forms are quaternary structure active intermediates or tetramer degradation products. Likewise the overall yield of active enzyme obtained upon refolding has not been determined. Since activity with benzoylcholine is very low, the efficiency of the refolding reaction is presumably poor. This is not surprising since disulfide-bonded proteins and moreover oligomeric proteins have proven difficult to refold. So far, attempts to increase the recovery of active BuChE by cosolvent assisted enzyme refolding failed : incorporation of PEG 4000 (3mg/ml), PEG 3350-bis (p-nitrophenyl carbonate)[13] or PEG 3350-bis (procainamide) did not enhance enzyme refolding. Study of other folding helpers is in progress.

CONCLUSION

This work demonstrates that recombinant BuChE can be produced in *E. coli* and renatured to generate active tetramer. However, for a practical interest, the conditions of refolding have to be optimized. Nevertheless, two conclusions can be drawn : a) sugars are not essential for BuChE activity (but sugars are important for biological stability and circulatory lifetime of the plasma enzyme since they protect it against proteases) ; b) though BuChE is a disulfide-linked dimer of dimers composed of three-disulfide looped single domain subunits, its folding to the functionally active state is thermodynamically controlled.

ACKNOWLEDGMENTS

We wish to thank Dr. Jacques Grassi (CEN, Saclay, France) for the gift of monoclonal antibodies to human BuChE. This work was supported by US Army Medical Research and Development Command DAMD 17-91-Z-1003 to O.L.

REFERENCES

1. M. Arpagaus, M. Kott, K.P. Vatsis, C.F. Bartels, B.N. La Du and O . Lockridge, Structure of the gene for human butyrylcholinesterase. Evidence for a single copy, *Biochemistry*, 29: 124 (1990).
2. .P. W. Allderdice, H.A.R. Gardner, D. Galutira, O. Lockridge, B. N. La Du, and P.J. McAlpine. The cloned butyrylcholinesterase (BCHE) gene maps to a single chromosome site,3q26, *Genomics*, 11 : 452 (1991).
3. G. Gaughan, H. Park, J. Priddle, I. Craig, and S. Craig, Refinement of the localization of human butyrylcholinesterase to chromosome 3q 26.1-q26.2 using a PCR-derived probe. *Genomics*, 11 : 455 (1991).
4. O. Lockridge. Genetic variants of human serum cholinesterase influence metabolism of the muscle relaxant succinylcholine. *Pharmac.Ther.*, 47 : 35 (1990).
5. C. A. Broomfield, D.M. Maxwell, R.P. Solana, C.A. Castro, A.V. Finger and D.E. Lenz. Protection by butyrylcholinesterase against organophosphorus poisoning in nonhuman primates. *J. Pharm. Exp. Ther.*, 259 : 633 (1991)
6. C. McTiernan, S. Adkins, A. Chatonnet, T. A. Vaughan, C.F. Bartels, M. Kott, T.L. Rosenberry, B.N.La Du, and O.Lockridge. Brain cDNA clone for human cholinesterase. *Proc. Natl. Acad. Sci USA*, 84 : 6682 (1987).
7. J. Sambrook, E.F. Fritsch, and T.Maniatis, Expression of cloned genes in *Escherichia coli* (17.32), in "Molecular cloning", Cold Spring Harbor Laboratory Press, NY. (1989).
8. P. Masson, A naturally occuring molecular form of human plasma cholinesterase is an albumin conjugate. *Biochim. Biophys. Acta*, 988 : 258 (1989).
9. M. J. Karnovsky, and L. Roots, A "direct-coloring" thiocholine method for cholinesterase. *J. Histochem.Cytochem.* 12 : 219 (1964).
10. R. Halenbeck, E. Kawasaki, J. Wrin, and K.Koths, Renaturation and purification of biologically active recombinant human macrophage colony-stimulating factor expressed in *E. coli. Bio/Technol.*, 7 : 710 (1989).
11. T. E. Creighton, Protein folding. *Biochem. J.*, 240 : 1 (1990).
12. O. Lockridge, C. F. Bartels, T. A. Vaughan, C. K. Wong., S. E. Norton,and L. L. Johnson, Complete aminoacid sequence of human serum cholinesterase. *J. Biol. Chem.*, 262 : 549 (1987).
13. J.L. Cleland, and D.I. Wang, Cosolvent assisted protein refolding. *Bio/Technol.*, 8 : 1274 (1990).

PART 1: GENETIC VARIANT OF HUMAN ACETYLCHOLINESTERASE
PART 2: SV-40 TRANSFORMED CELL LINES, FOR EXAMPLE COS-1, BUT NOT PARENTAL UNTRANSFORMED CELL LINES, EXPRESS BUTYRYLCHOLINESTERASE (BCHE)

Oksana Lockridge[1], Cynthia F. Bartels[1], Teresa Zelinski[2], Omar Jbilo[1], and Morena Kris[1]

[1]Eppley Institute, University of Nebraska Med. Ctr.
600 S. 42nd St., Omaha, NE 68198-6805, USA
[2]Rh Laboratory, 735 Notre Dame Ave., Winnipeg, Manitoba
R3E OL8 Canada

INTRODUCTION

Until now no genetic variant of human acetylcholinesterase has been reported. This enzyme is considered essential to life and it was thought that genetic variants of acetylcholinesterase were incompatible with life. However, we have found a common polymorphism in human acetylcholinesterase, histidine 322 being changed to asparagine, in 5% of ACHE alleles of European and American populations. Furthermore, this genetic variation is associated with the YT blood group. We conclude that the YT blood group antigen is located on red blood cell acetylcholinesterase.

The literature shows only one cultured cell line that has BChE activity, the HuH-7 hepatoma cell line (Hada et al., 1987). We have found that SV40 transformed cell lines including COS-1 and COS-7 monkey kidney cell lines, and WI38 VA13 and MRC-5 SV40 human lung embryonal cells have significant levels of BChE both in the cell lysate and secreted into the culture medium. In contrast, the nontransformed parental cell lines CV-1, WI38, and MRC-5 have little or no BChE. The enzyme produced by the SV40 transformed cells was identified as BChE on the basis of the following: 1) the enzyme hydrolyzed benzoylcholine, butyrylthiocholine, and acetylthiocholine, 2) activity was inhibited by diisopropylfluorophosphate, iso-OMPA and eserine, but not by BW284C51, 3) metabolic labeling with S35-methionine followed by immunoprecipitation yielded an intense band of 85 kD, the correct size for the subunit monomer of BChE, 4) polyA+mRNA from the SV40 transformed cells hybridized with a BCHE cDNA probe on Northern blots. The mRNA band size of 2.5 kb was the same as the most intense BCHE band in human liver. The nontransformed parental cell lines showed no BCHE mRNA. Transfection of COS-1 cells with a plasmid containing human BCHE cDNA increased the activity of secreted BChE from 0.001 units/ml (5 ng/ml) to 0.002 units/ml (10 ng/ml). This suggests that the COS-1 system is not useful for studying transient expression of BChE mutants whose activity is low compared to wild-type BChE. The presence of BChE activity in SV40 transformed cells is reminiscent of the correlation found by Layer (1991) between rapidly proliferating cells and high BChE activity.

Multidisciplinary Approaches to Cholinesterase Functions, Edited by
A. Shafferman and B. Velan, Plenum Press, New York, 1992

53

METHODS

Amplification and Sequencing of ACHE

Genomic DNA was purified from blood that had been phenotyped for the YT blood group (Zelinski et al., 1991) as well as from random blood. Portions of the human ACHE gene were amplified by the polymerase chain reaction using oligonucleotide primers based on the cDNA sequence of human ACHE (Soreq et al., 1990). The human ACHE gene is 4.3 kb in length (Li et al., 1991) with introns of 350 bp, 1 kb, and 78 bp. For the purpose of determining the sequence of exon/intron junctions we amplified across introns and sequenced the amplified DNA. Primers based on the sequence of introns near exon/intron junctions were used to amplify the complete coding regions of individual exons. Each coding region of the ACHE gene was completely sequenced.

Analysis of BCHE in Cell Lines

Secreted BChE was collected into serumless culture medium to avoid complications by acetylcholinesterase in fetal calf serum when activity assays were to be done with acetylthiocholine or butyrylthiocholine as substrate. BChE was collected into culture medium containing 10% fetal calf serum, when activity assays were to be done only with benzoylcholine. Washed cells were lysed by addition of 1 ml of 1 M NaCl, freeze-thawed 3 times, and centrifuged to remove debris. BChE activity was measured by hydrolysis of benzoylcholine, butyrylthiocholine, and acetylthiocholine (Arpagaus et al., 1991). PolyA+mRNA was prepared with the FastTrack mRNA isolation system from Invitrogen Corp. (San Diego, CA). Probes for Northern blots were antisense cDNA synthesized by PCR (Bednarczuk et al., 1991) using as templates exons 2 and 4 of human BCHE (Arpagaus et al., 1990). Immunoprecipitation of S35-labeled BChE from COS-1 cells was with commercial polyclonal antibodies made against human BChE (AXL-155, Accurate Chemical and Scientific Corp.). The immunoprecipitated proteins were released from Protein-A Sepharose beads by boiling in buffer containing SDS and mercaptoethanol and loaded on 10% polyacrylamide gel containing 0.1% SDS.

RESULTS AND DISCUSSION

Acetylcholinesterase Genetic Variant

A nucleotide substitution in codon 322 was found in all samples phenotyped as carrying the rare YT2 blood group. The substitution changed histidine 322 (CAC) to asparagine (AAC). Table 1 shows results for samples that have been sequenced to date. The two people phenotyped as YT 1/1 (YT 1 is called Yta in earlier literature) were homozygous at codon 322 of ACHE, showing His 322 in both alleles. The two people phenotyped as YT 2/2 (YT 2 is called Ytb in earlier literature) were homozygous at codon 322 of ACHE, showing Asn 322 at both alleles. The two people phenotyped as YT 1/2, that is heterozygous for the YT blood group, were also heterozygous at codon 322 of ACHE, showing His 322 at one allele and Asn 322 at the second allele.

Table 1. Sequence of the human ACHE gene in 6 samples that have been phenotyped for the YT blood group, and in 9 random samples.

phenotype	number	ACHE sequence
YT 1/1	2	His 322
YT 1/2	2	His 322/Asn 322
YT 2/2	2	Asn 322
nd	8	His 322
nd	1	His 322/Asn 322

nd = not determined

Of the nine random samples we have sequenced, 8 were homozygous at codon 322 of ACHE, showing His 322 at both alleles, and one was heterozygous, showing both His and Asn at codon 322. Though the number of samples sequenced to date is small, the frequency of the Asn 322 allele in the above random samples is a very good match with the known frequency of the YT2 allele. Table 2 shows that 8 to 10% of Europeans and Americans have the heterozygous YT 1/2 phenotype.

The frequency of the YT blood group antigens has been determined for several populations and is shown in Table 2. These data suggest that the genetic variation in ACHE at codon 322 is a common polymorphism. Carriers of the rare allele constitute 26% of Israelis, 23% of Arabs, and 8 to 10% of Europeans, American Negroes, and Canadians.

Table 2. Frequency of YT blood group phenotypes in various populations.

Phenotype	Israelis[a]	Arabs[a]	Europeans[b]	American Negroes[c]	Canadians[d]
YT 1/1	0.7386	0.7647	0.9191	0.9160	0.8966
YT 1/2	0.2462	0.2117	0.0792	0.0822	0.1006
YT 2/2	0.0151	0.0235	0.0017	0.0018	0.0028

a. Levene et al (1987)
b. Giles et al (1967)
c. Wurzel and Haesler (1968)
d. Lewis et al (1987)

The association between the rare YT2 blood group phenotype and the ACHE genetic variation suggests that the YT antigen is located on red cell acetylcholinesterase. Furthermore, it suggests that the YT1 antigen includes His 322, while the YT2 antigen includes Asn 322.

We looked for a genetic variation in ACHE samples phenotyped as YT2 because a report by Spring and Anstee (1991) concluded that the YT blood group antigens are located on human erythrocyte AChE. This conclusion was based on the following experimental results. Immunoprecipitation with anti-YT antibodies from intact radioiodinated red cells yielded bands of 160,000 (nonreducing) and 72,000 (reducing conditions) on SDS gel electrophoresis. Immunoprecipitation with monoclonal anti-AChE (AE-1 and AE-2) also gave bands of 160,000 and 72,000. Furthermore, anti-YT immune precipitates contained AChE activity.

From the work of Spring and Anstee (1991) it was not known whether the YT antigen was a sugar on AChE, or a part of the glycolipid anchor of AChE, or a portion of the amino acid sequence. Our sequencing results show that amino acid 322 of human acetylcholinesterase is asparagine in the rare YT2 blood group, but is histidine in the common YT1 blood group. This suggests that amino acid 322 of AChE is a component of the YT blood group antigen.

Additional evidence supporting the conclusion that AChE contains the YT blood group antigen is the fact that both the YT blood group (Zelinski et al., 1991) and ACHE (P. McAlpine, personal communication; P. Taylor, personal communication) map to human chromosome 7q22. Furthermore, in paroxysmal nocturnal hemoglobinuria, a disease wherein the red cells are depleted of glycolipid anchored proteins, both AChE and the YT blood group are depleted (Telen et al., 1990). A final argument is the observation that red cells treated with papain lose the YT antigen (Eaton et al., 1956; Vengelen-Tyler and Morel, 1983). This is consistent with the observation that papain both releases and destroys red cell AChE (Dutta-Choudhury and Rosenberry, 1984).

The antigen in many blood groups includes a carbohydrate. In the example of the MN and Ss system, the antigen includes both an amino acid change in glycophorin and a carbohydrate (Anstee, 1990). Our results do not rule out the possibility of carbohydrate involvement in the YT blood group antigen. However, the His 322 to Asn mutation does not directly affect carbohydrates, since it neither deletes nor adds a carbohydrate addition site. The amino acid sequence His-Gly-Leu becomes Asn-Gly-Leu, and neither sequence corresponds to the consensus sequence for Asn linked sugars, Asn-X-Ser/Thr.

The significance of the conclusion that AChE contains the YT blood group antigen lies in the fact that anti-YT blood group antibodies are anti-AChE antibodies. The risk to health of having anti-AChE antibodies is not known, but could include neurologic effects. Anti-YT antibodies are found in people who have undergone multiple blood transfusions (Eaton et al., 1956; Ikin et al., 1965). Thus, the population to study for effects on health of anti-AChE antibodies is the group that has received multiple transfusions.

Our finding has implications for the use of recombinant AChE to treat intoxication by organophosphate esters. If the genotypes of the recombinant AChE and the recipient AChE do not match, then the recipient will develop anti-AChE antibodies. The presence of these antibodies will lead to agglutination and lysis of donor red cells (Bettigole et al., 1968; Dobbs et al., 1968; Gobel et al., 1974; Ballas and Sherwood, 1977; Davey and Simpkins, 1981; Mohandas et al., 1985; Levy et al., 1988), should the recipient require a blood transfusion at a later time.

Butyrylcholinesterase in SV40 Transformed Cell Lines

When we were doing transient expression studies of human BChE in COS-1 cells, we noticed that untransfected COS-1 cells had a significant background level of BChE activity, hydrolyzing 0.0003 to 0.004 µmoles benzoylcholine per min per ml of culture medium. This was surprising since no established cell lines were known that expressed BChE, with the exception of the HuH-7 hepatoma cell line which secreted so little activity that the culture medium had to be concentrated 1000 fold before activity was detectable (Hada et al., 1987). Another cell line that had been reported to have BChE activity, an Epstein Barr virus transformed B lymphoblastoid line (Rubinstein et al., 1984), is no longer available either from the investigators or from the American Type Culture Collection. The observation of background levels of BChE activity in untransfected COS-1 cells ruled out the use of the COS cell expression system for studying BChE mutants having low activity. Transfection of COS-1 cells with a plasmid encoding wild-type BCHE only doubled the activity secreted into the culture medium.

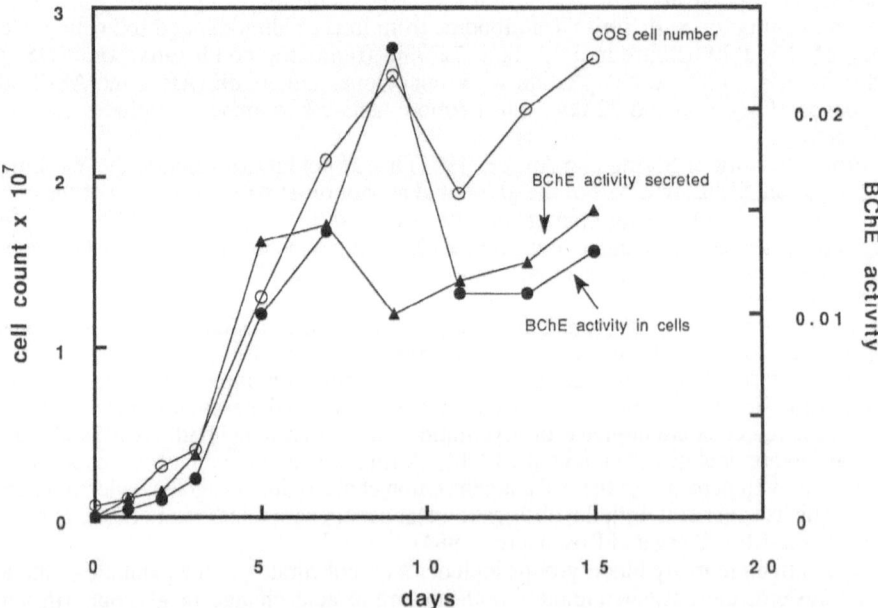

Fig. 1. Butyrylcholinesterase activity as a function of number of COS-1 cells.
COS-1 cells were seeded into twenty 60 mm dishes, at 700,000 cells per dish. Each dish received 5 ml culture medium containing 10% fetal calf serum. The cells were confluent on day 2 when there were 3 million cells. Culture medium was replaced on days 3, 5, 7, 9, 11, and 13. Two dishes were harvested on day 0, 1, 2, 3, 5, 7, 9, 11, 13, and 15 for counting of cells, and for determination of rate of benzoylcholine hydrolysis. Cell counts are for the total cells in one 60 mm dish. Activity in culture medium is the total for 5 ml medium. Activity in cell lysate is the total for the cells in one dish.

In transient expression studies the culture medium and COS-1 cells are harvested 48 to 72 hours after transfection. Fig. 1 shows that on days 2 and 3 the background level of BChE activity from untransfected COS-1 cells was approximately 0.0013 units per 5 ml culture medium. The activity increased ten fold by day 7. The number of cells also increased approximately ten fold in this time period to 30 million cells per 60 mm dish. BChE activity was distributed equally in the culture medium and in the cell lysate. Fig. 1 suggests that the amount of BChE activity is directly proportional to the number of cells. Based on these results we harvested cells for mRNA isolation when the cell density was approximately 30 million per dish, 7 to 15 days after cells had become confluent.

COS cells are African green monkey kidney cells that have been transformed with SV40 large T antigen. We wondered whether transformation with SV40 large T antigen was generally associated with BChE expression. For this purpose we obtained two other SV40 transformed cell lines, MRC-5 SV40 and WI38 VA13, from human embryonal lung. We also obtained the nontransformed parental cell lines, CV-1 from African green monkey kidney cells, and MRC-5 and WI38 from human lung cells. All cell lines were tested for BChE activity and for the presence of BChE mRNA. Table 3 shows that the SV40 transformed cell lines had BChE activity, but the parental nontransformed cell lines had no BChE activity in the case of MRC-5 and WI38, and a very low level of activity in CV-1.

Table 3. SV40 transformed cell lines have BCHE activity

Cell line	Description	BChE activity* secreted	in cells
COS-1	Monkey kidney transformed with SV40	0.04 units/ 10 ml	0.04 units per confluent dish
CV-1	Parental, nontransformed monkey kidney	0.001 units/10 ml	0.0002 units per confluent dish
WI38 VA 13	Human embryonic lung transformed with SV40	0.05 units/ 10 ml	0.05 units per confluent dish
WI38	Parental, nontransformed embryonic lung	0	0
MRC-5 SV40	Human embryonic lung fibroblast transformed with SV40	0.05 units/ 10 ml	0.05 units per confluent dish
MRC-5	Parental, nontransformed embryonic lung	0	0

* Assayed with benzoylcholine: 0.05 units = 250 nanograms BChE protein
The lower limit of detection is about 0.0003 units/10 ml.

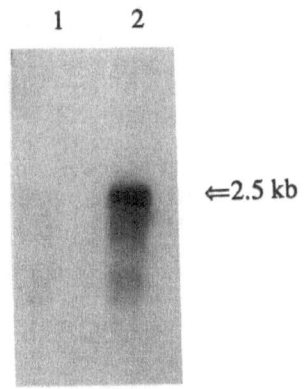

1   2

⇐2.5 kb

Fig. 2. Northern blot analysis of BCHE in COS-1 and CV-1 cells. Lane 1, 12 μg poly(A)+mRNA from CV-1 cells; Lane 2, 12 μg poly(A)+mRNA from COS-1 cells. The blot was hybridized with a BCHE cDNA probe for exon 2.

Northern blots probed with human BCHE cDNA showed a band of 2.5 kb for polyA+mRNA from the SV40 transformed cell lines, COS-1, MRC5-SV, and WI38 SV, but no band for poly A+mRNA from the nontransformed parental cell lines, CV-1, MRC-5, and WI38. See Fig. 2. Human liver mRNA also gave a band at 2.5 kb. The presence of BCHE mRNA in the SV40 transformed cells supports the conclusion that these cell lines express BChE.

Our finding that BChE is expressed in cells that have the SV40 large T antigen is intriguing in relation to Paul Layer's hypothesis that BChE has a role in cell proliferation (Layer, 1991). Layer measured cholinesterase activity in the developing chicken nervous system by histological staining and measured DNA synthesis by 3H-thymidine uptake. He found a strong correlation between BChE activity and cellular proliferation (Layer, 1991; Layer and Sporns, 1987). SV40 transformed cells are constantly dividing as they have lost normal growth control. Thus we conclude that SV40 transformed cells show the same correlation between BChE activity and cell proliferation that has already been noted by Layer in the avian nervous system.

ACKNOWLEDGEMENT

Supported by US Army Medical Research Development Command DAMD17-91-Z-1003 (to O.L.), Grant MT3391 from MRC Canada (to T.Z), American Cancer Society SIG16, and Laboratory Cancer Research Center Support Grant P30 CA36727

REFERENCES

Anstee, D.J., 1990, Blood group-active surface molecules of the human red blood cell. *Vox Sang.* 58: 1-20.

Arpagaus, M., Kott, M., Vatsis, K.P., Bartels, C.F., La Du, B.N., and Lockridge, O., 1990, Structure of the gene for human butyrylcholinesterase. Evidence for a single copy. *Biochemistry* 29: 124-131.

Arpagaus, M., Chatonnet, A., Masson, P., Newton, M., Vaughan, T.A., Bartels, C.F., Nogueira, C.P., La Du, B.N., and Lockridge, O., 1991, Use of the polymerase chain reaction for homology probing of butyrylcholinesterase from several vertebrates. *J. Biol. Chem.* 266: 6966-6974.

Ballas, S.K., and Sherwood, W.C., 1977, Rapid *in vivo* destruction of Yt(a+) erythrocytes in a recipient with anti-Yta. *Transfusion* 17: 65-66.

Bednarczuk, T.A., Wiggins, R.C., and Konat, G.W., 1991, Generation of high efficiency, single-stranded DNA hybridization probes by PCR. *BioTechniques* 10: 478.

Bettigole, R., Harris, J.P., Tegoli, J., and Issitt, P.D., 1968, Rapid *in vivo* destruction of Yt(a+) red cells in a patient with anti-Yta. *Vox Sang* 14: 143-146.

Davey, R.J., and Simpkins, S.S., 1981, 51Chromium survival of Yt(a+) red cells as a determinant of the *in vivo* significance of anti-Yta. *Transfusion* 21: 702-705.

Dobbs, J.V., Prutting, D.L., Adebahr, M.E., Allen, F.H., and Alter, A.A., 1968, Clinical experience with three examples of anti-Yta. *Vox Sang* 15: 216-221.

Dutta-Choudhury, T.A. and Rosenberry, T.L., 1984, Human erythrocyte acetylcholinesterase is an amphipathic protein whose short membrane-binding domain is removed by papain digestion. J. Biol. Chem. 259: 5653-5660.

Eaton, B.R., Morton, J.A., Pickles, M.M., and White, K.E., 1956, A new antibody, anti-Yta, characterizing a blood-group antigen of high incidence. *Brit. J. Haemat.* 2: 333-341.

Giles, C.M., Metaxas-Buhler, M., Romanski, Y., and Metaxas, M.N., 1967, Studies on the Yt blood group system. *Vox Sang* 13: 171-180.

Gobel, U., Drescher, K.H., Pottgen, W., and Lehr, H.J., 1974, A second example of anti-Ytq with rapid *in vivo* destruction of Yt(a+) red cells. *Vox Sang* 27: 171-175.

Hada, T., Yamamoto, T., Imanishi, H., Takahashi, S., Amuro, Y., Higashino, K. and Sato, J., 1987, Novel cholinesterase expression in the HuH-7 cell line. *Tumor Biology* 8: 3-8.

Ikin, E.W., Giles, C.M., and Plaut, G., 1965, A second example of anti-Ytb. *Vox Sang* 10: 212-213.

Layer, P., 1991, Cholinesterases during development of the avian nervous system. *Cell. Molec. Neurobiol.* 11: 7-33.

Layer, P.G., and Sporns, O., 1987, Spatiotemporal relationship of embryonic cholinesterases with cell proliferation in chicken brain and eye. *Proc. Natl. Acad. Sci. USA* 84: 284-288.

Levene, C., Bar-Shany, S., Manny, N., Moulds, J.J., and Cohen, T., 1987, The Yt blood groups in Israeli Jews, Arabs, and Druse. *Transfusion* 27: 471-474.

Levy, G.J., Selset, G., McQuiston, D., Nance, S.J., Garratty, G., Smith, L.E., and Goldfinger, D., 1988, Clinical significance of anti-YTb. Report of a case using a 51Chromium red cell survival study. *Transfusion* 28: 265-267.

Lewis, M., Kaita, H., Philipps, S., McAlpine, P.J., Wong, P., Giblett, E.R., and Anderson, J., 1987, The Yt blood group system (ISBT No. 011). Genetic studies. *Vox Sang* 53: 52-56.

Li, Y., Camp, S., Rachinsky, T.L., Getman, D., and Taylor, P., 1991, Gene structure of mammalian acetylcholinesterase. Alternative exons dictate tissue-specific expression. *J. Biol. Chem.* 266: 23083-23090.

Mohandas, K., Spivack, M., and Delehanty, C.L., 1985, Management of patients with anti-Cartwright (Yta). *Transfusion* 25: 381-384.

Rubinstein, H.M., Lubrano, T., Mathews, H.L., Lange, C.F., Silberman, S., Adams, E.M., and Minowada, J., 1984, A lymphocyte cell line that makes serum cholinesterase instead of acetylcholinesterase. *Biochem. Genet.* 22: 1171-1175.

Soreq, H., Ben-Aziz, R., Prody, C.A., Seidman, S., Gnatt, A., Neville, L., Lieman-Hurwitz, J., Lev-Lehman, E., Ginzberg, D., Lapidot-Lifson, Y., and Zakut, H., 1990, Molecular cloning and construction of the coding region for human acetylcholinesterase reveals a G+C-rich attenuating structure. *Proc. Natl. Acad. Sci. USA* 87: 9688-9692.

Spring, F.A., and Anstee, D.J., 1991, Evidence that the Yt blood group antigens are located on human erythrocyte acetylcholinesterase (AChE). *Transfusion Medicine* 1: Suppl. 2: 42.

Telen, M.J., Rosse, W.F., Parker, C.J., Moulds, M.K., and Moulds, J.J., 1990, Evidence that several high-frequency human blood group antigens reside on phosphatidylinositol-linked erythrocyte membrane proteins. *Blood* 75: 1404-1407.

Vengelen-Tyler, V., and Morel, P.A., 1983, Serologic and IgG subclass characterization of Cartwright (Yt) and Gerbich (Ge) antibodies. *Transfusion* 23: 114-116.

Wurzel, H.A., and Haesler, W.E., 1968, The Yt blood groups in American Negroes. *Vox Sang* 15: 304-305.

Zelinski, T., White, L., Coghlan, G., and Philipps, S., 1991, Assignment of the YT blood group locus to chromosome 7q. *Genomics* 11: 165-167.

KEYWORDS

genetic variant of acetylcholinesterase
histidine 322 of acetylcholinesterase
YT blood group
blood group
red cell acetylcholinesterase, agglutination
COS-1 cells
WI38 SV cells
MRC-5 SV40 cells
CV-1 cells
SV40 large T antigen
SV40 transformed cells
butyrylcholinesterase

HETEROGENEITY OF THE SILENT PHENOTYPE OF HUMAN BUTYRYLCHOLINESTERASE - IDENTIFICATION OF EIGHT NEW MUTATIONS

Sergio L. Primo-Parmo[1], Cynthia F. Bartels[2], Harold Lightstone[3], Abraham F.L. van der Spek[1], and Bert N. La Du[1]

[1]Departments of Pharmacology and Anesthesiology, University of Michigan, Ann Arbor, MI; [2]Eppley Institute, University of Nebraska, Omaha, NE; [3]Department of Anesthesiology, Albert Einstein Medical Center, Philadelphia, PA (USA)

INTRODUCTION

The silent phenotype of human butyrylcholinesterase (BChE; E.C. 3.1.1.8) is characterized by the absence or very low levels of enzyme activity. Heterogeneity of this phenotype was recognized soon after the first apparently homozygotes were identified (Goedde and Altland, 1968).

Immunological studies have suggested the existence of at least four different silent alleles (Whittaker et al., 1991). The serum samples studied differed both in the level of enzyme activity and the amount of immunoreactive protein. Studies at the DNA level have already established the structural alterations in four alleles. Nogueira et al.(1990) identified a frameshift mutation at Gly 117 (GGT ->GGAG). Muratani et al. (1991) described an insertion of an *Alu* element between positions nt 1062 and nt 1076 in exon 2 of the *BCHE* gene. Hidaka et al.(pers. commun.) identified two other silent alleles: one of them is characterized by a frameshift mutation at Thr 315 (ACC -> AACC), and the other shows a point mutation at amino acid 365 substituting arginine for glycine. No enzyme activity as well as no immunoreactive protein could be detected in the serum of individuals homozygous for the frameshift mutations and the *Alu* insertion. Very low levels of activity and protein were found in the serum of homozygotes for the other allele.

In order to further investigate the genetic heterogeneity of the silent phenotype we have sequenced the four exons of the *BCHE* gene (Arpagaus et al., 1990) of additional individuals carrying at least one silent allele. In this study we describe eight new mutations associated with this phenotype.

MATERIAL AND METHODS

Phenotype Determinations

Enzyme activity, dibucaine (DN) and fluoride (FN) numbers were determined in plasma according to Kalow and Genest (1957) and Harris and Whittaker (1961) using benzoylcholine as substrate. Rocket immunoelectrophoresis was performed according to Laurell (1966) in 1% agar gel containing 0.075% rabbit polyclonal antibody against

Multidisciplinary Approaches to Cholinesterase Functions, Edited by
A. Shafferman and B. Velan, Plenum Press, New York, 1992

human BChE. The samples showing non detectable enzymatic activity were mixed in equal proportions with a pool made of 10 individuals of usual phenotype (BCHE U). Some samples were diluted with gel buffer in order to achieve the level of activity shown by the least active sample. This procedure allowed us to load the same amount of enzyme units in the same volume. After electrophoresis, the heights of the rockets were compared with the one shown by the U pool. Rockets higher than this were assumed to contain immunoreactive protein coded by the silent allele.

DNA Samples

Individuals carrying at least one silent allele were selected from our laboratory collection from patients with increased sensitivity to succinylcholine. Whenever possible fresh blood samples were collected from the patients and their relatives. When this was not possible DNA was isolated directly from the original plasma samples. In both cases the DNA was extracted according to Muellenbach et al. (1989).

DNA Amplification and Sequencing

Genomic DNA was amplified by PCR according to McGuire et al. (1989). The amplification primers used allowed the amplification of the four *BCHE* exons as well as the exon-intron boundaries (Arpagaus et al., 1990). Direct sequencing of the amplified double-stranded DNA was performed using internal primers by the dideoxy method of Sanger et al. (1977).

RESULTS AND DISCUSSION

Individuals were assumed to carry at least one silent allele if: a) their enzyme activity levels were zero or extremely low; b) their phenotype as determined by the dibucaine and fluoride numbers was atypical but pedigree analysis showed that they could not be homozygous A/A; c) their phenotype was atypical but the DNA sequencing showed heterozygosity at the atypical variant site.

Samples corresponding to sixteen apparently unrelated probands were selected by the criteria above. The DNA structural alterations found in their silent *BCHE* alleles are listed in Table 1. In Table 2 are shown the phenotype characteristics of the probands.

Complete absence of enzymatic activity clearly would be expected for the polypeptide products coded by S-2 and S-11. The frameshift mutation in S-2 alters almost all the usual polypeptide chain sequence, whereas S-11 codes for a variant that lacks the essential serine residue at the active center. The stop codon caused by the nucleotide change in S-3 determines the synthesis of a polypeptide chain that is 13% shorter than the usual one. The missing C-terminus peptide is involved in two disulfide bridges, one intrachain (Cys 519) and one intersubunits (Cys 571) (Lockridge et al., 1987). All the other alleles show mutations that result in changes of single amino acids. These substitutions range from very conservative (S-9), to changes in charge (S-7, S-8), in hydrophobicity (S-10) to changes in the polypeptide secondary structure (S-4).

The S-7 allele codes for an unstable form of BChE. When the plasma sample from patient M.F. (homozygous S-7/S-7) was assayed in our laboratory no BChE activity could be detected although the same plasma sample was reported to show very low levels of activity when previously assayed at another laboratory. Two great-grandsons of patient M.F. (J.P. and T.P.) are heterozygous for S-7 and the atypical (A) allele. When their plasma samples were first assayed their DN and FN were 35 and 51 (T.P.) and 37 and 40 (J.P.), respectively. After two cycles of freezing and thawing, however, these values were 20 and 31 (T.P.) and 24 and 34 (J.P.) respectively. The changes in DN and FN, unusual for variants of BChE, clearly show the unstable character of this new variant. This is the first report of an unstable variant of BChE and it requires the examination of fresh plasma samples for its detection. Protein instability might also be the result of the mutation in S-4. Small amounts of immunoreactive protein could be detected in patient M.C. (heterozygous A/S-4) but not in H.B. (homozygous S-4/S-4). Unfortunately we were unable to get fresh blood samples from these individuals to test this hypothesis.

Table 1. DNA structural alterations found in the silent alleles.

Trivial Name	Alleles	DNA changes			Affected Pedigrees
S-1	BCHE*FS117[b]	nt 351	GGT → GGAG		1
S-2	BCHE*FS6	nt 16	ATT → _TT		2
S-3	BCHE*500STOP	nt 1500	TAT → TAA	(Tyr → STOP)	1
S-4	BCHE*37S	nt 109	CCT → TCT	(Pro → Ser)	2
S-5	BCHE*365R[c]	nt 1093	GGA → CGA	(Gly → Arg)	-
S-6	BCHE*FS315[c]	nt 944	ACC → AACC		-
S-7	BCHE*115D	nt 344	GGT → GAT	(Gly → Asp)	3
S-8	BCHE*471R	nt 1411	TGG → CGG	(Trp → Arg)	1
S-9	BCHE*170E	nt 509	GAT → GGT	(Asp → Glu)	2
S-10	BCHE*518L	nt 1553	CAA → CTA	(Gln → Leu)	1
S-11	BCHE*198G	nt 592	AGT → GGT	(Ser → Gly)	1
S-12	BCHE*INS1077[d]	nt 1077	*Alu* insert		-

[a] according to La Du et al. (1991); [b] Nogueira et al. (1990);
[c] Hidaka et al (pers. commun.); [d] Muratani et al. (1991).

Table 2. Phenotypic characteristics of the probands studied.

Probands	Enzyme activity[a]	DN	FN	Immunoreactive protein[b]	Genotype
E.V.	0.14	14	10	not detected	A/S-1
Y.S.	zero	-	-	not detected	S-2/S-3
J.D.	0.38	81	68	not detected	K/S-2
H.B.	zero	-	-	not detected	S-4/S-4
M.C.	0.12	8	10	++	A/S-4
M.F.	zero	-	-	+	S-7/S-7
M.P.	0.08	36	51	+	A/S-7
J.B.	0.19	17	20	not tested	A/S-7
J.G.	0.22	23	28	+	A/S-8
R.W.	zero	-	-	not detected	S-9/ S-?
P.P.	0.10	25	31	not detected	A/S-9
W.B.	0.07	78	59	++	U/S-10; ?/S-10
C.A.F.	0.16	30	25	+++	A/S-11
J.S.Jr	0.61	80	57	not tested	U/S-?
D.A.	0.32	26	-	not detected	A/S-?
A.D.	0.60	79	60	not tested	U/S-?

[a] normal values for the usual phenotype = 0.8 - 1.2 μmoles/min/ml; [b] the results express amount of immunoreactive protein in excess of what was expected from the BChE activity levels.

The S-8, S-9 and S-10 alleles were only detected in heterozygous state with the atypical allele. Immunoreactive protein in levels higher than expected by the enzyme activity could be clearly demonstrated in patients carrying the S-8 and S-10 alleles . Whether these variants present very low levels of BChE activity or none at all is difficult to determine due to the relatively broad ranges for the enzyme activity and DN and FN for the atypical phenotype.The complete absence of immunoreactive protein and enzyme activity in patient R.W. (heterozygous for S-9 and another unidentified silent allele) is not expected for a variant that shows a very conservative amino acid substitution. It is possible that this allele presents another mutation, probably in the regulatory regions of the BCHE gene not sequenced by us. This is also expected to occur in the BCHE alleles of three other patients in whom no mutations could be detected.

We intend to express the BCHE silent alleles described in the present study in order to characterize their effects on the structure of the enzyme. The large number of different mutations found in this small number of patients examined to date seems to indicate that the heterogeneity of the molecular basis of the silent phenotype is much greater than had been anticipated.

REFERENCES

Arpagaus, M., Kott, M., Vatsis, K.P., Bartels, C.F., La Du, B.N. and Lockridge, O., 1990, Structure of the gene for human butyrylcholinesterase. Evidence for a single copy, *Biochemistry* 29:124.

Goedde, H.W. and Altland, K., 1968, Evidence for different silent genes in the human serum cholinesterase polymorphism, *Ann. N.Y. Acad. Sci.* 151:540.

Harris, H. and Whittaker, M., 1961, Differential inhibition of human serum cholinesterase with fluoride:recognition of two new phenotypes, *Nature* 191:496.

Kalow, W. and Genest, K., 1957, A method for the detection of atypical forms of human cholinesterase:determination of dibucaine numbers, *Can. J. Biochem.* 35:339.

La Du, B.N., Bartels, C.F., Nogueira, C.D., Arpagaus, M. and Lockridge, O., 1991, Proposed nomenclature for human butyrylcholinesterase genetic variants identified by DNA sequencing, *Cell. Mol. Neurobiol.* 11:79.

Laurell, C.B., 1966, Quantitative estimation of protein by electrophoresis in agarose gel containing antibodies, *Anal. Biochem.* 15:45.

Lockridge, O., Bartels, C.F., Vaughan, T.A., Wong, C.K., Norton, S.E. and Johnson, L.L., 1987, Complete amino acid sequence of human serum cholinesterase, *J. Biol. Chem.* 262:549.

McGuire, M.C., Nogueira, C.P., Bartels, C.F., Lightstone, H., Hajra, A., van der Spek, A.F.L., Lockridge, O. and La Du, B.N., 1989, Identification of the structural mutation responsible for the dibucaine-resistant (atypical) variant form of human serum cholinesterase, *Proc. Natl. Acad. Sci. USA* 86:953.

Muellenbach, R., Lagoda, P.J.L. and Welter, C., 1989, An efficient salt-chlorophorm extraction of DNA from blood and tissues. *T.J.G.* 5:391.

Muratani, K., Hada, T., Yamamoto, Y., Kaneko, T., Shigeto, Y., Ohue, T., Furuyama, J. and Higashino, K., 1991, Inactivation of the cholinesterase gene by *Alu* insertion: possible mechanism for human gene transposition, *Proc. Natl. Acad. Sci. USA* 88:11315.

Nogueira, C.P., McGuire, M.C., Graeser, C., Bartels, C.F., Arpagaus, M., van der Spek, A.F.L., Lightstone, H., Lockridge, O. and La Du, B.N., 1990, Identification of a frameshift mutation responsible for the silent phenotype of human serum cholinesterase, Gly 117 (GCT → GGAG), *Am. J. Hum. Genet.* 46:934.

Sanger, F., Nicklen, S. and Coulson, A.R., 1977, DNA sequencing with chain termination inhibitors, *Proc. Natl. Acad. Sci. USA* 74:5463.

Whittaker, M., Jones, J. and Braven, J., 1991, Heterogeneity of the silent gene for plasma cholinesterase: immunological studies, *Hum. Hered.* 41:77.

NEMATODE ACETYLCHOLINESTERASES:

SEVERAL GENES AND MOLECULAR FORMS OF THEIR PRODUCTS

Martine Arpagaus,[1,2] Patricia Richier,[1] Yann L'Hermite,[1] Florence Le Roy,[1] Jean Bergé,[2] Danielle Thierry-Mieg,[3] and Jean-Pierre Toutant[1]

[1]Différenciation cellulaire et Croissance, INRA Montpellier
[2]Biologie des invertébrés, INRA Antibes
[3]Centre de Recherche en Biologie Moléculaire, CNRS, Montpellier

INTRODUCTION

In vertebrates, acetylcholinesterase (AChE) and butyrylcholinesterase (BChE) are polymorphic enzymes presenting both globular and asymmetric forms[1]. In invertebrates, only AChE has been characterized so far that presents a reduced molecular diversity. In insects for example the major molecular form of AChE is an amphiphilic dimeric form[2,3] attached to the membrane through a glycolipid covalently linked at the C-terminus of each catalytic subunit[4,5,6]. This AChE has a substrate specificity intermediate to those of mammalian AChE and BChE[3,4]. A glycolipid-anchored 7.5S form has also been observed in the trematode *Schistosoma mansoni*[7]. Asymmetric forms have never been convincingly reported in invertebrates except in the more evolved animals such as *Amphioxius*[8]. In the latter case also there is no BChE but AChE presents catalytic properties intermediate to those of vertebrate AChE and BChE[8].

We are now interested in nematode AChE(s) for the following reasons:
-several species are agricultural pests and it is important to get further informations on the target of potential nematicides
-it has been shown that at least three different genes code for AChEs in *Caenorhabditis elegans*[9,10,11]. It is therefore interesting to see whether the presence of multiple genes results in an increased molecular diversity, to define what are the structural characteristics of each gene product and finally to clone and sequence the three genes for evolutionary relationships with the other members of the cholinesterase family.

MATERIALS AND METHODS

Steinernema carpocapsae were reared on their natural hosts *Galleria melonella. Caenorhabditis elegans* (N2 strain or wild type) as well as the

Multidisciplinary Approaches to Cholinesterase Functions, Edited by
A. Shafferman and B. Velan, Plenum Press, New York, 1992

65

mutants *ace-1*, *ace-2*, *ace-3* and *ace-1/ace-2* were reared on agar plates seeded with bacterial strain OP50.

For the study of molecular forms of AChEs, we used a combination of sucrose gradient centrifugation and non-denaturing electrophoreses as previously described[12]. The amphiphilic forms were characterized by using Brij 96 as well as Triton X100 in centrifugation and negatively-charged sodium deoxycholate in ND-PAGE[2,12]. Molecules anchored to the membranes by a glycosylphosphatidylinositol were detected by their sensitivity to PI-PLC from *Bacillus thuringiensis*[13]. The identification of different gene products was based on the differential sensitivity of AChE to eserine (Sigma). Labeling with ^{3}H-DFP (NEN) was achieved on crude extracts incubated for three hours at a final concentration of 10^{-6} mol/l (20 µCi of ^{3}H-DFP in 2ml of extract). Isolated fractions of sucrose gradients were also labeled in the same conditions. After extensive dialysis the labeled samples were analyzed by SDS-PAGE in reducing or non-reducing conditions.

RESULTS

The eserine sensitivity of acetylthiocholine-splitting activity in a crude extract of *Steinernema* is shown on figure 1A. The curve presents a plateau between 10^{-6}M and 5×10^{-4}M indicating the presence of at least two types of enzyme active on this substrate and that differ strikingly in their eserine susceptibility. The two components were isolated (see below) and their substrate specificity compared: both of them are most active on acetylthiocholine than on other thiocholine esters (1B). They are refered to as classes A and B AChEs.

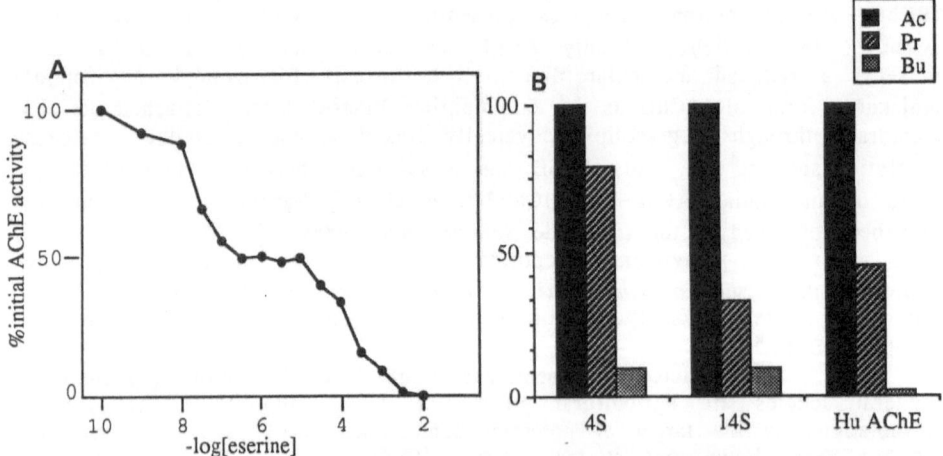

Figure 1. A: Inhibitory effect of eserine on the hydrolysis of acetylthiocholine by a direct DS extract of *Steinernema carpocapsae*. B: substrate specificity of class A AChE (14s form) and class B AChE (4s form). Ac, Pr, Bu: acetyl-, propionyl- and butyrylthiocholine iodides. Substrate specificity of human erythrocyte AChE is shown for comparison.

Molecular Forms of AChEs in *Steinernema*

Washed nematodes (1g) were sequentially extracted (as described previously[1,2]) in 5ml of low salt buffer, then in 5ml of low salt buffer containing 1% of Triton X100 and finally in 5 ml of high-salt buffer (1M NaCl). The solubilized fractions were refered respectively to as Low Salt-Soluble (LSS),

Detergent-Soluble (DS) and High Salt-Soluble (HSS). These extracts were first analyzed by sucrose gradient centrifugation in the presence of Triton X100 in the gradient and the peak fractions were run on non-denaturing PAGE containing either 0.5% Triton X100, or no detergent, or 0.2%Triton X100+0.2% sodium deoxycholate. Figure 2 shows the results of such an analysis of both LSS and DS extracts.

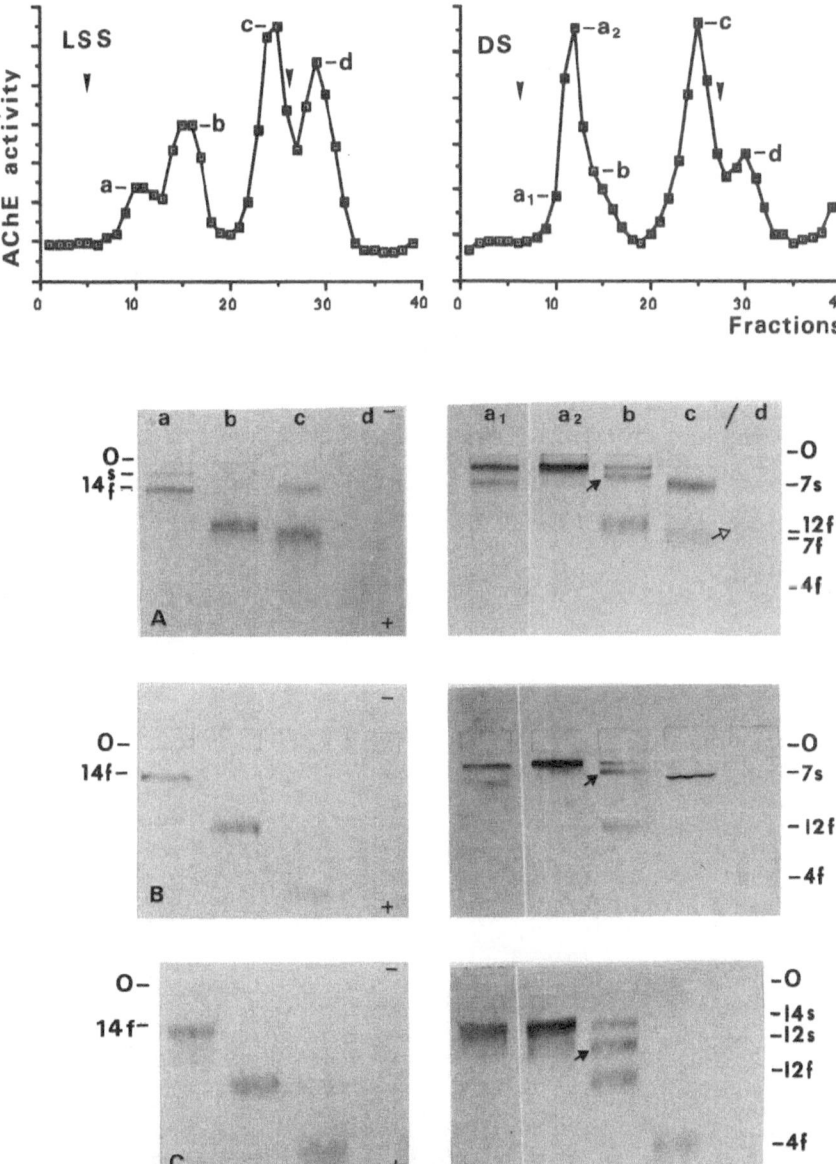

Figure 2. LSS and DS extracts were run on sucrose gradients containing 1% Triton X100. Black arrowheads indicate the position of β-galactosidase (16S, left) and alkaline phosphatase (6.1S, right). In each extract, the fractions labeled a to d were electrophoresed in non-denaturing PAGE containing 0.5% Triton X100 (A), no detergent (B) or 0.2% Triton X100+ 0.2% deoxycholate (C). Eight electromorphs were identified (1) according to the S value of the fraction in which they are found and (2) as s (slow-migrating) or f (fast-migrating). Black arrows show the 12s component in A, B and C and the open arrow shows a faint 4s band only observed in A.

Among the eight electromorphs observed, all slow components were amphiphilic. This is apparent on figure 2 for the 14s and 12s bands that present enhanced migrations in the presence of deoxycholate (C) and for the 7s band that shows a typical precipitation line in the absence of detergent (B). The amphiphilic nature of 14s, 7s and 4s forms was further demonstrated by the existence of a shift to lower S values in sedimentation in sucrose gradients containing Brij-96 (figure 3). The fast bands were hydrophilic (no interaction with non-denaturing detergents). It should be noted that HSS extracts contained

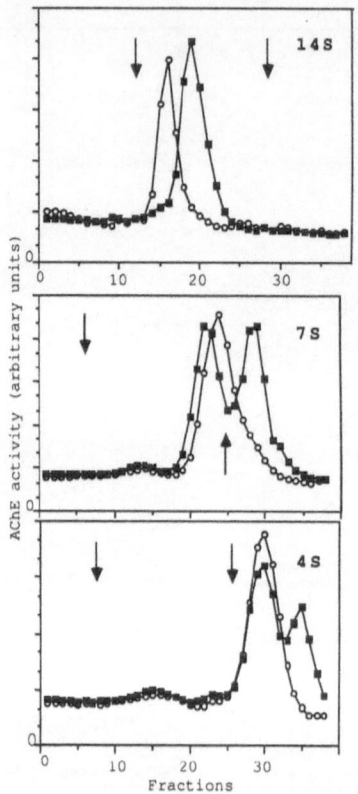

Figure 3. Analysis of isolated peak fractions on Triton X100- (O) and Brij 96-containing gradients (■). As previously reported[2], amphiphilic components shift to lower S values in the presence of Brij 96 than in Triton X100. This results from the lower density of Brij 96 compared to Triton X100. Hydrophilic molecules show identical S values in both conditions. The 14S peak was composed almost entirely of an amphiphilic form shifted from 13.7S to 11.8S. 7S and 4S peaks were mixtures of both amphiphilic and hydrophilic forms. 7s (amphiphilic) form was shifted from 6.5S to 3.8S and 4s form from 4.0S to 1.5S. Vertical arrows show the position of β-galactosidase (16S) and alkaline phosphatase (6.1S).

no extra forms that could be related to the asymmetric, "collagen-tailed" forms of vertebrate cholinesterases.

Eserine sensitivity of isolated sedimentation peaks. Both 14S and 12S peaks (14s, 14f, 12s, 12f components) were highly sensitive to eserine ($I50 = 5 \times 10^{-8}$M, monophasic curve, class A AChE) whereas the 4S peak (4s and 4f) was less sensitive ($I50 = 10^{-4}$M, monophasic curve, class B AChE). The inhibition of the 7S peak presented a plateau similar to that in figure 1A. Further analyses showed that the native 7f form was from class A and the native 7s form from class B. By comparison to *C. elegans* AChEs, class A and class B were assumed to be the products of two different genes: ace-1 and ace-2 respectively.

Treatments by proteases and PI-PLC. A mild treatment by proteinase K converted the 14s form into the 14f component. *Bacillus thuringiensis* PI-PLC as well as mild proteolytic treatments converted the 7s form into a 7f hydrophilic counterpart indicating that the 7s form was associated to the membrane through a glycolipid anchor.

Quaternary structure of molecular forms in class A AChE. Reduction of 14s, 14f, and 12f forms by DTT (10^{-2}M) dissociated these forms into a hydrophilic 7f form and a 4f component (this 4f component of class A migrated like the 4f form of class B). This indicates that (1) the three forms (14s, 14f and 12s) are oligomers of the 4f hydrophilic subunit and (2) the amphiphilic properties of the 14s form result from the association of a non catalytic hydrophobic element to the hydrophilic subunits. S values suggest that 12f, 7f and 4f forms are likely tetramers, dimers and monomers. The 14s and 14f are also likely tetramers of catalytic subunits, the shift from 12S to 14S being due to the presence of the structural element. The dissociating effect of DTT on oligomeric forms is likely due to the cleavage of interchain disulfide bonds and it indicates that the catalytic subunits in the tetramer are not bound by additional hydrophobic interactions as usually reported for vertebrate G4 forms of both AChE[14,15] and BChE[16].

Class B AChE. The 7s form is a dimer of glycolipid-anchored catalytic subunits. This form was inhibited by 10^{-4}M DTT even in the presence of edrophonium chloride. We thus failed to characterize the type of association of the monomers in the dimer. The nature of the hydrophobic domain in the 4s monomeric form is unknown.

[3]H-DFP labeling of catalytic subunits of classes A and B AChEs. A DS extract was incubated with ^{3}H-DFP, dialyzed, concentrated and run on SDS-PAGE in reducing conditions. Two bands were observed at 65 kDa and 90 kDa.

Only the 65 kDa band was observed when an isolated 4S peak fraction was DFP-labeled whereas the 90 kDa peptide was present in isolated 12 and 14S peaks. We thus suggest that these peptides represent the products of ace-1 (90 kDa, class A AChE) and ace-2 (65 kDa, class B AChE).

Conclusion. A working hypothesis on the quaternary structures of the molecular forms in the two classes A and B of *Steinernema* AChE is presented in figure 4. In class A, linkage of catalytic subunits in oligomeric forms (14S and 12S) as well as the association of the structural element in 14s and 14 f forms likely involve disulfide bonds. Note that the minor 12s form is not interpreted in this scheme. It may represent a hydrophobic tetramer with a different glycosylation than the major 14s from. In class B AChE, the nature of the link between two catalytic subunits is unknown.

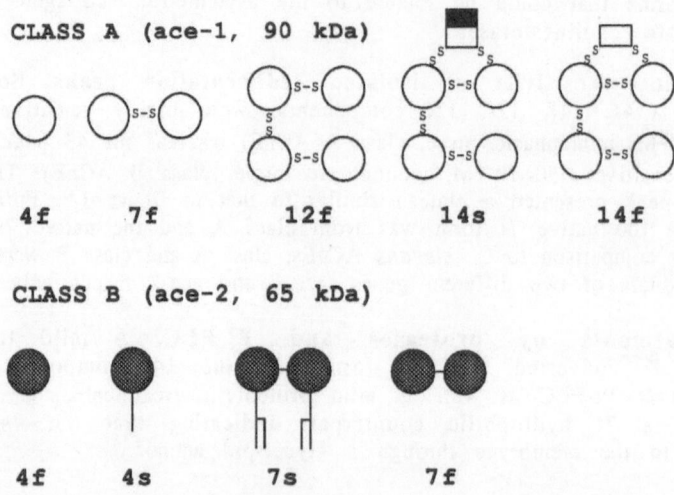

CLASS A (ace-1, 90 kDa)

4f   7f   12f   14s   14f

CLASS B (ace-2, 65 kDa)

4f   4s   7s   7f

Figure 4. Quaternary structure of molecular forms in class A and class B AChEs from *Steinernema* (ace-1 and ace-2 products). In 14s form, the hydrophobic domain of the non-catalytic subunit that mediates the membrane attachment is drawn in black. In class B, hydrophobic domains of the 7s form are glycolipid anchors.

Molecular forms of *Caenorhabditis* AChEs and the effect of *ace* mutations

Analysis of *C. elegans* extracts by sucrose gradient sedimentation showed molecular forms comparable to those of *Steinernema*. As previously mentioned by C. Johnson and R. Russell[17] the study is rendered difficult however 1) by the low level of AChE activity and 2) by the strong inhibition exerted by Triton X100 on certain molecular forms. We used Brij 96 as a detergent in both extraction and sucrose gradient media.

Three major peaks were usually observed in sedimentation (5S, 7S and 13S) with a shoulder of the 13S peak at 11.5S. Preliminary results indicate that most of the structures found in *Steinernema* AChE molecular forms were also observed in *C. elegans*. We studied the AChE activity in the wild strain N2 and the following mutants *ace-1, ace-2, ace-3* and *ace-1/ace-2*. The proportions of AChE activity recovered in mutant animals relative to the N2 strain were 63% (*ace-1*), 44% (*ace-2*), 97% (*ace-3*). These values are in good agreement with those reported by C. Johnson, R. Russell and colleagues[9,10,11,17]. No AChE activity was detected in *ace-1/ace-2* mutants. The sedimentation profiles of AChE in N2, *ace-1, ace-2* and *ace-3* were compared (figure 5). N2 and *ace-3* profiles were very similar. *Ace-1* mutants were devoid of fast sedimenting forms (13 and 11.5S) whereas these forms were present in *ace-2* along with a dramatic reduction of the light components (7 and 5S). AChE components of *C. elegans* were labeled with 3H-DFP. Two major bands were observed in N2 strain at 90 and 65kDa as in *Steinernema*. In *ace-1* and *ace-2* animals both bands were also present indicating that in these mutants a peptide was produced that contained an active serine but was devoid of AChE activity. This is quite possible since mutants are obtained through ethylmethanesulfonate treatment that generates point mutations. Such inactive AChE peptides binding DFP have already been described in *Torpedo*[18].

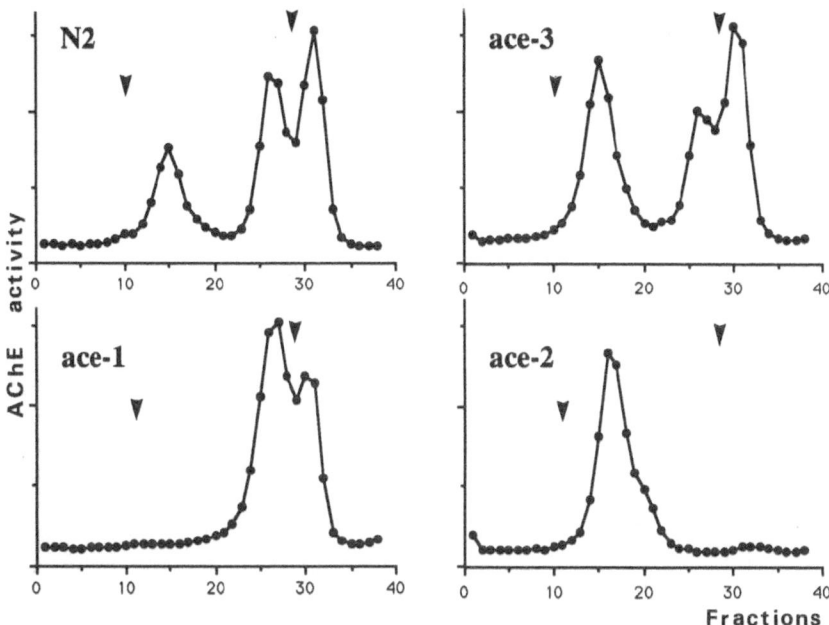

Figure 5. Sedimentation analysis of molecular forms of *Caenorhabditis elegans* AChEs in wild type animals (N2) and homozygous *ace* mutants. The different strains were first extracted twice in low salt buffer containing Brij 96, then in high salt buffer (HSS fraction). HSS fractions analyzed on HS gradients containing 0.5% Brij 96 are shown. Arrowheads indicate the position of β-galactosidase (16S, left) and alkaline phosphatase (6.1S, right). AChE activity is expressed in arbitrary units.

Cloning of *ace-3* by Transformation Rescue in *C. elegans*

Although we had very few informations about *ace-3* product, we started the study of this gene because it was best located on the genetic map.

A very precise physical map of *C. elegans* genome has been established by cloning the entire genome (100 Mb) in cosmid and YAC vectors. Sets of overlapping cosmids ("contigs") have been aligned with the genetic map (figure 6). This is critical to allow use of the genetic map for cloning genes defined through mutations (see Coulson et al.[19] for a review). *Ace-3* has been mapped on chromosome II[20] close to unc-52 a cloned gene that lies in the center of a contig. This contig however was not oriented relatively to the chromosome map and in this region the correlation between the genetic and the physical metrics had not yet been established.

We first measured precisely the genetic distance between *ace-3* and *unc-52* and found 0.3 map unit. This allowed the selection of 16 cosmids covering 600 kb which should contain *ace-3*. These cosmids were provided by the Cambridge Mapping Project Laboratory (UK). We then microinjected individual cosmid DNA together with a marker into the germ line of adult hermaphrodites homozygous for *ace-3*. *Ace-3* mutants have no visible phenotype but are sensitive to 5 µM trichlorfon: wild type animals behave normally when reared on plates containing the anticholinesterase agent whereas *ace-3* animals are shortened and uncoordinated. Animals injected with the cosmid containing the wild type allele of *ace-3* should generate transgenic animals insensitive to trichlorfon. We found one cosmid that rescued *ace-3*. We are now subcloning

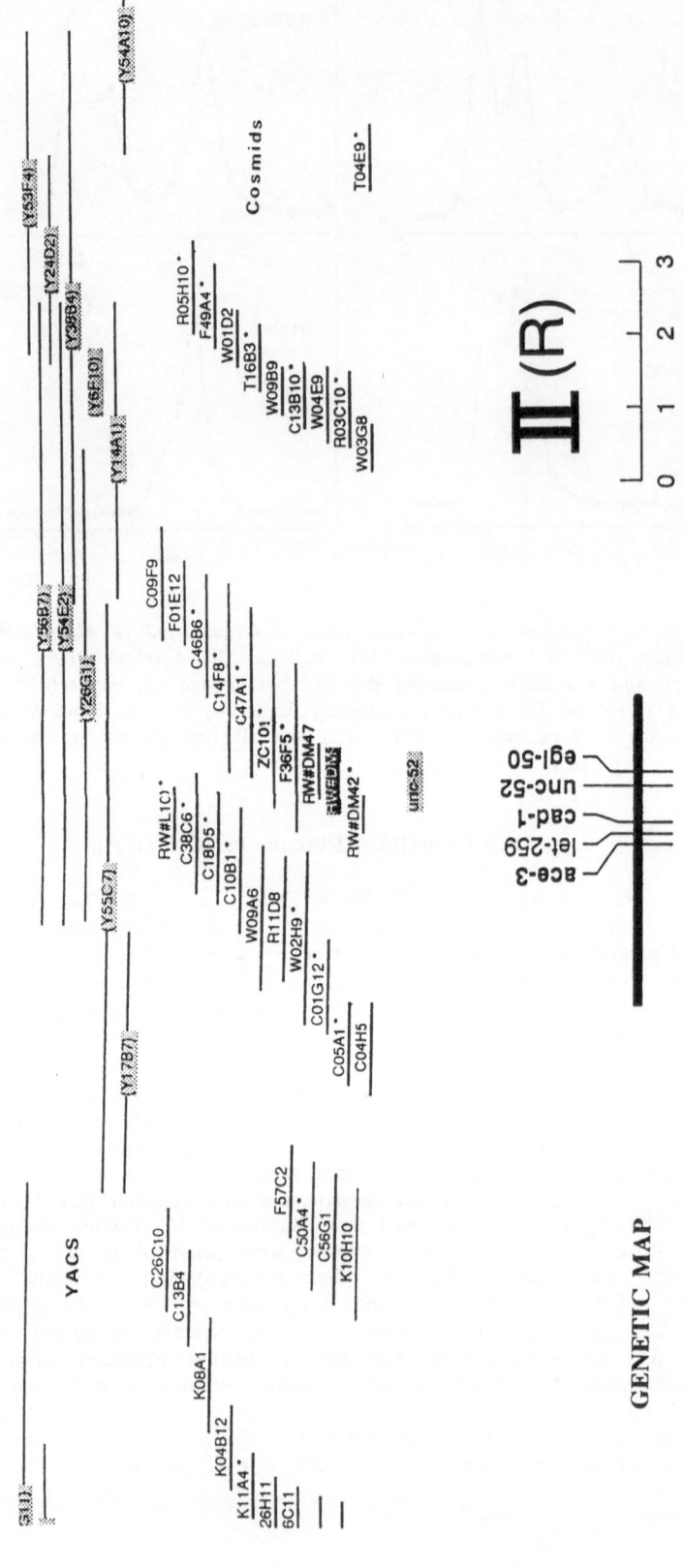

Figure 6. Correspondence between the physical and genetic maps in the region of *ace-3* on chromosome II in *C. elegans*. The positions of YACs (top, long horizontal bars), cosmids (small horizontal bars) and genetic loci are indicated. We found that cosmid WO2H9 could rescue *ace-3* mutants on trichlorfon plates.

this cosmid in order to isolate (by transformation rescue as above) a fragment of DNA suitable for sequencing.

DISCUSSION

The presence of several genes coding for AChE in *Steinernema* results in an increased molecular polymorphism as compared to the very simple situation found in insects for example. As in other invertebrates however there are no asymmetric forms similar to the collagen-tailed oligomers present in vertebrate cholinesterases.

Ace-3 product participates marginally to the hydrolysis of ACh in *Steinernema* and *Caenorhabditis*. It is however essential for survival since triple mutants *ace-1/ace-2/ace-3* but not double mutants *ace-1/ace-2* are lethal[11]. Due to its very low activity, we have been unable so far to characterize *ace-3* product. Expression of the cloned *ace-3* gene should allow an easier study of the biochemical properties of this enzyme.

The products of the two genes *ace-1* and *ace-2* were characterized. We found that the structure of their molecular forms resembled those found in other animals. In particular, the two types of amphiphilic forms (likely amphiphilic dimers and tetramers) presented two modes of membrane association already characterized in other phyla: a hydrophobic structural element linked to hydrophilic catalytic subunits as in the membrane bound G4 form of vertebrate brains[21] and a dimeric form attached to the membrane through glycolipid anchors as in *Drosophila* AChE[5,6], dimers of class I in *Torpedo* cholinesterases[22] and mammalian erythrocyte membranes[23]. At variance from the situation in vertebrates (where only one gene codes for the two types of catalytic subunits by the splicing of an alternative exon[24,25]) two distinct genes are required to give rise to this polymorphism in nematodes. Thus the comparison of the structure and sequences of *ace-1* and *ace-2* genes with vertebrate AChE genes is very promising.

Acknowledgments

This work was supported by grants from the *Institut National de la Recherche Agronomique.*

REFERENCES

1. J. Massoulié and S. Bon, The molecular forms of cholinesterases in vertebrates, *Ann. Rev. Neurosci.* 5:57 (1982).
2. J.P. Toutant, M. Arpagaus and D. Fournier, Native molecular forms of head acetylcholinesterase from adult *Drosophila melanogaster*: Quaternary structure and hydrophobic character, *J. Neurochem.* 50:209 (1988).
3. J.P. Toutant, Insect acetylcholinesterase. Catalytic properties, tissue distribution and molecular forms, *Progr. Neurobiol.* 32:423 (1989).
4. A.L. Gnagey, M. Forte and T.L. Rosenberry, Isolation and characterization of acetylcholinesterase from *Drosophila, J. Biol. Chem.* 262:13290 (1987).
5. D. Fournier, J.B. Bergé, M.L. Cardoso de Almeida and C. Bordier, Acetylcholinesterase from *Musca domestica* and *Drosophila melanogaster* are linked to membranes by a glycophospholipid anchor sensitive to an endogenous phospholipase, *J. Neurochem.* 50:1158 (1988).
6. R. Haas, T.L. Marshall and T.L. Rosenberry, *Drosophila* acetylcholinesterase demonstration of a glycoinositol anchor and an endogenous proteolytic cleavage, *Biochemistry* 27:6453 (1988).

7. B. Espinosa, R. Tarrab-Hazdai, I. Silman and R. Arnon, Acetylcholinesterase in *Schistosoma mansoni* is anchored to the membrane via covalently attached phosphatidylinositol, *Mol. Biochem. Parasitol.* 29:171 (1988).

8. L. Pezzementi, M. Sanders, T. Jenkins, D. Holliman, J.P. Toutant and R. Bradley, Structure and function of cholinesterase from *Amphioxus*, *in*: "Cholinesterases", J. Massoulié *et al.*, eds. ACS, Washington (1991).

9. C.D. Johnson, J.G. Duckett, J.G. Culotti, R.K. Herman, P.M. Meneely and R.L. Russell, An acetylcholinesterase-deficient mutant of the nematode *Caenorhabditis elegans*, *Genetics* 97:261 (1981).

10. J.G. Culotti, G. Von Ehrenstein, M.R. Culotti, and R.L. Russell, A second class of acetylcholinesterase-deficient mutants of the nematode *Caenorhabditis elegans*, *Genetics* 97:281 (1981).

11. C.D. Johnson, J.R. Rand, R.K. Herman, B.D. Stern, and R.L. Russell, The acetylcholinesterase genes of *C. elegans*: identification of a third gene (*ace-3*) and mosaic mapping of a synthetic lethal phenotype, *Neuron* 1:165 (1988).

12. M. Arpagaus and J.P. Toutant, Polymorphism of acetylcholinesterase in adult *Pieris brassicae* heads. Evidence for detergent-insensitive and Triton X-100-interacting forms, *Neurochem. Int.* 7: 793 (1985).

13. T.L. Rosenberry, J.P. Toutant, R. Haas and W.L. Roberts, Identification and analysis of glycoinositol phospholipid anchors in membrane proteins, *Meth. Cell Biol.* 32: 231 (1989).

14. M. Vigny, S. Bon, J. Massoulié and V. Gisiger, The subunit structure of mammalian acetylcholinesterase: catalytic subunits, dissociating effect of proteolysis and disulphide reduction on the polymeric forms, *J. Neurochem.* 33:559 (1979).

15. H. Heider and U. Brodbeck, Monomerization of tetrameric bovine caudate nucleus acetylcholinesterase. Implications for hydrophobic assembly and membrane anchor attachment site, *Biochem. J.* 281:279 (1992).

16. O. Lockridge, H.W. Eckerson and B.N. La Du, Interchain disulfide bonds and subunit organization in human cholinesterase, *J. Biol. Chem.* 254:8324 (1979).

17. C.D. Johnson, and R.L. Russell, Multiple molecular forms of acetylcholinesterase in the nematode *Caenorhabditis elegans*, *J. Neurochem.* 41:30 (1983).

18. S. Stieger, U. Brodbeck and V. Witzemann, Inactive monomeric acetylcholinesterase in the low-salt-soluble extract of the electric organ from *Torpedo marmorata*, *J. Neurochem.* 49:460 (1987).

19. A. Coulson, Y. Kosono, B. Lutterbach, R. Shownkeen, J. Sulston and R. Waterston, YACs and the *C. elegans* genome, BioEssays 13:413 (1991).

20. D.L. Kolson and R.L. Russell, A novel class of acetylcholinesterase, revealed by mutations, in the nematode *Caenorhabditis elegans*, *J. Neurogenet.* 2:93 (1985).

21. N.C. Inestrosa, W.L. Roberts, T.L. Marshall and T.L. Rosenberry, Acetylcholinesterase from bovine caudate nucleus is attached to membranes by a novel subunit distinct from those of acetylcholinesterase in other tissues, *J. Biol. Chem.* 262:4441 (1987).

22. S. Bon, J.P. Toutant, K. Méflah and J. Massoulié, Amphiphilic and nonamphiphilic forms of *Torpedo* cholinesterases: II. Electrophoretic variants and phosphatidylinositol phospholipase C-sensitive and -insensitive forms, *J. Neurochem.* 51:786 (1988).

23. I. Silman and A.H. Futerman, Modes of attachment of acetylcholinesterase to the surface membrane, *Eur. J. Biochem.* 170:11 (1987).

24. J.L. Sikorav, N. Duval, A. Anselmet, S. Bon, E. Krejci, C. Legay, M. Osterlund, B. Reimund and J. Massoulié, Complex alternative splicing of acetylcholinesterase transcripts in *Torpedo* electric organ; primary structure of the precursor of the glycolipid-anchored dimeric form, *EMBO J.* 7:2983 (1988).

25. Y. Li, S. Camp, T.L. Rachinsky, D. Getman and P. Taylor, Gene structure of mammalian acetylcholinesterase. Alternative exons dictate tissue-specific expression, *J. Biol. Chem.* 266:23083 (1991).

DROSOPHILA ACETYLCHOLINESTERASE: ANALYSIS OF STRUCTURE AND SENSITIVITY TO INSECTICIDES BY *IN VITRO* MUTAGENESIS AND EXPRESSION

Didier Fournier, Annick Mutero, Madeleine Pralavorio and Jean Marc Bride

INRA, Laboratoire de Biologie des Invertébrés, 06606, BP 2078, Antibes Cedex, France

INTRODUCTION

Mutations in *Drosophila* hold promise for studying effects of protein alterations on development, neurophysiology and behavior of the organism. For that purpose, several mutations which affect acetylcholinesterase (AChE) were isolated (Greenspan *et al.*, 1980; Morton and Singh, 1984; Pralavorio and Fournier, 1992). These mutations are either lethal, temperature-sensitive or modify the catalytic properties of the enzyme. Lethal and conditional mutations were obtained by mutagenesis experiments and mutations affecting the catalytic properties of the enzyme were found in insects resistant to insecticides which have an altered AChE less sensitive to inhibition by organophosphate and carbamate compounds.

The gene coding for AChE in *Drosophila* was cloned by chromosome walking on the previously localized *Ace* locus (Bender *et al.*, 1983). The coding region was sequenced and the protein sequence deduced (Hall and Spierer, 1986). The gene structure was established and transient expression was obtained in *Xenopus* oocyte (Fournier *et al.*, 1989; 1992). In contrast with vertebrate cholinesterases which are highly polymorphic in their quaternary structure, AChE is found as one main form in insects: a globular disulfide-linked dimer of 150 kDa, attached to the membrane via a glycosyl-phosphatidylinositol anchor (Gnagey *et al.*, 1987; Fournier *et al.*, 1988a; Haas *et al.*, 1988). A particularity of AChE in *Drosophila* is that each subunit of the protein is composed of two non-covalently linked polypeptides of 18 and 55 kDa which arise from the processing of a 75 kDa precursor (Fournier *et al.*, 1988b).

Here we investigated the role of amino acids involved in post-translational modifications using site directed-mutagenesis and expression in *Xenopus* oocytes. We mapped mutations in temperature sensitive and in insecticide resistant mutant strains by sequencing their AChE

Multidisciplinary Approaches to Cholinesterase Functions, Edited by
A. Shafferman and B. Velan, Plenum Press, New York, 1992

coding regions after PCR amplification. Importance of these mutations were analyzed following expression in *Xenopus* oocytes.

AMINO ACIDS INVOLVED IN POST-TRANSLATIONAL MODIFICATIONS OF AChE

Mapping of Amino Acids Important for Post-Translational Modifications

We investigated three post-translational modifications: dimerization, proteolytic cut of the precursor into two polypeptides, and glycosylation. Homology with vertebrate cholinesterases suggested that two cysteines at positions 328 and 615 (Fig. 1) were not involved in intrachain disulfide bonds and were therefore candidate for the intersubunit disulfide bond involved in the formation of dimers. Only the mutation of cysteine 615 affected the apparent molecular weight of the protein indicating that this cysteine is involved in the intersubunit disulfide bond. *Drosophila* AChE differs from vertebrate cholinesterases in two aspects. First, the 75 kDa precursor polypeptide is split into two polypeptides of 18 and 55 kDa; second, it exists an additional hydrophilic peptide in the primary sequence deduced from the cDNA (Arg 147 to Pro 180). These two aspects are related since deletion of the hydrophilic peptide inhibits the cut. The cut occurs in several sites in the hydrophilic peptide leading to an heterogenous population of polypeptides. Among the five potential sites of carbohydrate chain linkage on asparagine residues, four are actually glycosylated at positions 126, 174, 331 and 531 and the carbohydrates contribute to 10% of the molecular mass of the protein (Mutero and Fournier, 1992).

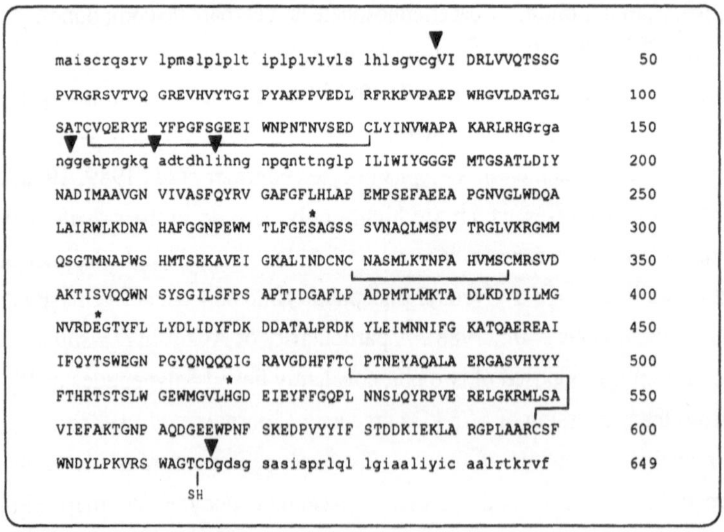

```
        maiscrqsrv lpmslplplt iplplvlvls lhlsgvcgVI DRLVVQTSSG      50

        PVRGRSVTVQ GREVHVYTGI PYAKPPVEDL RFRKPVPAEP WHGVLDATGL     100

        SATCVQERYE YFPGFSGEEI WNPNTNVSED CLYINVWAPA KARLRHGrga     150

        nggehpngkq adtdhlihng npqnttnglp ILIWIYGGGF MTGSATLDIY     200

        NADIMAAVGN VIVASFQYRV GAFGFLHLAP EMPSEFAEEA PGNVGLWDQA     250

        LAIRWLKDNA HAFGGNPEWM TLFGESAGSS SVNAQLMSPV TRGLVKRGMM     300

        QSGTMNAPWS HMTSEKAVEI GKALINDCNC NASMLKTNPA HVMSCMRSVD     350

        AKTISVQQWN SYSGILSFPS APTIDGAFLP ADPMTLMKTA DLKDYDILMG     400

        NVRDEGTYFL LYDLIDYFDK DDATALPRDK YLEIMNNIFG KATQAEREAI     450

        IFQYTSWEGN PGYQNQQQIG RAVGDHFFTC PTNEYAQALA ERGASVHYYY     500

        FTHRTSTSLW GEWMGVLHGD EIEYFFGQPL NNSLQYRPVE RELGKRMLSA     550

        VIEFAKTGNP AQDGEEWPNF SKEDPVYYIF STDDKIEKLA RGPLAARCSF     600

        WNDYLPKVRS WAGTCDgdsg sasisprlql lgiaaliyic aalrtkrvf      649
                         |
                         SH
```

Figure 1. Primary structure of Drosophila AChE. The arrows indicate proteolytic cuts ant the stars mark the residues of the catalytic triad.

Role of Post-Translational Modifications

The three mutated, monomeric, uncut, or unglycosylated, proteins were active. We did not detect any significant difference in their affinity towards acetylcholine (Km) or in their catalytic efficiency (kcat). It appears that dimerization, proteolytic cut and glycosylation are not necessary for the catalytic function of the AChE in *Drosophila*. The mutated enzymes are secreted outside the cell. Thus, the three post-translational modifications do not seem to be important for the targeting of the protein. We found only one difference between the mutated and the wild type protein: unglycosylated and monomeric mutants were about four-fold more sensitive to pronase. Thus dimerization and glycosylation seem to protect the enzyme from proteolysis while monomer and unglycosylated enzymes displaying regions susceptible to proteases which are masked in the wild type protein. In contrast, we did not find any role for the hydrophilic peptide. A similar hydrophilic peptide, although not conserved, has been found at the same location in the other known insect AChE sequences (Malcolm and Hall, 1990). Thus this peptide is likely to play a physiological role.

AMINO ACIDS INVOLVED IN THE FOLDING OF ACHE

We found mutations involved in the folding of the enzyme in two different experiments. First, by mutagenizing some amino acids highly conserved in cholinesterases and related sequences, we found that one of them is essential for the folding of the protein. Second, by analyzing thermosensitive mutant flies, it appeared that mutations found in their enzymes affect the folding of the protein in a thermosensitive manner.

Non Conditional Mutations

In order to search for amino acids important for the enzyme function, we mutated the aspartate 248 to an asparagine (Fig. 1). This mutation had a drastic effect on the activity and on the secretion of the protein. *Xenopus* oocytes injected with a gene bearing this mutation produced an inactive protein which remained sequestered inside the cell. This protein was visible on gel performed in denaturing and reducing conditions, but not in the absence of reducing agent. In native conditions, the AChE produced by the oocytes injected with the wild type gene migrates as a single band, in contrast the mutated protein could never be observed in non denaturing conditions. This differential comportment on gels indicates that the mutated protein is misfolded, these forms aggregates with other proteins maybe by erroneous bonding and / or exposition of abnormal hydrophobic regions.

This aspartate is highly conserved in cholinesterases as well as in cholinesterase-like proteins devoid of esterase activity (Krejci *et al.*, 1991). This conservation seems to be due to the essential role of this amino acid in the folding of the protein. Conversely, we did not find any alteration of the biosynthesis or any effect on the activity when we mutated another conserved aspartate at position 130. Thus, high conservation does not always imply an importance for the folding, the structure or the catalysis of the neurotransmitter. This suggests that these amino acids may have another role.

Temperature Sensitive Mutations

Ace^{J40} is a heat sensitive mutant isolated by Greenspan et al. (1980). In homozygous conditions, this mutation is lethal when flies are raised above 25°C. At permissive temperature, flies have 30% of wild type activity, but the mutation does not disturb the overall structure of the enzyme (Toutant and Arpagaus, 1989). Mutated enzyme displays a slightly increased sensitivity to heat inactivation at 50°C, but this alteration does not explain the lethality of the flies at restrictive temperature. We PCR amplified the exons of the gene encoding AChE from DNA extracted from homozygous flies and we found one mutation, the proline 75 is changed to a leucine (Fig. 1). We expressed the mutated protein in Xenopus oocytes. The secreted enzyme was not modified in its catalytic or structural properties, but secretion was temperature dependent, higher at 20°C than at 25°C. Specific activity of the enzyme found inside the oocyte, en route to the external medium was also lower, suggesting that a part of the protein was misfolded and remained sequestered in the secretory pathway. We tested this hypothesis by comparing patterns obtained in gels run in denaturing and native conditions. We observed a difference between the two gels suggesting that thermosensitive mutation results in the misfolding of a part of the protein, which is inactive and not secreted. Proline 75 is highly conserved in cholinesterases and related proteins and might be important for the folding due to its cyclic structure.

Ace^{J29} is a cold sensitive mutant (Greenspan et al., 1980) lethal when flies are raised under 23°C. As for the Ace^{J40} mutant, no enzyme alteration could explain the lethality of the flies at restrictive temperature. We PCR amplified the exons of the gene and we found one mutation : the serine 314 is changed to a phenylalanine. We expressed the mutated protein in Xenopus oocytes. Excretion was temperature dependent (higher at 25°C than at 20°C) due to the misfolding of a part of the protein.

AMINO ACIDS INVOLVED IN THE RESISTANCE TO INSECTICIDE

Mutation of Tyr 109

Tyrosine 109 in the Drosophila AChE corresponds to an aspartate in vertebrate sequences. Mutation of this amino acid to glycine in the human butyrylcholinesterase gives rise to the "atypic" phenotype characterized by a reduced activity for charged compounds (Lockridge and

La Du, 1978). We investigated the importance of tyrosine 109 by substituting a glycine, an aspartate, or a lysine using *in vitro* mutagenesis, we then expressed the mutant proteins in *Xenopus* oocytes (Mutero *et al.*, 1992). These mutations affected some catalytic properties of the enzyme and its sensitivity to insecticides. The mutated enzymes were slightly different from the wild type, either more susceptible or more resistant.

Mutations Found in Resistant Strains

Since Smissaert (1964), numerous pest insect strains were shown to display an AChE with a reduced sensitivity to insecticides. Among these species, a *Drosophila* strain (MH19) resistant to malathion was isolated by Morton and Singh (1982). Following PCR amplification and sequencing, we found only one non-silent mutation resulting in the replacement of a phenylalanine to a tyrosine at position 368. To verify that this change is indeed responsible for the resistance, we introduced the MH19 mutation in a minigene and transformed AChE deficient flies. Rescued flies were more resistant than their counterparts transformed with the wild type minigene. The mutant protein expressed in *Xenopus* oocytes was also resistant. This phenylalanine 368 is conserved in the Torpedo sequence in which it corresponds to Phe 290 (Krejci *et al.*, 1991). This amino-acid belongs to the active site pocket and is supposed to be involved in the substrate guidance (Sussman *et al.*, 1991). It is tempting to speculate that mutation of phenylalanine to tyrosine disturbs the aromatic guidance of both substrate and insecticides.

In order to find other mutations, we selected field populations originating from several countries. Thirteen among the twenty two strains selected were resistant to parathion due to an altered AChE. The enzymes with the higher resistance levels have been characterized with respect to their cross-resistance towards several insecticides. The patterns obtained allowed us to distinguish several modified proteins (Pralavorio and Fournier, 1992). We determined the sequences of the AChE coding regions in these strains after PCR amplification. Three mutations have been identified: Phe115 to Ser, Gly303 to Ala, and Ileu199 to Val. Mutated proteins expressed in *Xenopus* oocytes were assayed for their sensitivity towards insecticides. Each of these mutations led to a different pattern of resistance but resistance was too low to permit the fly to escape usual field pesticide treatment. We noted that several mutations were often present in the same gene in most resistant flies and we hypothesized that high levels of resistance originate from the combination of several mutations in the same gene.

Combination of Several Mutations

We tested this hypothesis by expressing several combinations of mutations affecting the sensitivity to organophosphates. It appears that associations of several mutations result in high resistance levels that reach one thousand-fold resistance to carbaryl for the protein bearing mutations Ala303 and Ser368.

Until now it was thought that high levels of resistance allowing the insect to escape to insecticide treatments originate from the appearance of a new point mutation. This hypothesis does not explain several observations. For example, it does not explain the high frequency of appearance of resistant insects or why some insect species become resistant whereas other do not. If high resistance originates from the intracistronic recombination between point mutations responsible for weak resistances, we better understand the high frequency of appearance of resistance in some species. Indeed, recombination rate depends of several biological factors as for example the length and number of introns which are different from species to species.

REFERENCES

Bender, W., Spierer, P., and Hogness, D.S., 1983, Chromosomal walking and jumping to isolate DNA from the *Ace* and *rosy* loci and the bithorax complex in *Drosophila melanogaster. J. Mol. Biol.*, 163: 17.

Fournier, D., Bergé, J.B., Cardoso de Almeida, M.L., and Bordier, C., 1988a, Acetylcholinesterase from *Musca domestica* and *Drosophila melanogaster* brain are linked to membranes by a glycophospholipid anchor sensitive to an endogenous phospholipase. *J. Neurochem.*, 50: 1158.

Fournier, D., Bride, J.M., Karch, F., and Bergé, J.B., 1988b, Acetylcholinesterase from Drosophila melanogaster : identification of two subunits encoded by the same gene. *FEBS L.*, 238: 333.

Fournier, D., Karch, F., Bride, J.M., Hall, L.M.C., Bergé, J.B., and Spierer, P., 1989, The *Drosophila melanogaster* gene : structure, evolution and mutations. *J. Mol. Biol.*, 210: 15.

Fournier, D., Mutero, A., and Rungger, D., 1992, *Drosophila* acetylcholinesterase : expression of a functional precursor in *Xenopus* oocytes. *Eur. J. Biochem.*, 203: 513.

Gnagey, A.L., Forte, M., and Rosenberry, T.L., 1987, Isolation and characterization of acetylcholinesterase from *Drosophila. J. Biol. Chem.*, 262: 13290.

Greenspan, R.J., Finns, J.A., and Hall, J., 1980, Acetylcholinesterase mutants in *Drosophila* and their effects on the structure and function of the central nervous system. *J. Comp. Neurol.* 189: 741.

Haas, R., Marshall, T.L., and Rosenberry, T.L., 1988, *Drosophila* acetylcholinesterase: demonstration of a glycoinositol phospholipid anchor and an endogenous proteolytic cleavage. *Biochemistry*, 27: 6453.

Hall, R., and Spierer, P., 1986, The *Ace* locus of *Drosophila melanogaster*: structural gene for acetylcholinesterase with an unusual 5' leader. *EMBO J.*, 5: 2949.

Krejci, E., Duval, N., Chatonnet, A., Vincens, P., and Massoulié, J., 1991, Cholinesterase-like domains in enzymes and structural proteins: Functional and evolutionary relationships and identification of a catalytically essential aspartic acid. *Proc. Natl. Acad. Sci. USA.*, 88: 6647.

Lockridge, O., and La Du, B.N., 1978, Comparison of atypical and usual human serum cholinesterase. *J. Biol. Chem.*, 253: 361.

Malcolm, C.A., and Hall, L.M.C., 1990, Cloning and characterisation of a mosquito acetylcholinesterase gene. *in* :"Molecular Insect Science" Hagedorn, H.H. et al. eds., pp. 57-65, Plenum Publishing Corp., New York.

Morton, R.A., and Singh, R.S., 1982, The assosciation between malathion resistance and acetylcholinesterase in *Drosophila melanogaster. Biochem. Genet.*, 20: 179.

Mutero, A., and Fournier, D., 1992, Post-translational modifications of *Drosophila* acetylcholinesterase, *in vitro* mutagenesis and expression in *Xenopus* oocytes. *J. Biol. Chem.*, 267:1695.

Mutero, A., Pralavorio, M., Simeon, V., and Fournier, D., 1992, Catalytic properties of cholinesterases: importance of tyrosine 109 in *Drosophila* protein. *Neuroreport*, 3: 39.

Pralavorio, M., and Fournier, D., 1992, *Drosophila* acetylcholinesterase: characterization of different mutants resistant to insecticides. *Biochem. Genet.*, 30:77.

Smissaert, H.R., 1964, Cholinesterase inhibition in spider mites susceptible and resistant to organophosphate. *Science*, 143: 129.

Sussman, J.L., Harel, M., Frolow, F., Oefner, C., Coldman, A., Toker, L., and Silman, I., 1991, Atomic structure of acetylcholinesterase from Torpedo californica: a prototypic acetylcholine-binding protein. *Science*, 253: 872.

Toutant, J.P., and Arpagaus, M., 1989, Quaternary structure and hydrophobic interactions of *Drosophila* acetylcholinesterase in wild type flies and in mutants of the Ace locus. *in* "Insect Neurochemistry and Neurophysiology" Borkovec A.B. and Masler E.P. eds, pp. 239-242, The Humana Press Inc.

ALTERED FORMS OF ACETYLCHOLINESTERASE IN INSECTICIDE-RESISTANT HOUSEFLIES (*MUSCA DOMESTICA*)

Martin S. Williamson, Graham D. Moores, and Alan L. Devonshire

AFRC Institute of Arable Crops Research
Rothamsted Experimental Station
Harpenden, Herts, AL5 2JQ, United Kingdom

INTRODUCTION

Widespread use of insecticides has led to the selection of resistance in many insect species. The biochemical mechanisms responsible are of two broad classes: enhanced insecticide degradation by various enzyme groups, and decreased sensitivity of the target protein to the toxicant.[1] As the target for organophosphorus (OP) and carbamate insecticides, acetylcholinesterase (AChE) forms showing insensitivity to these inhibitors have been selected in various pests.

Detailed studies on these enzyme forms require their isolation in strains homozygous for a particular AChE allele and this has traditionally been achieved by intensive insecticide selection pressure in the laboratory. However, this can mask the genetic diversity in field populations since it biasses towards genotypes with the most marked insensitivity to the selectant. This can now be overcome by simplified studies of inhibition kinetics to identify both homozygotes and heterozygotes in large numbers of individual insects taken from field populations. Since several insecticidal inhibitors can be used for each individual, an 'insensitivity profile' can be established.

BIOCHEMICAL ANALYSIS OF AChE INSENSITIVITY

AChE assays are done with the Ellman technique; initial work used conventional spectrophotometry but the advent of kinetic microplate readers operating on all 96 wells simultaneously now enables much greater sensitivity and throughput; several hundred insects as small as a whitefly (20 μg) can be characterized in a day. For this, the activity in a portion of each homogenate without inhibitor is compared with mean activity in the presence of each diagnostic inhibitor concentration.[2] This form of inhibition assay in the presence of substrate gives reaction curves conforming to first order kinetics, but the kinetic constants cannot be determined reliably or simply, especially when the insect is heterozygous for the AChE gene. Instead, the mean activity over the assay period, determined by the on-line microcomputer, is expressed as a percentage of the uninhibited rate, and this gives a rapid and robust way of distinguishing insects carrying different AChE genes.

Multidisciplinary Approaches to Cholinesterase Functions, Edited by
A. Shafferman and B. Velan, Plenum Press, New York, 1992

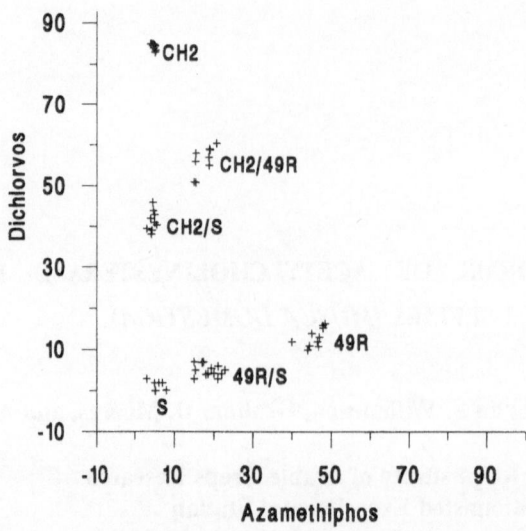

Figure 1. Bivariate plot of mean percentage activity remaining during inhibition of AChE by dichlorvos and azamethiphos for all six AChE genotypes derived from strains CH2, 49R and S.

We have exploited this technique to identify multiple forms of AChE in the housefly. An example of the analysis is shown in figure 1 for two well characterized strains, 49R and CH2, showing converse patterns of insensitivity to azamethiphos and dichlorvos. A bivariate plot of mean percentage remaining activity for these inhibitors not only distinguishes the susceptible, 49R and CH2 homozygotes, but also resolves flies heterozygous for these alleles (figure 1). This technique has been extended to field populations using a wider range of inhibitors, to resolve up to 10 genotypes within a single farm population of houseflies (Denholm, unpublished).

To gain an insight into the genetic diversity which underlies these altered forms of AChE, we are now using gene cloning techniques to isolate the housefly gene sequences encoding these variant AChE proteins.

CLONING OF HOUSEFLY AChE GENE SEQUENCES

Our strategy for cloning the housefly AChE gene was based initially on the polymerase chain reaction (PCR) homology probing technique.[3] Cholinesterase gene sequences have now been reported for *Torpedo*, *Drosophila*, human and mosquito (see ref[4] and references therein) and comparison of their predicted amino acid sequences has revealed a number of short, conserved regions within these proteins despite their diverse origins. The PCR homology probing technique exploits these conserved sequences using degenerate oligonucleotide primers designed from them to selectively amplify the intervening regions of the gene from other species.

We selected two regions from the C terminal end of the ChE sequence for the design of oligonucleotide primers (figure 2). PCR reactions with these primers using housefly genomic DNA resulted in the amplification of two DNA fragments of 400 bp and 600 bp (not shown). The coding sequence between these primers requires a fragment of only 285 bp; however, since both the *Drosophila* and mosquito AChE

```
512                           519         600                      606
E   W   M   G   V   L   H   G             F   W   N   D   Y   L   P
                I                                     Q   F   F
```

```
5'-GAR TGG ATG GGN GTN BTN CAY GG-3'

                              3'-AAR ACC TTR XTN AZR RAN GG-5'

    23mer (degeneracy=768)              20mer (degeneracy=1024)
```

Figure 2.   Conserved regions of the cholinesterase sequence used to design degenerate oligonucleotide primers for PCR. The peptides correspond to residues 512-519 and 600-606 of the *Drosophila* sequence.[5] The primers contain all possible DNA sequences which could encode these peptides.

R = A or G; X = C or G; Z = T or A; B = A, C or T; N = A, C, G or T.

```
             10               30               50          ↓  70
D.mel  VIDRLVVQTSSGPVRGRSVTVQGREVHVYTGIPYAKPPVEDLRFRKPVPAEPWHGVLDATGLSATCVQER
       ...........................................................

             90  ↓            110              130
D.mel  YEYFPGFSGEEIWNPNTNVSEDCLYINVWAPAKARLRHGRGANGGEHPNGKQADTDHLIHNGNPQNTTNG
                                   ||  ||||||    |  |||||||   |||||||
M.dom  ..................................RGTNGGEH..SSKTDQDHLIHSATPQNTTNG

             150              170              190              210
D.mel  LPILIWIYGGGFMTGSATLDIYNADIMAAVGNVIVASFQYRVGAFGFLHLAPEMPSEFAEEAPGNVGLWD
       ||||||||||||||||||||||||||  ||  ||||||||||||||||||||||  |  |||  |  |||||||||||
M.dom  LPILIWIYGGGFMTGSATLDIYNAEIMSAVGNVIVASFQYRVGAFGFLHLSPVMPG.FEEEAPGNVGLWV

             230         *    250              270
D.mel  QALAIRWLKDNAHAFGGNPEWMTLFGESAGSSSVNAQLMSPVTRGLVKRGMMQSGTMNAPWSHMTSEKAV
       ||||  ||||  ||  |||||||||||||||||||||||||||||||||||||||||||||||||||||||
M.dom  QALALRWLKENARAFGGNPEWMTLFGESAGSSSVNAQLMSPVTRGLVKRGMMQSGTMNAPWSHMTSEKAV

             290  ↓        ↓310              330              350
D.mel  EIGKALINDCNCNASMLKTNPAHVMSCMRSVDAKTISVQQWNSYSGILSFPSAPTIDGAFLPADPMTLMK
       ||||||  ||||||||  |     ||  |||  |||||||||||||||||||||||||||||||||||||||||||  |
M.dom  EIGKALVNDCNCNASLLPENPQAVMACMRQVDAKTISVQQWNSYSGILSFPSAPTIDGAFLPADPMTLLK

             *370             390              410
D.mel  TADLKDYDILMGNVRDEGTYFLLYDFIDYFDKDDATALPRDKYLEIMNNIFGKATQAEREAIIFQYTSWE
       ||||    ||||   |||  ||||||||||||||||||||||||   |||||||||||||||  ||  ||||||||||||||||
M.dom  TADLSGYDILIGNVKDEGTYFLLYDFIDYFDKDDATSLPRDKYLEIMNNIFQKASQAEREAIIFQYTSWE

             430       ↓      450              470         *    490
D.mel  GNPGYQNQQQIGRAVGDHFFTCPTNEYAQALAERGASVHYYYFTHRTSTSLWGEWMGVLHGDEIEYFFGQ
       ||||||||||||||||||||||||||||||||||||||||||||||||||||||||||||||||||||||||||||||||||||||||||||||
M.dom  GNPGYQNQQQIGRAVGDHFFTCPTNEYAQALAERGASVHYYYFTHRTSTSLWGEWMGVLHGDEIEYFFGQ

             510              530              550         ↓
D.mel  PLNNSLQYRPVERELGKRMLSAVIEFAKTGNPAQDGEEWPNFSKEDPVYYIFSTDDKIEKLARGPLAARC
       ||||||||||||||||||||||   |||||||  ||||   ||||||||||||||||||   |||||   ||||||   |||||||  ||
M.dom  PLNNSLQYRPVERELGKRMLNSVIEFAKSGNPAVDGEEWPNFSKEDPVYYVFSTDEKIEKLQRGPLAKRC

             570              590              610
D.mel  SFWNDYLPKVRSWAGTCDGDSGSASISPRLQLLGIAALIYICAALRTKRVF
       ||||||||||||||   |        |     |     |
M.dom  SFWNDYLPKVRSWIGSECENKSSTSFAAIYEMKMQQLTLLAVAIILTMVNSIFQ
```

Figure 3.   Comparison of the partial housefly AChE sequence encoded by pACE 1 (*M. dom.*) with the mature protein sequence of *Drosophila melanogaster*.[5] Identical sequences are marked by vertical lines. Based on the structure of the *Torpedo* enzyme,[7] the 3 residues which probably form the catalytic triad (Ser, Glu and His) are indicated by asterisks and the 6 cysteines which form intra-subunit disulphide bonds are arrowed.

genes contain an intron of approximately 100 bp within this sequence,[4,6] the amplified fragment of 400 bp was chosen as the most likely to contain the corresponding housefly sequence.

Attempts to clone this 400 bp PCR fragment proved unsuccessful and so the fragment was instead radiolabelled and used to screen a housefly cDNA library. The library was constructed in λZapII (Stratagene) using oligo dT primed adult head mRNA from a reference susceptible strain as the template. Around 3×10^4 plaques were screened using duplicate 'lifts' from the library and one clone, named pACE1, was identified showing positive hybridisation on both filters. Preliminary sequencing of one end of this cDNA revealed a close similarity to the *Drosophila* AChE sequence and so it was fully sequenced.

The pACE1 cDNA is 2077 bp in length with 1506 bp of coding sequence followed by a 3' non-coding region of 571 bp. The 502 amino acid housefly AChE sequence predicted from this cDNA is not full length, lacking the 109 N terminal amino acids of the corresponding *Drosophila* protein. An alignment of the housefly AChE sequence to that of *Drosophila* (figure 3) illustrates the high degree of similarity between the two sequences with an overall amino acid homology of 85%. It contains conserved features characteristic of all ChE sequences, such as the region around the active site serine, the catalytic triad residues and cysteines involved in disulphide bonding (figure 3). The housefly sequence also encodes a short hydrophilic region (residues 1-30 in predicted sequence of pACE1) similar to that found in *Drosophila* and mosquito. This region is characteristic of insect AChEs and is thought to contain the proteolytic site in *Drosophila* which generates the large (55kD) and small (16kD) polypeptides of the mature protein.[8,9] A similar pattern of cleavage is seen for housefly AChE.[10] The major region of divergence between the sequences is in the final 30 residues at the C termini. The predominant form of *Drosophila* and housefly AChE consists of an amphiphilic dimer,[9,11] analogous to the G2 form of vertebrates, and these hydrophobic 'tail' sequences are removed from the mature protein following the attachment of a glycophospholipid which anchors the protein to the membrane.

The cDNA clone described here contains 80% of the coding region of the housefly AChE gene. We are now using oligonucleotide primers derived from this sequence to clone upstream gene sequences which encode the N terminal region of the protein, and to amplify the corresponding AChE sequences from the insecticide-resistant housefly strains. This will allow us to identify the changes responsible for insensitivity and so provide detailed information on the critical residues influencing how these insecticides interact with their target enzyme.

REFERENCES

1. F.J. Oppenoorth. Comprehensive Insect Physiology Biochemistry and Pharmacology, Vol. 12, ed. G.A. Kerkut & L.I. Gilbert. Pergamon, Oxford, 713-773 (1985).
2. G.D. Moores, A.L. Devonshire and I. Denholm. *Bull. Ent. Res.* 78:537 (1988).
3. S.J. Gould, S. Subramani and I.E. Scheffler. *Proc. Natl. Acad. Sci. USA* 86:1934 (1989).
4. L.M.C. Hall and C.A. Malcolm. *Cell Mol. Neurobiol.* 11:131 (1991).
5. L.M.C. Hall and P. Spierer. *EMBO J.* 5:2949 (1986).
6. D. Fournier, F. Karch, J.M. Bride, L.M.C. Hall, J.B. Berge and P. Spierer. *J. Mol. Biol.* 210:15 (1989).
7. J.L. Sussman, M. Harel, F. Frolow, C. Oefner, A. Goldman, L. Toker and I. Silman. *Science* 253:872 (1991).
8. D. Fournier, J.M. Bride, F. Karch and J.B. Berge. *FEBS Lett.* 238:333 (1988).
9. R. Haas, T.L. Marshall and T.L. Rosenberry. *Biochemistry* 27:6453 (1988).
10. R.W. Steele and B.N. Smallman. *Biochim. Biophys. Acta.* 445:147 (1976).
11. D. Fournier, J.B. Berge, M.L. Cardoso de Almeida and C. Bordier. *J. Neurochem.* 50:1158 (1988).

PCR GENERATED HOMOLOGOUS DNA PROBES AND SEQUENCE FOR ACETYLCHOLINESTERASE GENES IN INSECT PESTS

C.A. Malcolm[1], S. Rooker[1], A. Edwards[2], D. Heckel[3] and L.M.C. Hall[2]

[1]School of Biological Sciences, Queen Mary and Westfield College
University of London, Mile End Road, London E1 4NS, UK

[2]Department of Medical Microbiology, London Hospital Medical School
University of London, Turner Street, London E1 2AD, UK

[3]Department of University, 132 Long Hall, Clemson, South Carolina
29634-1903, USA

INTRODUCTION

Insect acetylcholinesterase (AChE) is the target for organophosphate and carbamate insecticides. A wide range of medically and agriculturally important insect pests have developed insecticide resistance involving an altered AChE with reduced sensitivity to insecticide inhibition. This has prompted studies to determine the nature of the genetic mutations causing resistance.

The first insect AChE gene to be cloned and sequenced was from *Drosophila melanogaster* (Hall and Spierer 1986). A *D. melanogaster* AChE gene cDNA clone was used as a heterologous probe in an attempt to clone AChE genes from mosquitoes. Success was only obtained with the Indo-Pakistan malaria vector *Anopheles stephensi*. The complex genomic organisation of the AChE gene in *D. melanogaster* (Fournier et al. 1989) was also found in *A. stephensi*, but the average intron size was about seven times smaller. The complexity of the gene appears likely to have contributed to the problems of using a heterologous probe. Lack of sequence homology towards the 3' end of the gene was another factor which will also have contributed to problems in cloning from cDNA. Increased homology towards the 5' end of the gene also includes regions of homology with other insect esterase genes. (Malcolm and Hall 1990, Hall and Malcolm 1991).

In this paper an approach using the Polymerase chain reaction (PCR) method to generate homologous DNA probes from genomic DNA is described, first working with other Dipteran insects and then with non-Dipterans.

Multidisciplinary Approaches to Cholinesterase Functions, Edited by
A. Shafferman and B. Velan, Plenum Press, New York, 1992

METHODS

DNA from mosquito (except *Culex pipiens*) and moth species was prepared from colonies maintained at QMW, London, and at Clemson University, South Carolina, respectively. DNA from *C. pipiens* was supplied by Dr. N. Pasteur (University of Montpellier, France). DNA and insects from aphid and beetle species were supplied by Dr. A. Devonshire (Rothamsted Experimental Station, Harpenden, UK) and Dr. P. Mason (CSL, Slough, UK) respectively. Frozen medflies were supplied by Dr. R. Wood, (University of Manchester, UK).

Genomic DNA libraries were prepared in the bacteriophage vector Lambda Dash™ for insecticide resistant *C. pipiens* and insecticide susceptible *Anopheles gambiae* as previously published (Malcolm and Hall 1990). Polymerase chain reaction was performed following standard protocols (Innis et al. 1990).

RESULTS AND DISCUSSION

No examples of insecticide insensitive AChE (iAChE) have been reported in *An. stephensi*, but homozygous iAChE strains are available of two other important mosquito disease vectors, *An. albimanus* and *C. pipiens*. A third species, *An. gambiae*, does not have a reported iAChE, but was included because of its importance as the African malaria vector.

PCR primers were designed from an alignment of the predicted protein sequences for the two insect AChEs and available vertebrate cholinesterase sequences. A large number have now been prepared and used on genomic DNA from all three mosquito species. One pair termed F and G were most successful and were selected to extend the studies to other species. Primers F to G cover amino acid residues 109 to 175 and encompass intron 2. The region also includes the short 35 to 41 amino acid hydrophilic sequence, where proteolysis occurs to produce the two polypeptide subunits of the insect AChE monomer.

The mosquito FG PCR fragments were partially sequenced and compared to the predicted protein sequence of *An. stephensi*. The splice site for intron 2 was conserved in all three species. All of the 16 predicted amino acids residues for *An. gambiae*, and 14 out of 15 for *An. albimanus*, and 9 out of 10 for *C. pipiens*, showed identity with *An. stephensi* AChE sequence. The biggest difference between the species was the larger intron 2 in *C. pipiens*, this can be seen in the size of the FG PCR fragments obtained (Table 1.).

The FG PCR fragments were used to screen genomic DNA libraries. Eight positive clones were obtained from screening approximately 80,000 recombinants of the *C. pipiens* library and two positives from approximately 30,000 recombinants of the *An. albimanus* library. This contrasts with one positive out of approximately 150,000 *An. stephensi* clones using the heterologous *D. melanogaster* probe. Four of the *C. pipiens* clones were restriction mapped. The smallest contained a fragment of 16kb which is contained within the inserts of two of the other three and overlaps substantially with the third. The two *An. albimanus* clones were also very similar. Direct sequencing of PCR fragments from these clones has provided most of the coding sequence for the *An. albimanus* AChE gene, which shows about 95% homology with *An. stephensi* based on predicted protein sequence. Across exon 3, one of the most conserved regions of the gene, nucleotide identity was about 85%. The average intron lengths were similar, but nucleotide identity drops to about 53%. Sequencing of the *C. pipiens* AChE gene has been slower due to larger introns, results to date show extensive homology to the anopheline AChE genes, thus reflecting the results from FG PCR fragments.

Based on the results obtained new primers are being designed to amplify the alternative insecticide sensitive or insensitive AChE alleles by PCR. To date fragments have been obtained for about one third of the alternative *C. pipiens* AChE allele. Also an FG PCR fragment from a second form of *C. pipiens*, *C. molestus*, has been obtained (Table 1.)

Table 1. AChE PCR fragment sizes (base pairs) obtained using primers F and G. (Sizes are based on measurements from PAGE or sequence))

Anopheles albimanus	(South American malaria vector)	295
Anopheles gambiae	(African malaria vector)	320
Culex pipiens	(Egyptian filariasis vector)	580
Culex molestus	(Form of *C. pipiens*)	580
Cerititus capitata	(Mediterranean fruitfly)	330
Heliothis virescens	(Tobacco budworm)	365
Anticarsia gemmatalis	(Velvetbean caterpillar)	349
Helicoverpea zea	(Corn earworm)	349
Eucles imperialis	(Imperial moth)	360
Plutella xylostella	(Diamondback moth)	301
Myzus persicae	(Peach potato aphid)	332
Phorodon hummuli	(Aphid)	197
Bemisia tabaci	(Sweet potato whitefly)	339
Aphis gossypii	(Aphid)	342
Tetramychus urticae	(Two spotted spider mite)	350
Oryzaephilus surinamensis	(Grain beetle)	346

A wide range of agricultural pests have also developed insecticide resistance due to iAChE. Given the success of the FG PCR primers on mosquito genomic DNA, it was decided to evaluate the potential to rapidly generate homologous AChE probes for a more diverse range of species. An inter-species comparison of mutations causing iAChE would be as valuable as an intra-species survey, in furthering understanding of this resistance mechanism. Through collaboration with other groups working on well characterised insect strains homozygous for particular iAChE variants, it should be possible to build a relatively extensive list of candidate insecticide insensitivity mutations.

The results obtained using the FG primers are listed in Table 1. Not all the species used were known to possess an iAChE variant. For example, in the Medfly, *Ceratitus capitata*, insecticide resistance due to iAChE has not been reported. This is in spite of extensive Medfly control programmes involving the use of organophosphate insecticides. Preliminary confirmation of the Medfly AChE PCR fragment was made by cross-hybridisation to a Southern blot of the restriction enzyme digested cDNA clone of the *D. melanogaster* AChE gene. This PCR fragment was used to obtain clones from a genomic DNA library, from which a part of the AChE gene has now been sequenced (see G.K. Banks this proceedings).

iAChE has been reported in *H. virescens*. The FG PCR fragment from this species produced clear cut positive results when hybridised to Southern blots of the *D. melanogaster* AChE cDNA clone and to *H. virescens* and *H. subflexa* (a close relative) genomic DNA. A

H. virescens juvenile hormone esterase gene probe hybridised to a replicate filter of the genomic DNA produced a different pattern of bands to the PCR fragment. The *H. virescens* PCR fragment was also hybridised to a Southern blot of primary FG PCRs from the other four moth species. Single bands were obtained in each lane for *Anticarsia gemmatalis*, *Helicoverpea zea*, and *Eucles imperialis* corresponding to the expected fragment sizes (Table 1). The fourth moth, *Plutella xylostella* which produces a smaller fragment did not hybridise. The *H. virescens* and the *P. xylostella* PCR fragments are now being sequenced and used to screen genomic DNA libraries.

The beetle, mite and aphid species produced fragment sizes similar to the moths, however small but distinct differences in size were observed between most on polyacrylamide gel electrophoresis (PAGE). The exception was *P. hummuli*, which produced a much smaller fragment, possibly consistent with the absence of intron 2. The results were all reproducible when repeated with different batches of reaction components. This also cross-checked possible contaminations. Where multiple bands were seen on electrophoresis, the most likely candidate was usually easily distinguished on the basis of intensity and size. In each case this fragment was eluted from the gel and reamplified.

The results accumulated, although not fully analysed by sequencing, demonstrate the potential to rapidly expand the collection of insect AChE sequence to a diverse range of species.

ACKNOWLEDGEMENTS

We are grateful to Nicole Pasteur and Michel Raymond (University of Montpellier, France); Alan Devonshire (Rothamsted Experimental Station, Harpenden, UK); Richard Brown (QMW, London, UK), and Phil Mason (CSL, Slough, UK) for the supply of DNA and insects and their collaboration.

REFERENCES

Fournier, D., Karch, F., Bride, J.M., Hall, L.M.C., Berge, J.B. and Spierer, P., 1989,The *Drosopila melanogaster* acetylcholinesterase gene: Structure, evolution and mutations, J. Mol. Biol. 210:15.

Hall, L.M.C. and Malcolm, C.A., 1991, The acetylcholinesterase gene of *Anopheles stephensi*, Cell. Mol. Neurobiol. 11:131.

Hall, L.M.C. and Spierer, P., 1986, The Ace locus of *Drosophila melanogaster*: Structural gene for acetylcholinesterase with an unusual 5' leader, EMBO J. 5:2949

Innis, M.A., Gelfand, D.H., Sninsky, J.J. and White, T.J., 1990, "PCR Protocols," Academic Press, Inc. San Diego.

Malcolm, C.A. and Hall, L.M.C., 1990, Cloning and characterisation of a mosquito acetylcholinesterase gene, in: "Molecular Insect Science," H.H. Hagedorn, J.G. Hildebrand, M.G. Kidwell, and J.H. Law, eds., Plenum, New York.

TESTICULAR GENE AMPLIFICATION AND IMPAIRED BCHE TRANSCRIPTION INDUCED IN TRANSGENIC MICE BY THE HUMAN BCHE CODING SEQUENCE

Rachel Beeri [1,] Averell Gnatt[1], Yaron Lapidot-
Lifson[1,2], Dalia Ginzberg[1], Moshe Shani[3,]
Haim Zakut[2] and Hermona Soreq[1]

[1]Dept. of Biol. Chem., The Life Sciences Inst., The Hebrew
Univ., Jerusalem 91904, Israel. [2]Dept of Obs/ Gyn., The
Edith Wolfson Med. Ctr., Holon, The Sackler Faculty of Med.
Tel-Aviv Univ., Israel. [3]Dept. of Genet. Eng., The Inst.
of Animal Sci., Agric. Res., 906, Beit Dagan 50250. Israel

INTRODUCTION

Multiple findings implicate acetylcholine with sperm functioning [1,2] and acetyl- and butyrylcholinesterase activities (ACHE, BCHE) were observed in mammalian sperm cells and during oocyte development [1-3.] *In vivo* amplification of the human BCHE gene was first found in a father and son exposed to cholinesterase inhibitors [4], but it remained unclear whether the amplified DNA was transmitted as such from father to son or whether the amplification phenomenon re-occurred in germ cells, particularly during male meiosis or sperm differentiation.

Amplified genes in tumors and in cultured cells often cause, through selection advantage, resistance to inhibitory agents as a result of the amplified genes being over expressed [5]. Selection advantage may also result from over expression of genes encoding growth factors or of oncogenes [6.] In certain cases, the amplified sequences may undergo rearrangements and mutagenesis, lose the ability to be expressed and get lost during cell division. In other cases, additional gene amplifications subsequently appear [7.] Co-amplification of BCHE, ACHE and certain oncogenes was found in DNA from leukemic patients, in ovarian carcinomas and in noncancerous blood cells from a patient with defective hemopoiesis due to Lupus Erythematosus ,[8,9,10.] To reveal whether the human BCHE coding region includes domains which suffice for its amplification, whether disruption of cellular mechanisms induced by this process may cause amplification of other genes, and to what extent this amplification can be transferred to future generations, we constructed and studied transgenic mice carrying the BCHEcDNA sequence.

METHODOLOGY

Transgenic mice carrying the pSVL-CHE DNA (Fig.1) were constructed [11,12.] Examination of structure and copy number of the amplified transgene and endogenous sequences in mouse testes and somatic DNA was performed by DNA blot

Multidisciplinary Approaches to Cholinesterase Functions, Edited by
A. Shafferman and B. Velan, Plenum Press, New York, 1992

91

hybridizations[8,12]. Transcription leved of BCHE and smooth muscle T actin (SMGA)[13] genes in mouse testes was examined by reverse transcription and direct PCR amplification[14] of total RNA extracted from tissues[15].

RESULTS AND DISCUSSION

Three out of 17 founders and several FI transgenic mice carrying the human BCHE coding region in an SV-40 vector carried 1-3 copies of BCHE cDNA hybridizing sequences. Two independent FII pedigrees were subsequently established and tail DNA hybridization demonstrated wide variability in BCHE DNA copy numbers . Altogether 31% of tail DNA samples from FII mice carried 5-200 copies of the BCHE DNA sequence but only 4% of FIII generation mice displayed this phenomenon (Fig.1).

Since the BCHE gene co-amplifies with ACHE in various systems, we also examined the endogenous ACHE coding sequence. None of the FII mice displayed somatic ACHE DNA amplification, but tail DNA from the two FIII mice carrying amplified BCHE DNA also carried co-amplified ACHE sequences. Independent amplification events were suggested by the variability in restriction patterns observed for the amplified BCHE DNA in different FII mice from the same litter. Re-hybridization with SV-40 DNA revealed fragments of the same size as the BCHE labeled ones, indicating that the amplified BCHE DNA is the transgene. Since transgene integrations occur as random events, the appearance of BCHE amplifications in two pedigrees suggests that this sequence could contain elements which control its own amplification or cis-activate randomly localized origins of replication in the mouse genome.

To explain the loss of amplification between generations, we examined DNA from tails, testes and epididymis of male transgenic mice by hybridization with cDNA probes for ACHE, BCHE, the amplifiable oncogene c-raf and the tissue specific haptoglobin gene. Tail DNA from all these examined FII and FIII mice did not carry amplified genes, however DNA from testes and epididymis of all of these mice displayed BCHE, c-raf and SV-40 amplifications, and most of the mice also showed amplified endogenous ACHE and cFES(fps) DNA in similar levels (Fig. 1). In all the tissues examined, haptoglobin DNA was found in 1-3 copies. In view of our previous observations of ACHE, BCHE and c-raf co-amplifications in humans [8-10], we suggest that the CHE genes belong to a family of genes which can undergo amplifications in various biosystems, and may further disrupt an equilibrium which normally prevents uncontrolled DNA replication.

In 4 weeks old mice the hybridization signals in testes, which at this age is poorer in mature sperm cells, were lower than those observed with epididymis DNA, implicating the amplification with late stages of sperm development. The loss of amplification could be due to functional defects in sperm cells in which the amplification occurred. Therefore, we examined mRNA levels in testes by direct reverse transcription and PCR amplification. Total RNA from testes of two control mice and six transgenic mice carrying between 5 and 15 copies of the amplified genes displayed similar levels of actin mRNA. Drastic reductions in BCHE mRNA levels were, however, observed in the testes of transgenic mice carrying amplifications versus control mice. Tail BCHE mRNA levels were identical in control and transgenic mice carrying no amplifications,focusing the transcriptional damage to the testes tissues. This decrease in BCHE gene expression may be implicated with the impaired fertilizing capacity of sperm cells carrying the amplifications, perhaps due to the amplification of the ACHE gene which is inversely regulated with the BCHEgene in multiple systems [16].

Sperm motility defects, with which the cholinergic system has been implicated [2,3] belong to the primary factors causing human male infertility. Our current findings suggest that exposure to cholinesterase inhibitors could be causally involved with human infertility. The apparently inheritable BCHE gene amplification which we previously observed [4] could thus reflect repetitive responses to environmental exposure to such inhibitors.

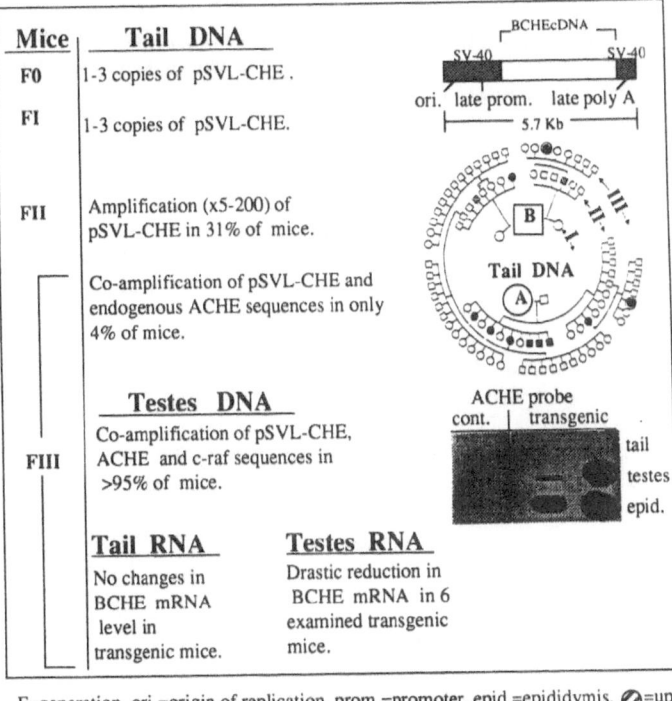

Mice	Tail DNA	
F0	1-3 copies of pSVL-CHE .	
FI	1-3 copies of pSVL-CHE.	
FII	Amplification (x5-200) of pSVL-CHE in 31% of mice.	
	Co-amplification of pSVL-CHE and endogenous ACHE sequences in only 4% of mice.	

Testes DNA

| FIII | Co-amplification of pSVL-CHE, ACHE and c-raf sequences in >95% of mice. | |

Tail RNA — **Testes RNA**

No changes in BCHE mRNA level in transgenic mice.

Drastic reduction in BCHE mRNA in 6 examined transgenic mice.

F=generation, ori.=origin of replication, prom.=promoter, epid.=epididymis, ⊘=up to 20 copies of pSVL-CHE, ●=up to 200 copies, ◉= up to 200 copies of pSVL-CHE and ACHE.

Figure 1. Genomic DNA amplifications and Transcriptional aberrations in pSVL-CHE transgenic mice.

REFERENCES

1. B.V. Rama Sastry, V.E. Janson and A.K. Chaturvedi. Inhibition of human sperm motility by inhibitors of choline acetyltransferase. J. Pharmacol. Exp. Ther. 216: 378-384 (1981).
2. C.F. Ibanez, M. Pelto-Huikko, o. Soder, et al. Expression of choline acetyltransferase mRNA in spermatogenic cells results in an accumulation of the enzyme in the postacrosomal region of mature spermatozoa. Proc. Natl. Acad. Sci. USA. 88: 3676-3680 (1991).
3. G. Malinger , H. Zakut and H. Soreq. Cholinoceptive properties of human primordial, pre-antral and mature oocytes: In-situ hyridization and biochemical evidence for expression of cholinesterase genes. J. Mol. Neuroscience 1: 77-84 (1989).
4. C.A. Prody, P.Dreyfus, R. Zamir, et al. *De-novo* amplification within a "silent" human cholinesterase gene in a family subjected to prolonged exposure to organophosphorous insecticides. Proc. Natl. Acad. Sci. USA. 86: 690-694 (1989).
5. R.T. Schimke, Gene amplification in cultured animal cells. Cell 37:705-713 (1984).
6. J.M. Bishop, The molecular genetics of cancer. Science 235:305-311 (1987).
7. P.G. Pauw, M.D. Johnoson, P. Moore , et al. Stable gene amplification and overexpression of sodium and potassium activated ATPase in Hela cells. Mol. Cell. Biol. 6: 1164-1171 (1986).
8. Y. Lapidot-lifson, C.A. Prody, D. Ginzberg, et al. Co-amplification of human acetylcholinesterase and butyrylcholin-esterase genes in blood cells: correlation with various leukemias and abnormal megakaryocytopoiesis. Proc. Natl. Sci. USA. 86:4715-4719 (1989).

9. H. Zakut , G.Ehrlich , A.Ayalon , et al. Acetylcholinesterase and butyrylcholinesterase genes coamplify in primary ovarian carcinomas. J. Clin. Invest. 86: 900-908 (1990).

10. H. Zakut, , Y. Lapidot-Lifson, , R. Beeri, et al. In-vivo gene amplification in non cancerous cells: cholinesterase genes and oncogenes amplify in thrombocytopenia associated with Lupus erythematosus. Mutation Research, in press. (1992).

11. M. Shani. Tissue specific expression of rat myosin light chain 3 gene in transgenic mice. Nature 314: 283-286 (1985).

12. R. Beeri, , A. Gnatt, , Y. Lapidot-Lifson, et al. Gene amplification and its impaired transmission studied in transgenic mice carrying human butyrylcholinesterase cDNA. submitted (1992).

13. E. Kim, S.H. Waters ,L.E. Hake, et al. Identification and developmental expression of a smooth-muscle gamma-actin in postmeiotic male germ cells of mice. Mol. Cell Biol. 1875 1881 (1989).

14. Y. Lapidot-Lifson, D. Patinkin, C.A. Prody, et al. Cloning and antisense oligodeoxynucleotide inhibition of a human homolog of cdc2 required in hematopoiesis. Proc. Natl. Acad. Sci. USA, 89:579-583 (1992).

15. C.A. Prody , D. Zevin-Sonkin , A. Gnatt , et al. Isolation and characterization of full-length cDNA clones coding for cholinesterase from fetal human tissues. Proc. Natl. Acad. Sci. USA, 84:3555-3559 (1987).

16. P.G. Layer , Cholinesterases during development of the avian nervous system. Cell. Mol. Neurobiol. 11:7-33. (1991).

THREE DIMENSIONAL STRUCTURE OF ACETYLCHOLINESTERASE

J.L. Sussman[1], M. Harel[1], and I. Silman[2]

[1]Department of Structural Biology
[2]Department of Neurobiology
The Weizmann Institute of Science
Rehovot 76100 Israel

INTRODUCTION

Acetylcholinesterase (AChE, acetylcholine hydrolase, EC 3.1.1.7) plays a key role in cholinergic neurotransmission. By rapid hydrolysis of the transmitter acetylcholine (ACh), the enzyme terminates the chemical impulse, thereby permitting rapid repetitive responses and allowing re-uptake of choline (Katz, 1966; Barnard, 1974):

$$CH_3COOCH_2CH_2N^+(CH_3)_3 + AChE$$
$$\downarrow$$
$$CH_3CO-AChE + HOCH_2CH_2N^+(CH_3)_3$$
$$\downarrow$$
$$CH_3COO^- + H^+ + AChE$$

In keeping with this requirement, AChE possesses a remarkably high specific activity, especially for a serine hydrolase (Quinn, 1987), functioning at a rate approaching that of a diffusion-controlled reaction (Hasinoff, 1982; Bazelyansky *et al.*, 1986). The powerful acute toxicity of organophosphorus poisons is primarily because they are potent inhibitors of AChE (Koelle, 1963). They inhibit by forming a covalent bond to a serine residue in the active site (Quinn, 1987). AChE inhibitors are used in treatment of various disorders such as myasthenia gravis and glaucoma (Taylor, 1990), and their use has been proposed as a possible therapeutic approach in the management of Alzheimer's disease (Hallak and Giacobini, 1989), which is known to be associated with a depletion of levels of ACh. Consequently, a variety of AChE inhibitors have been synthesized and characterized pharmacologically. Knowledge of the three-dimensional structure of AChE is, therefore, essential for understanding its remarkable catalytic efficacy, for rational drug design, and for developing therapeutic approaches to organophosphate poisoning. Furthermore,

Multidisciplinary Approaches to Cholinesterase Functions, Edited by
A. Shafferman and B. Velan, Plenum Press, New York, 1992

information about the ACh-binding site of AChE will help us understand the molecular basis for the recognition of ACh by other ACh-binding proteins such as the various ACh receptors (Dougherty and Stauffer, 1990).

We present here a brief overview of the three-dimensional structure of AChE followed by a description of the structure of complexes of the enzyme with two anticholinesterase agents of pharmacological interest.

METHODS

In *Torpedo*, a major form of AChE is a homodimer bound to the plasma membrane via covalently attached phosphatidylinositol (Silman and Futerman, 1987). The purification procedure, crystallization conditions, structure determination and refinement of *T. californica* AChE have been described (Sussman *et al.*, 1988; Sussman *et al.*, 1991) and the coordinates are available from the authors as well as from the Brookhaven Protein Data Bank (code 1ACE) (Bernstein *et al.*, 1977).

Crystals of complexes of AChE with edrophonium (ethyl(3-hydroxyphenyl)dimethylammonium) (EDR) and with tacrine (1,2,3,4-tetrahydro-9-aminoacridine) (THA) were obtained by soaking native crystals in solutions of the drugs (10mM EDR chloride and saturated THA, respectively). X-ray intensity data for each of the two conjugates were collected in the same way as for the native crystal (Sussman *et al.*, 1991). The structures were determined by difference Fourier techniques and refined using simulated annealing and restrained refinement in a way similar to that of the native enzyme (Sussman *et al.*, 1991).

RESULTS

Structure of AChE

The molecule has an ellipsoidal shape with dimensions ~45 x 60 x 65 Å. It belongs to the class of α/β proteins (Levitt and Chothia, 1976; Richardson, 1985) (see Figs. 1,2 and Table 1). The AChE homodimer, whose subunits are related by a crystallographic two-fold axis, appears to be held together by a four-helix bundle composed of two helices from each subunit (Sussman *et al.*, 1991).

One of the most surprising results that emerged from the structure determination of AChE was the identification of a new protein structural motif, the α/β hydrolase fold, which is common to several hydrolytic enzymes of widely differing phylogenetic origin and catalytic function (Ollis *et al.*, 1992). These enzymes consist of AChE from *Torpedo californica* (Sussman *et al.*, 1991), carboxypeptidase II (CPW) from wheat (Liao and Remington, 1990), dienelactone hydrolase (DLH) from *Pseudomonas sp.* B13 (Pathak *et al.*, 1988; Pathak and Ollis, 1990), haloalkane dehalogenase (HAL) from *Xanthobacter autotrophicus* (Franken *et al.*, 1991), and a lipase (GLP) from *Geotrichum candidum* (Schrag *et al.*, 1991). They have very different sequences*, substrates and physical properties. The core of each enzyme is similar: an α/β-sheet (not barrel) of eight β-strands connected by α-helices (Fig. 3). These enzymes have evidently diverged from a common ancestor so as to preserve the active site geometry, but not the binding site. They all have a catalytic triad the elements of which are borne on loops which are the best-conserved structural features in the fold. Only the histidine in the nucleophile-histidine-acid catalytic

---

*The only exception is that AChE & GLP have ~25% identical amino acids (Slabas *et al.*, 1990).

triad is completely conserved, with the nucleophile and acid loops accommodating more than one type of amino acid. There are now four groups of enzymes which contain catalytic triads and which are related by convergent evolution towards a stable, useful active site: the eukaryotic serine proteases, the cysteine proteases, subtilisins, and the α/β hydrolase fold enzymes. A more thorough discussion of the fascinating similarity of these hydrolase fold enzymes has been presented recently (Ollis *et al.*, 1992).

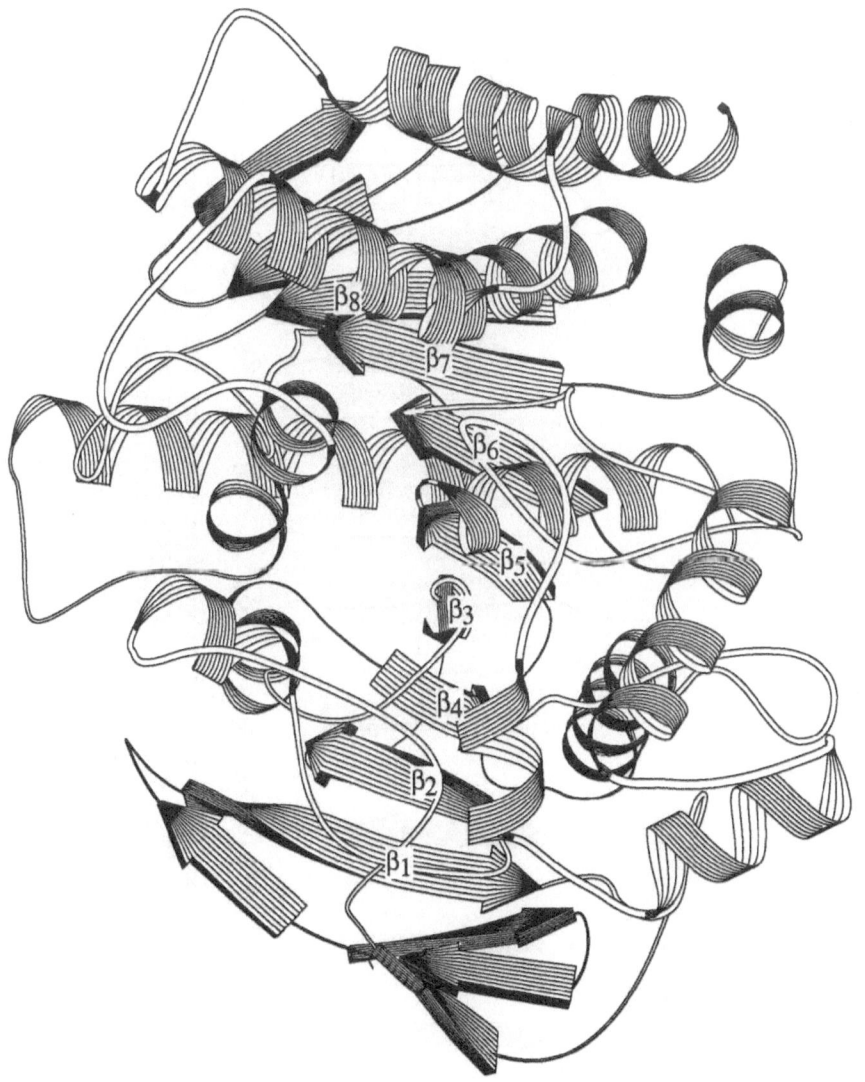

Figure 1. Schematic ribbon diagram of the 3-D fold of *T. californica* AChE. The N-terminus is located at the bottom left, and the C-terminus at the top right. The structure consists of an 11-stranded central mixed β-sheet surrounded by 15 α-helices, and a short 3 stranded β-sheet at the N-terminus which is not hydrogen bonded to the central sheet*. The 8 strands of the central β-sheet which are found in all the α/β hydrolase fold proteins are labeled.

---

*This description of the secondary structure of AChE, i.e. the separation of the central β-sheet from the short N-terminus β-sheet, is slightly different from the way it was originally described (Sussman *et al.*, 1991), after carefully comparing its three-dimensional structure with that of GLP (Cygler *et al.*, 1992).

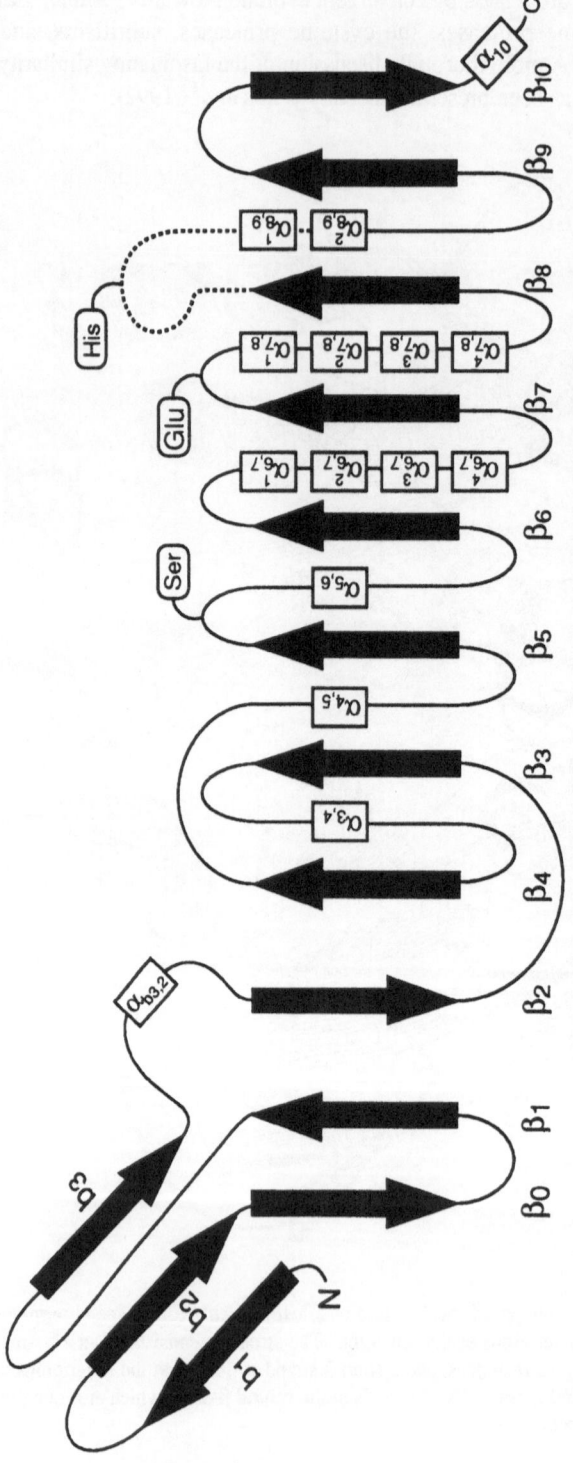

Figure 2. Secondary-structure cartoon showing the topology of AChE, with the β-sheets represented by black arrows and the α-helices by small open rectangles. The numbering of the central β-sheet corresponds to that of the eight-stranded β-sheet of the α/β hydrolase fold of enzymes (Ollis *et al.*, 1992) and the numbering of the α-helices is that of the cholinesterase family of enzymes (Massoulié *et al.*, 1992). The positions of the residues of the catalytic triad (S[200], E[327], H[440]) are indicated.

Table 1. Secondary structure of *T. californica* AChE

Segment Name	Residue Numbers	Comments
b_1	6-10	Small N-terminal β-sheet
b_2	13-16	"
β_0	18-21	Large central β-sheet
β_1	26-34	"
b_3	57-60	Small N-terminal β–sheet
$\alpha_{b3,2}$	79-85	α-helix
β_2	96-102	Large central β-sheet
β_3	109-116	"
$\alpha_{3,4}$	132-139	α-helix
β_4	142-147	Large central β-sheet
$\alpha_{4,5}$	168-183	α-helix
β_5	193-199	Large central β-sheet
$\alpha_{5,6}$	200-211	α-helix
β_6	220-226	Large central β-sheet
$\alpha_{6,7}^1$	238-252	α-helix
$\alpha_{6,7}^2$	259-268	α-helix
$\alpha_{6,7}^3$	271-278	α-helix
$\alpha_{6,7}^4$	305-311	α-helix
β_7	318-324	Large central β-sheet
$\alpha_{7,8}^1$	329-335	α-helix
$\alpha_{7,8}^2$	349-360	α-helix
$\alpha_{7,8}^3$	365-376	α-helix
$\alpha_{7,8}^4$	384-411	α-helix (*kink* at residue 400)
β_8	417-423	Large central β-sheet
$\alpha_{8,9}^1$	443-448	α-helix
$\alpha_{8,9}^2$	460-479	α-helix
β_9	502-505	Large central β-sheet
β_{10}	510-514	Large central β-sheet
α_{10}	518-534	α-helix

The existence of a catalytic triad in AChE has been the subject of controversy (Quinn, 1987). The earlier identification of S^{200} as the active-site serine of *T. californica* AChE (MacPhee-Quigley *et al.*, 1985) has recently been supplemented by the designation of H^{440} as the catalytic histidine residue on the basis of sequence comparison (Doctor *et al.*, 1989; Gentry and Doctor, 1991) and site-directed mutagenesis (Gibney *et al.*, 1990). Our chain tracing clearly supports this assignment by placing H^{440} close to S^{200} and E^{327} (Sussman *et al.*, 1991) and the participation of E^{327} has also recently been verified by site directed mutagenesis (Shafferman *et al.*, 1992a). The three residues form a planar array which resembles the catalytic triad of chymotrypsin (Cht) and other serine proteases (Steitz and Shulman, 1982). There are, however, two important differences:

1) AChE, together with the above mentioned GLP (Schrag *et al.*, 1991), are, to the best of our knowledge, the first published cases of Glu occurring instead of Asp in a catalytic triad.

2) As in CPW, DLH and the neutral lipases, which have a similar fold to AChE, this triad is of the opposite 'handedness' to that of Cht (Sussman *et al.*, 1991), hence, changes the direction of the polypeptide backbone around the His and Ser residues. This suggests that the oxyanion hole, which is formed by the amide NH of the active site Ser in the serine proteases, would be formed instead by the amide NH of the *following* residue in AChE, A^{201}, as appears to be the case for human pancreatic lipase (Winkler *et al.*, 1990; Gubernator *et al.*, 1991) and for the other structurally related hydrolases (Ollis *et al.*, 1992). All three residues of the triad occur within highly conserved regions of the sequence (Sussman *et al.*, 1991) and, as is typical of active sites in α/β-proteins (Richardson, 1981), are in loops following the C-termini of β-strands.

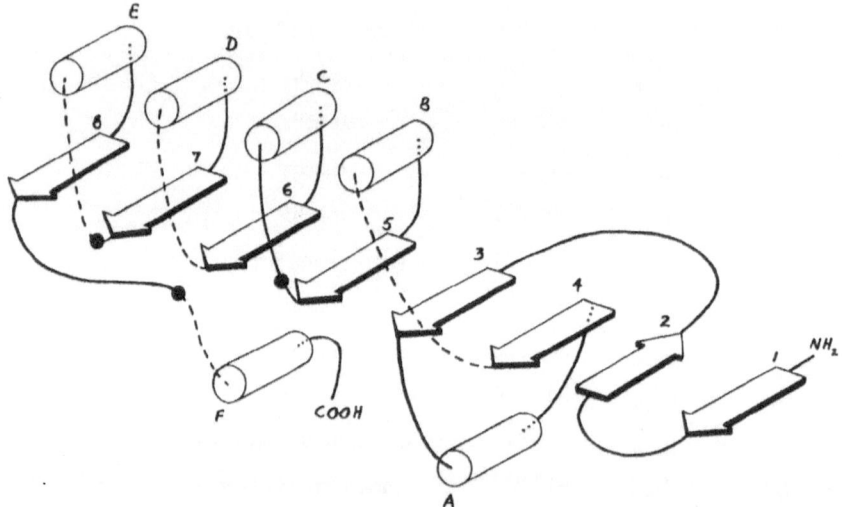

Figure 3. A schematic drawing of the hydrolase fold, which consists of an 8 stranded β-sheet surrounded by helices. The residues of the catalytic triad are indicated with black circles. The broken lines indicate places where some of the structures have additional excursions.

The most remarkable feature of the structure is a deep and narrow gorge, about 20 Å long, which penetrates halfway into the enzyme and widens out close to its base. We have named this cavity the 'active site gorge' (Sussman *et al.*, 1991) because it contains the AChE catalytic triad. S^{200} Oγ, which can be seen from the surface of the enzyme, is about 4 Å above the base of the gorge. 14 aromatic residues line a substantial portion of the surface of the gorge (~40%) (Y[70], W[84], W[114], Y[121], Y[130], W[233], W[279], F[288], F[290], F[330], F[331], Y[334], W[432], Y[442]). These residues and their flanking sequences are highly conserved in AChE's from widely different species (Sussman *et al.*, 1991), and are located primarily in loops between β-strands. It should be noted that the gorge contains only a few acidic residues, which include D[285] and E[273] at the very top, D[72], hydrogen-bonded to Y[334], about half way down, and E[199] close to the bottom.

The presence of tryptophan in the active site of AChE was predicted by spectroscopic and chemical modification studies (Shinitzky *et al.*, 1973; Blumberg and Silman, 1978; Goeldner and Hirth, 1980). A recent affinity labelling study (Weise *et al.*, 1990) in fact identified W[84] as part of the putative 'anionic' (choline) binding site. Recent site directed mutagenesis studies of the human enzyme, also indicated the importance of W[84] for catalytic activity (Shafferman *et al.*, 1992b). An earlier photoaffinity labelling study implicated a peptide in electric eel AChE, homologous to *Torpedo* G[328]-S[329]-F[330]-F[331], as part of the binding site (Kieffer *et al.*, 1986). The observation of tyrosine residues close to the catalytic site agrees with chemical modification studies (Fuchs *et al.*, 1974; Blumberg and Silman, 1978; Page and Wilson, 1985).

Despite the structural complexity of the gorge and the flexibility of the natural substrate, ACh (Chothia and Pauling, 1968), a good fit of the extended, all-trans conformation of ACh was obtained by manual docking (Sussman *et al.*, 1991). Specifically, the acyl group was positioned to make a tetrahedral bond with the Oγ of S[200] while the quaternary group of the choline moiety was placed within van der Waals distance (~3.5Å) of W[84]. The model suggests that the 'oxyanion hole' (Steitz and Shulman, 1982) would be formed by the main chain nitrogens of G[118], G[119] and A[201] interacting with the carbonyl oxygen, and that the ester oxygen may interact with the imidazole of H[440]. G[118] and G[119] are part of a 10-residue conserved sequence which contains three glycines in a row; this may make the chain flexible enough to allow amide nitrogens from both G[118] and G[119] to be part of the oxyanion hole. E[199], which might serve as an anionic component of the substrate-binding site, appears in our model to make close contacts (~3Å) both to one of the quaternary methyl groups and to the α-carbon of the choline moiety, although it has been reported that mutating it to glutamine had little effect on the enzymic kinetic parameters (Gibney *et al.*, 1990). E[199] appears, however, to be hydrogen-bonded, either directly or through a water molecule, to E[443]. As both carboxylic acid sidechains are in a hydrophobic environment in the interior of AChE, it seems likely that one or both of them are protonated. This might explain the result of the mutagenesis experiment.

The high aromatic content of the walls and floor of the active site gorge, together with its dimensions, may help explain why biochemical studies have revealed a variety of hydrophobic and 'anionic' binding sites distinct from, or overlapping, the active site. For instance, chemical modification by various reagents (Purdie and McIvor, 1966; Meunier and Changeux, 1969; O'Brien, 1969; Fuchs *et al.*, 1974) greatly reduces enzymic activity towards ACh either without affecting, or sometimes actually enhancing, activity towards various neutral esters. This supports the existence of hydrophobic areas distinct from the binding site for ACh. Other evidence for hydrophobic sites extending beyond or distinct from the anionic site comes from studies on the affinities and reaction rates of homologous series of organophosphate inhibitors (Kabachnik *et al.*, 1970), on the affinities of various acridine derivatives (Steinberg *et al.*, 1975) and from studies employing resolved enantiomeric methylphosphonothioates (Berman and Decker, 1989; Berman and Leonard, 1989). The complexity of the array of aromatic residues also provides candidates for a binding site for aromatic cations, the existence of which, closer to the esteratic site than the 'anionic' site, was recently proposed (Berman and Leonard, 1990). All these results are consistent with the characteristics of the deep gorge extending up from the active site of AChE.

Two reports have used photolabelling (Amitai and Taylor, 1991) and affinity labelling (Weise *et al.*, 1990) to identify peptide sequences (residues 251-264 and 270-278 respectively) as part of the 'peripheral' binding site(s) for ACh and other quaternary ligands. These two neighboring peptides on the surface of the protein are close to the rim of the gorge. The complex and varied inhibitory effects of different peripheral site ligands (Bergmann *et al.*, 1950; Changeux, 1966; Belleau *et al.*, 1970; Rosenberry, 1975; Quinn, 1987) may be better understood taking into account the complex geometry of the gorge. Certain ligands may be too bulky to penetrate it, but still could partially block its entrance. Long bisquaternary compounds, which serve as potent inhibitors, might attach at one end to the peripheral site(s) and at the other end to any one of the various aromatic residues lining the walls of the gorge. However, because of its depth, shorter bisquaternary inhibitors and oxime reactivators might bind wholly within the gorge itself.

Crystal Structures of AChE - Inhibitor Complexes

Edrophonium (Fig. 4a) is a powerful competitive inhibitor of AChE (Hobbiger, 1952; Wilson and Quan, 1958) which is used clinically in the diagnosis of myasthenia gravis (Osserman and Genkins, 1971). Due to its quaternary character, it does not penetrate cell membranes or the blood-brain barrier and thus acts primarily at peripheral sites such as the muscle endplate (Taylor, 1990). It is an analog of the carbamate, neostigmine (Fig. 4b), which is employed clinically in the management of myasthenia gravis (Drachman, 1987). This latter compound contains a quaternary moiety almost identical to EDR, but since it serves as a carbamylating agent of the active-site serine of AChE, has a considerably longer duration of action (Taylor, 1990). Tacrine (Fig. 4c) is also a powerful competitive inhibitor of AChE (Heilbronn, 1961). Due to its tertiary character, it can penetrate the blood-brain barrier, and is currently under active consideration for the management of Alzheimer's disease (Summers *et al.*, 1986; Gauthier and Gauthier, 1991).

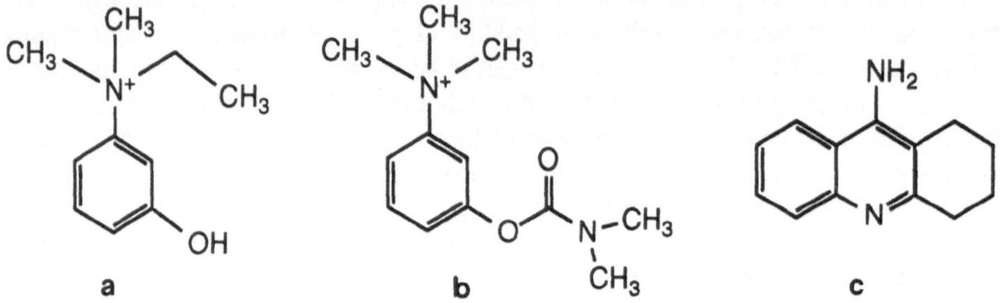

Figure 4. Chemical formulae of the anticholinesterase agents (a) EDR; (b) neostigmine; (c) THA.

The structure of the AChE:EDR complex (with 53 water molecules included) was refined to an R-factor of 18.4% (6-2.8Å data), while the structure of AChE:THA complex (with 75 water molecules included) was refined to an R-factor of 18.2%. In the course of the latter refinement it became clear that the side chain of W279 was disordered in 2 discrete conformations.

The overall conformations of native AChE, of the EDR and THA complexes are very similar. The quaternary nitrogen group of EDR nestles adjacent to the indole ring of W^{84}, as predicted for the quaternary group of ACh (Sussman *et al.*, 1991), with the three alkyl groups lying in a plane approximately parallel to the plane of the indole ring of W^{84}. The hydroxyl group, at the *meta* position in EDR, is positioned between $H^{440} N^{\epsilon 2}$ and $S^{200} O^{\gamma}$, making hydrogen bonds to these atoms of two of the three members of the catalytic triad (see Fig. 5a).

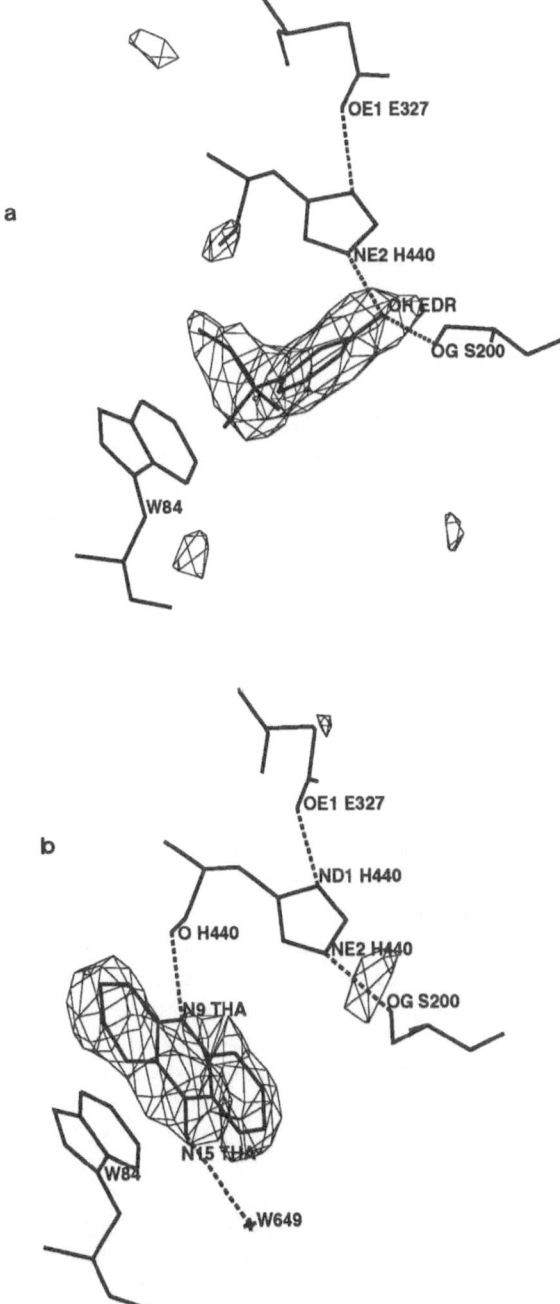

Figure 5. Omit (F_O-F_C) maps at 2.8 Å resolution, of drug-AChE complexes after 7 cycles of restrained least squares refinement (omitting the drug) (a) EDR-AChE (b) THA-AChE complex. Dashed lines indicate hydrogen bonds.

In the THA-AChE complex, the THA moiety is stacked against W[84], with its ring nitrogen forming a hydrogen bond with the main-chain carbonyl oxygen of H[440]; its amino nitrogen forms a hydrogen bond to a water molecule, which is found also in the EDR-AChE complex but *not* in the native enzyme, (see Fig. 5b). The most pronounced differences between the native AChE structure and that of the two complexes lie in the positioning of the aromatic rings of residues F[330] and W[279]. Both these residues belong to the set of highly conserved aromatic amino acids whose rings line the surface of the gorge leading to the active site (see above). In both complexes, the phenyl ring of F[330] swings away from its position in the native enzyme so as to to make a better aromatic-aromatic interaction (Burley and Petsko, 1985) with the ring(s) of the corresponding inhibitor. The indole moiety of W[279], which is located close to the entrance of the aromatic gorge, changes its orientation (rather differently in each complex) even though it is located ~8 Å away from the nearest atom of either of the inhibitor molecules.

DISCUSSION

The crystal structures of two complexes of *T. californica* AChE with ligands of clinical importance were determined and refined at 2.8Å resolution. As seen in the three-dimensional structure of native *Torpedo* AChE (Sussman *et al.*, 1991), the catalytic triad lies near the bottom of a long and narrow gorge, which is lined with the rings of 14 aromatic amino acid residues. Despite the complexity of this array of aromatic rings, we suggested, on the basis of modelling which involved docking of the ACh molecule in an all-*trans* conformation, that the quaternary group of the choline moiety makes close contact with the indole ring of W84 (Sussman *et al.*, 1991). This assignment is strongly supported by our crystallographic data for the EDR-AChE complex, which shows the aliphatic substituents of the quaternary nitrogen of EDR in a plane parallel to, and ~4Å from that of the W84 indole ring. This is the same residue which was covalently labelled by the aziridinium ion (Weise *et al.*, 1990), which is similar in structure to EDR; furthermore, EDR protects against labelling by aziridinium. Our data also demonstrate, therefore, a close correspondence between the crystal structure and the structure in solution.

The structure of the THA-AChE complex is also in agreement with earlier solution studies. Specifically, our finding that the 3-ring structure of THA is stacked opposite the indole ring of W84 is in full agreement with the spectroscopic observation that the competitive inhibitor, N-methylacridinium (MAC), which also possesses a 3-ring structure, forms a charge-transfer complex with a Trp residue in the active site of AChE in which the two ring structures face each other (Shinitzky *et al.*, 1973).

As pointed out above, the oxygen of the *m*-hydroxyl group of EDR is within hydrogen-bonding distance of two key atoms of residues of the catalytic triad, i.e., H[440] Nε2 and S[200] Oγ. This provides a structural basis for the observation that such *meta*-substituted anilinium ions are much more powerful competitive inhibitors of AChE than either the homologous non-substituted anilinium ions or ones in which the ring has been substituted at a different position (Hobbiger, 1952; Wilson and Quan, 1958).

Although both inhibitors make strong contacts with the indole ring of W[84], this is not the only aromatic ring in close contact with the ligand in the EDR-AChE and THA-AChE complexes. In both, the position of the phenyl ring of F[330] differs substantially from its position in the native enzyme, rotating about the first side chain torsion angle. In the EDR-AChE complex, the ring rotates so as to make close contact (4.2Å) with the quaternary group of EDR, while in the THA-AChE complex, it stacks against the THA molecule which thus becomes sandwiched between W[84] and F[330]. It should be noted that in this case, too, data obtained in solution support this observation (Kieffer *et al.*, 1986).

It is of particular interest that an additional aromatic residue, W^{279}, also displays substantial movement of its indole ring upon binding of the two drugs. This is surprising because the ring in question is at least 8Å away from the nearest atom of either of the two bound ligands. The role of the multiple aromatic rings in the upper part of the 'aromatic gorge', which cannot be involved directly in binding of substrate, is still obscure. We have suggested that they provide 'aromatic guidance', i.e. provide low affinity sites to which the quaternary group of choline can bind reversibly (Sussman et al., 1991). These sites might serve to assist both substrate (ACh) diffusion to, and product (choline) diffusion from the active site of this rapid enzyme, which functions at rates at which diffusion control may play a role (Quinn, 1987). They may also serve as candidates for the so-called 'peripheral' anionic site. The experimental evidence recently presented showed that this site may serve as a low-affinity ACh-binding site responsible for the phenomenon of substrate inhibition (Radic et al., 1991; Reiner et al., 1991), is not inconsistent with the 'aromatic guidance' role which we have put forward.

THA, while being considered for use in the management of Alzheimer's disease on the strength of its potent anticholinesterase action, has also been shown to exert diverse pharmacological effects, including the blockage of K^+ channels, as recently reviewed (Freeman and Dawson, 1991). Recent work on voltage-dependent K^+ channels has identified an 18-amino acid sequence, SS1-SS2, as providing the lining for the pore (Hartmann et al., 1991). In the model proposed (MacKinnon, 1991), residues 3-5 of the SS1-SS2 sequence, FWW, are close to the mouth of the pore, and it has been reported that site-directed mutagenesis of the phenylalanine residue affects pore selectivity (Yool and Schwartz, 1991). This aromatic motif could provides a plausible site of interaction for THA. Aromatic groups might also contribute to the site of external blockade of K^+ channels by tetraethylammonium. Both theoretical considerations and studies with model host compounds have been invoked recently to suggest participation of aromatic residues in binding sites both for ACh and for other quaternary compounds (Dougherty and Stauffer, 1990), as is directly supported by our crystallographic studies. The observation that mutation (Thr to Tyr) of the residue immediately following the C-terminus of the SS1-SS2 sequence strongly enhances sensitivity to external tetraethylammonium (MacKinnon, 1991) is in keeping with this notion. It is of interest to speculate whether conformational changes of the type induced in AChE by EDR and THA, binding might also occur upon ligand-binding in nicotinic or muscarinic ACh receptors or in receptors which are activated by biogenic amines bearing an aromatic group, such as adrenaline and serotonin. Photoaffinity labelling studies (Galzi et al., 1991) have shown that several aromatic rings are involved in the ACh-binding site of the nicotinic ACh receptor and that labelling of certain of these residues was increased in the desensitized state, suggesting that they might play a key role in the putative conformational changes involved in the regulation of ligand-gated ion channels.

Acknowledgments

We thank Miriam Lachever for help in various stages of this research and Lilly Toker for preparation of the AChE. This project was supported by the U.S. Army Medical Research and Development Command under Contract DAMD17-89-C9063, the Association Franco-Israélienne pour la Recherche Scientifique et Technologique, the Minerva Foundation, Munich, Germany and the Kimmelman Center for Biomolecular Structure and Assembly, Rehovot.

REFERENCES

Amitai, G., and Taylor, P., 1991, *in*: "Cholinesterases: Structure, Function, Mechanism, Genetics and Cell Biology," J. Massoulié, F. Bacou, E. Barnard, A. Chatonnet, B. P. Doctor, and D. M. Quinn, Eds, American Chemical Society, Washington, DC.

Barnard, E. A., 1974, *in*: "The Peripheral Nervous System," J. I. Hubbard, Eds, Plenum, New York.

Bazelyansky, M., Robey, C., and Kirsch, J. F., 1986, *Biochemistry*. 25:125.

Belleau, B., DiTullio, V., and Tsai, Y. H., 1970, *Mol. Pharmacol*. 6:41.

Bergmann, F., Wilson, I. B., and Nachmansohn, D., 1950, *Biochim. Biophys. Acta*. 6:217.

Berman, H. A., and Decker, M. M., 1989, *J. Biol. Chem*. 264:3951.

Berman, H. A., and Leonard, K., 1989, *J. Biol. Chem*. 264:3942.

Berman, H. A., and Leonard, K., 1990, *Biochemistry*. 29:10640.

Bernstein, F. C., Koetzel, T. F., Williams, G. J. B., Meyer, E. F., Jr., Brice, M. D., Rodgers, J. R., Kennard, O., Schimanouchi, T., and Tasunmi, M., 1977, *JMB*. 112:535.

Blumberg, S., and Silman, I., 1978, *Biochemistry*. 17:1125.

Burley, S. K., and Petsko, G. A., 1985, *Science*. 229:23.

Changeux, J. P., 1966, *Mol. Pharmacol*. 2:369.

Chothia, C., and Pauling, P., 1968, *Nature*. 219:1156.

Cygler, M., Schrag, J. D., Sussman, J. L., Harel, M., Silman, I., Gentry, M. K., and Doctor, B. P., 1992 (submitted).

Doctor, B. P., Smyth, K. K., Gentry, M. K., Ashani, Y., Christner, C. E., De La Hoz, D. M., Ogert, R. A., and Smith, S. W., 1989, *in*: "Computer-Assisted Modeling of Receptor-Ligand Interactions. Theoretical Aspects and Applications to Drug Design," R. Rein, and A. Golombek, Eds, A.R. Liss, New York.

Dougherty, D. A., and Stauffer, D. A., 1990, *Science*. 250:1558.

Drachman, D., Ed, 1987, "Myasthenia gravis: biology and treatment," vol. 555 of "Ann. N.Y. Acad. Sci.," New York.

Franken, S. M., Rozeboom, H. J., Kalk, K. H., and Dijkstra, B. W., 1991, *EMBO J*. 10:1297.

Freeman, S. E., and Dawson, R. M., 1991, *Prog. Neurobiol*. 36:257.

Fuchs, S., Gurari, D., and Silman, I., 1974, *Arch. Biochem. Biophys*. 165:90.

Galzi, J.-C., Revah, F., Bouet, F., Ménez, A., Goeldner, M., Hirth, C., and Changeux, J.-P., 1991, *Proc. Natl. Acad. Sci. USA*. 88:5051.

Gauthier, S., and Gauthier, L., 1991, *in*: "Cholinergic Basis of Alzheimer Therapy," R. Becker, and E. Giacobini, Eds, Birkhauser, Berlin.

Gentry, M. K., and Doctor, B. P., 1991, *in*: "Cholinesterases: Structure, Function, Mechanism, Genetics and Cell Biology," J. Massoulié, F. Bacou, E. Barnard, A. Chatonnet, B. P. Doctor, and D. M. Quinn, Eds, American Chemical Society, Washington, DC.

Gibney, G., Camp, S., Dionne, M., MacPhee-Quigley, K., and Taylor, P., 1990, *Proc. Natl. Acad. Sci. USA*. 87:7546.

Goeldner, M. P., and Hirth, C. G., 1980, *Proc. Natl. Acad. Sci. USA*. 77:6439.

Gubernator, K., Müller, K., and Winkler, F. K., 1991, *in*: "Lipases- Structure, Mechanism and Genetic Engineering," L. Alberghina, R. D. Schmid, and R. Verger, Eds, VCH, Weinheim.

Hallak, M., and Giacobini, E., 1989, *Neuropharmacology*. 28:199.

Hartmann, H. A., Kirsch, G. E., Drewe, J. A., Taglialatela, M., Joho, R. H., and Brown, A. M., 1991, *Science*. 251:942.

Hasinoff, B. B., 1982, *Biochim. Biophys. Acta*. 704:52.

Heilbronn, E., 1961, *Acta Chem. Scand*. 15:1386.

Hobbiger, F., 1952, *Brit. J. Pharmacol*. 7:223.

Kabachnik, M. I., Brestkin, A. P., Godovikov, N. N., Michelson, M. J., Rozengart, E. V., and Rozengart, V. I., 1970, *Pharmacol. Rev*. 22:355.

Katz, B., 1966, "Nerve, Muscle and Synapse," McGraw-Hill, New York.

Kieffer, B., Goeldner, M., Hirth, C., Aebersold, R., and Chang, J. Y., 1986, *FEBS Lett*. 202:91.

Koelle, G. B., Ed, 1963, "Cholinesterase and anti-cholinesterase agents," vol. 15 of "Handbuch der Experimentellen Pharmakologie," Springer-Verlag, Heidelberg.

Levitt, M., and Chothia, C., 1976, *Nature*. 261:552.

Liao, D., and Remington, S. J., 1990, *J. Biol. Chem*. 265:6528.

MacKinnon, R., 1991, *Current Opinion Neurobiol*. 1:14.

MacPhee-Quigley, K., Taylor, P., and Taylor, S., 1985, *J. Biol. Chem*. 260:12185.

Massoulié, J., Sussman, J. L., Doctor, B. P., Soreq, H., Velan, B., Cygler, M., Rotundo, R., Shafferman, A., Silman, I., and Taylor, P., 1992, *this volume*.

Meunier, J.-C., and Changeux, J.-P., 1969, *FEBS Lett*. 2:224.

O'Brien, R. D., 1969, *Biochem. J*. 113:713.

Ollis, D. L., Cheah, E., Cygler, M., Dijkstra, B., Frolow, F., Franken, S. M., Harel, M., Remington, S. J., Silman, I., Schrag, J., Sussman, J. L., Verschueren, K. H. G., and Goldman, A., 1992, *Protein Engineering*. 5:197.

Osserman, K. E., and Genkins, G., 1971, *Mt. Sinai J. Med.* 38:497.

Page, J. D., and Wilson, I. B., 1985, *J. Biol. Chem.* 260:1475.

Pathak, D., Ngai, K. L., and Ollis, D., 1988, *JMB.* 204:435.

Pathak, D., and Ollis, D., 1990, *JMB.* 214:497.

Purdie, J. E., and McIvor, R. A., 1966, *Biochim. Biophys. Acta.* 128:590.

Quinn, D. M., 1987, *Chem. Rev.* 87:955.

Radic, Z., Reiner, E., and Taylor, P., 1991, *Mol. Pharmacol.* 39:98.

Reiner, E., Aldridge, N., Simeon, V., Radic, Z., and Taylor, P., 1991, *in*: "Cholinesterases: Structure, Function, Mechanism, Genetics and Cell Biology," J. Massoulié, F. Bacou, E. Barnard, A. Chatonnet, B. P. Doctor, and D. M. Quinn, Eds, American Chemical Society, Washington, DC.

Richardson, J. S., 1981, *Adv. Protein Chem.* 34:167.

Richardson, J. S., 1985, *in*: "Diffraction Methods for Biological Macromolecules," H. W. Wyckoff, C. H. W. Hirs, and S. N. Timasheff, Eds, Academic Press, New York.

Rosenberry, T. L., 1975, *Adv. Enzymol.* 43:103.

Schrag, J. D., Li, Y., Wu, S., and Cygler, M., 1991, *Nature.* 351:761.

Shafferman, A., Kronman, C., Flashner, Y., Leitner, M., Grosfeld, H., Ordentlich, A., Gozes, Y., Cohen, S., Ariel, N., Barak, D., Harel, M., Silman, I., Sussman, J. L., and Velan, B., 1992a, *J. Biol. Chem.* (in press).

Shafferman, A., Velan, B., Ordentlich, A., Kronman, C., Grosfeld, H., Leitner, M., Flashner, Y., Cohen, S., Barak, D., and Ariel, N., 1992b, *this volume*.

Shinitzky, M., Dudai, Y., and Silman, I., 1973, *FEBS Lett.* 30:125.

Silman, I., and Futerman, A. H., 1987, *Eur. J. Biochem.* 170:11.

Slabas, A. R., Windust, J., and Sidebottom, C. M., 1990, *Biochem. J.* 269:279.

Steinberg, G. M., Mednick, M. L., Maddox, J., Rice, R., and Cramer, J., 1975, *J. Med. Chem.* 18:1056.

Steitz, T. A., and Shulman, R. G., 1982, *Annu. Rev. Biophys. Bioeng.* 11:419.

Summers, W. K., Majovski, L. V., Marsch, G. M., Tachiki, K., and Kling, A., 1986, *New England J. Med.* 315:1241.

Sussman, J. L., Harel, M., Frolow, F., Oefner, C., Goldman, A., Toker, L., and Silman, I., 1991, *Science.* 253:872.

Sussman, J. L., Harel, M., Frolow, F., Varon, L., Toker, L., Futerman, A. H., and Silman, I., 1988, *JMB.* 203:821.

Taylor, P., 1990, *in*: "The Pharmacological Basis of Therapeutics," A. G. Gilman, A. S. Nies, T. W. Rall, and P. Taylor, Eds, MacMillan, New York.

Weise, C., Kreienkamp, H.-J., Raba, R., Pedak, A., Aaviksaar, A., and Hucho, F., 1990, *EMBO J.* 9:3885.

Wilson, I. B., and Quan, C., 1958, *Arch. Biochem. Biophys.* 73:131.

Winkler, F. K., D'Arcy, A., and Hunziker, W., 1990, *Nature.* 343:771.

Yool, A. J., and Schwartz, T. L., 1991, *Nature.* 349:700.

SEQUENCE ALIGNMENT OF ESTERASES AND LIPASES BASED ON 3-D STRUCTURES OF TWO MEMBERS OF THIS FAMILY

Miroslaw Cygler¶, Joseph D. Schrag¶, Joel L. Sussman†, Michal Harel†, Israel‡ Silman, Mary K. Gentry§ and Bhupendra P. Doctor§

¶Biotechnology Research Institute, National Research Council of Canada Montreal, Quebec H4P 2R2, Canada

Department of Structural Biology† and Neurobiology‡ The Weizmann Institute of Science, Rehovot 76100, Israel

§Division of Biochemistry, Walter Reed Army Institute of Research Washington, DC 20307-5100, USA

INTRODUCTION

Acetylcholinesterases (AChE, EC 3.1.1.7) and butyrylcholinesterases (BChE, EC 3.1.1.8) from various organisms show a high degree of sequence similarity. They have been identified as members of a family of homologous proteins which includes other esterases (carboxylesterases and cholesterol esterases), lipases, and non-enzymatic domains of other proteins such as thyroglobulin (Krejci *et al.*, 1991; Gentry and Doctor, 1991).

Based on published sequence data, 30 proteins of this family have now been identified, three of which are devoid of enzymatic activity. The active sites of the enzymes contain a catalytic triad composed of Ser-His-Glu, with the exception of cholesterol esterases (DiPersio *et al.*, 1990, 1991, and Hui, personal communication) and *Drosophila* esterase P and esterase 6, which contain a Ser-His-Asp triad. Although the enzymes all cleave ester bonds, their substrates vary greatly (e.g. acetylcholine *vs.* triacylglycerols). Comparison of the sequences of these proteins shows that the N-terminal portions (~350 amino acids) are more conserved than the C-terminal parts (~200 amino acids), suggesting that the level of structural similarity in the latter parts will be low. The structures of two members of this family, namely AChE from *Torpedo californica* (Sussman *et al.*, 1991) and lipase (GCL) from *Geotrichum candidum* (Schrag *et al.*, 1991), have recently been determined. They showed an unexpectedly high level of structural similarity. Despite the weaker sequence similarity at the C-terminus, the structural similarity extends throughout the entire length of the sequences. Knowledge of these three-dimensional structures permits analysis of the positions and possible functional roles of highly conserved residues.

Multidisciplinary Approaches to Cholinesterase Functions, Edited by
A. Shafferman and B. Velan, Plenum Press, New York, 1992

STRUCTURAL ALIGNMENT AND TOPOLOGY

The three-dimensional structures of AChE and GCL were superimposed by an automatic procedure (Rossman and Argos, 1975) which overlapped 272 atoms, out of approximately 530 common Cα atoms, with a root-mean-square (rms) deviation of only 2.3 Å. This superposition is shown in Fig. 1. The best correspondence was found within the β-sheet and the α-helices which pack against the sheet. The greatest differences occur above the active site in the parts of the structures involved in substrate binding.

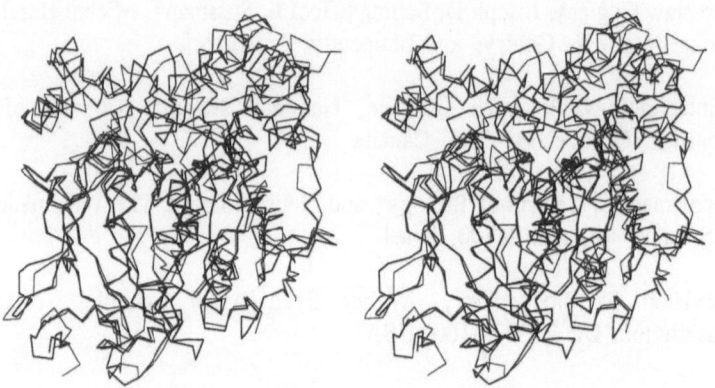

Figure 1. Superposition of GCL (thick line) and AChE (thin line). Catalytic triad residues are shown in full (bold lines). The best superposition is observed in the cores of the molecules, with deviations in atom positions increasing at the surface.

The topologies of AChE and GCL are nearly identical and a diagram of the topology of this enzyme family based on GCL is shown in Fig. 2. AChE and GCL are

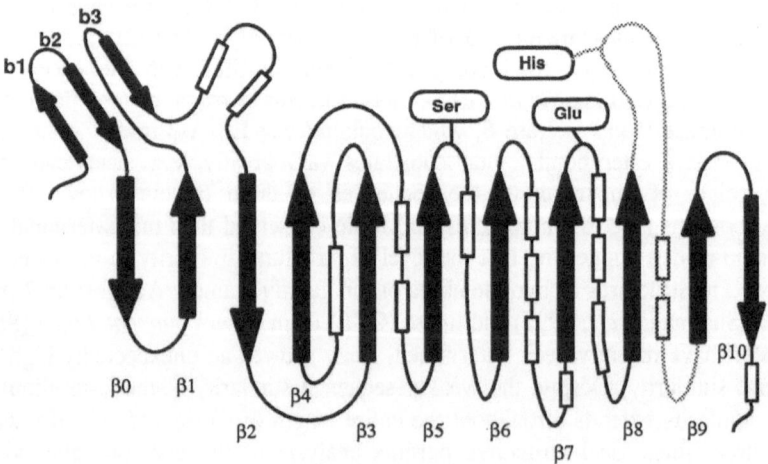

Figure 2. Schematic diagram showing the topology of GCL. This topology is nearly identical to that of AChE and is expected to be representative of the topology of this entire enzyme family.

folded around a highly twisted, 11 stranded, mixed β-sheet in which most strands are parallel. Five helices pack against this β-sheet, two on the concave side and three on the convex side. The remaining helices form a cap along the C-terminal ends of the β-strands (Fig. 1). There is also a small, three-stranded β-sheet at the N-terminus. This topology maintains all the conserved features of the α/β hydrolase fold, which was also identified in the three-dimensional structures of a number of enzymes having no sequence similarities to AChE or GCL (Ollis *et al.*, 1992).

The 3-D superposition of GCL and AChE formed the reference point for the alignment of their sequences and the sequences of other proteins from this family. This multiple sequence alignment shows that the level of identity between any pair from this set varies from 15% to 97%, with the average value of 25%. The graphical representation of sequence identities based on pairwise comparisons after alignment is shown in Fig. 3. This representation shows clearly a further division of these proteins into more closely related subfamilies, e.g. acetyl- and butyrylcholinesterases, carboxyl esterases, lipases, etc.

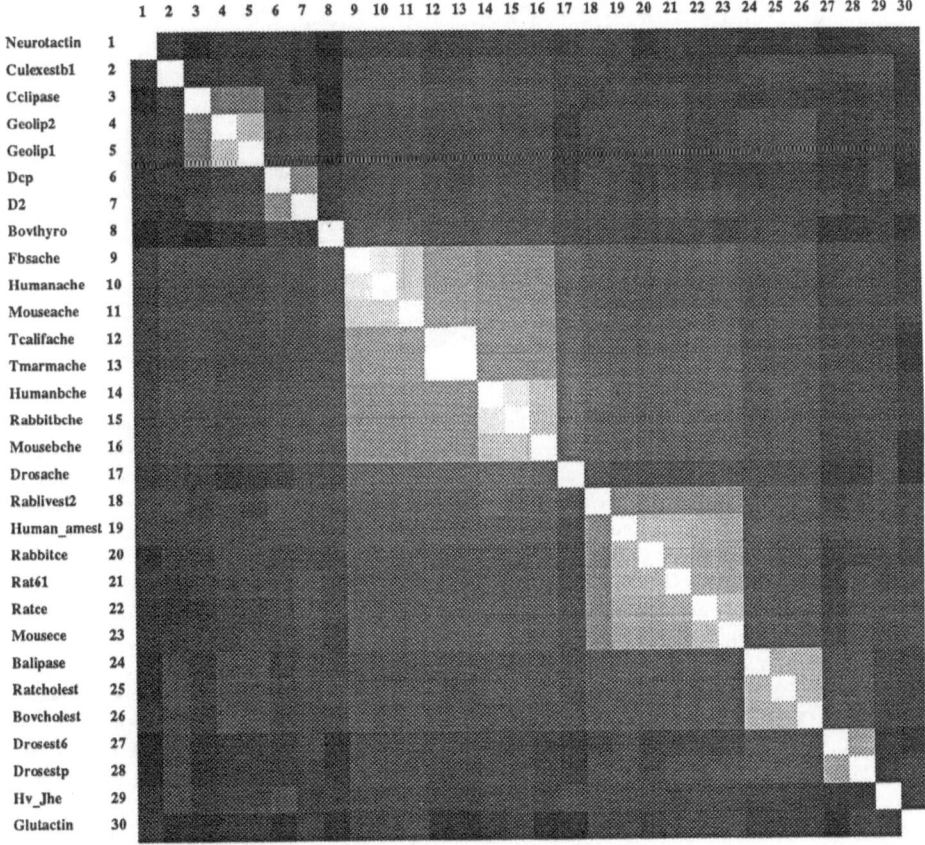

Figure 3. Pairwise comparison of sequences of the esterases/lipases family based on the multiple sequence alignment, as described in the text. Each box represents a percentage of identical residues between a pair of proteins. Greater similarity is represented by lighter boxes. References to sequences are given in the legend to Figure 4.

Examination of the aligned sequences shows that some residues are invariant and others show only low variability and are, generally, conservatively substituted. Analysis

```
Geolip1       AEGNTNAGLHDQRKGLEWVSDNIANFGGDPDKVMIFGESAGAMSVAHQLIAYGGD 233
Geolip2       AEGNTNAGLHDQRKGLEWVSDNIANFGGDPDKVMIFGESAGAMSVAHQLVAYGGD 233
Cclipase      AEGSANAGLKDQRLGMQWVADNIAAFGGDPTKVTIFGESAGSMSVMCHILWNDGD 225
Hv_Jhe        TKIPGNAGLRDQVTLLRWVQRNAKNFGGDPSDITIAGQSAGASAAHLLTLSKATE 219
Culexestbl    DGVPGNAGLKDQNLAIRWVLENIAAFGGDPKRVTLAGHSAGAASVQYHLISDASK 207
Drosestp      RHLPGNYGLKDQRLALQWIKKNIAHFGGMPDNIVLIGHSAGGASAHLQLLHEDFK 203
Drosest6      RDLPGNYGLKDQRLALKWIKQNIASFGGEPQNVLLVGHSAGGASVHLQMLREDFG 204
Ratcholest    ANLPGNFGLRDQHMAIAWVKRNIAAFGGDPDNITIFGESAGAASVSLQTLSPYNK 210
Bovcholest    SNLPGNYGLWDQHMAIAWVKRNIEAFGGDPDNITLFGESAGGASVSLQTLSPYNK 210
Balipase      ANLPGNYGLRDQHMAIAWVKRNIAAFGGDPNNITLFGESAGGASVSLQTLSPYNK 210
Rat61         EHSRGNWGHLDQVAALHWVQDNIANFGGNPGSVTIFGESAGGFSVSALVLSPLAK 219
Human_amest   EHSRGNWGHLDQVAALRWVQDNIASFGGNPGSVTIFGESAGGESVSVLVLSPLAK 220
Rabbitce      .....NIDELFLVAVNRWVQDNIANFGGDPGSVTIFGESAGGQSVSILLLSPLTK 211
Ratce         EHSRGNWAHLDQLAAALRWVQDNIANFGGNPDSVTIFGESAGGVSVSALVLSPLAK 219
Mousece       EHSPGNWAHLDQLAAALRWVQDNIANFGGNPDSVTIFGESSGGISVSVLVLSPLGK 219
Rablivest2    QHATGNHGYLDQVAALRWVQKNIAHFGGNPGRVTIFGESAGGTSVSSHVLSPMSQ 217
Humanbche     PEAPGNMGLFDQQLALQWVQKNIAAFGGNPKSVTLFGESAGAASVSLHLLSPGSH 214
Rabbitbche    PEAPGNMGLFDQQLALQWVQKNIAAFGGNPKSVTLFGESAGAASVSLHLLSPRSH 213
Mousebche     PDAPGNMGLFDQQLALQWVQRNIAAFGGNPKSITIFGESAGAASVSLHLLCPQSY 214
Tcalifache    QEAPGNVGLLDQRMALQWVHDNIQFFGGDPKTVTIFGESAGGASVGMHILSPGSR 216
Tmarmache     QEAPGNMGLLDQRMALQWVHDNIQFFGGDPKTVTLFGESAGGASVGMHILSPGSR 216
Fbsache       REAPGNVGLLDQRLALQSVQENVAAFGGDPTSVTLFGESAGAASVGMHLLSPPSR 219
Humanache     REAPGNVGLLDQRLALQWVQENVAAFGGDPTSVTLFGESAGAASVGMHLLSPPSR 219
Mouseache     REAPGNVGLLDQRLALQWVQENIAAFGGDPMSVTLFGESAGAASVGMHILSLPSR 219
Drosache      EEAPGNVGLWDQALAIRWLKDNAHAFGGNPEWMTLFGESAGSSSVNAQLMSPVTR 227
D2            DLMHGNYGFLDQIKALENVYNNIGSFGGNKEMITIWGESAGAFSVSAHLTFTYSR 209
Dcp           GLLSGNFGFLDQVMALDWVQENIEVFGGDKNQVTIYGESAGAFSVAAHLSSEKSE 212

Bovthyro      SELSGNWGLLDQVVALTWVQTHIQAFGGDPRRVTLAADRGGADIASIHLVTTRAA 194
Glutactin     DELPGNVALSDLQLALEWLQRNVVHFGGNAGQVTLVGQAGGATLAHALSLSGRAG 212
Neurotactin   PPTSGNYALTDIIAVLNWIKLNIVHFGGDPQSVTLLGHRAGATLVTLLVNSQKVK 214
```

Figure 4. Aligned sequences in the region of the active site Ser. The catalytic Ser is located in a G-X-S-X-G consensus sequence. The last three sequences are from non-hydrolytic proteins which lack this Ser. Black boxes mark residues which are highly conserved. The sources of the sequences listed are: Geolip1 - Shimada, Y., *et al., J. Biochem.* **107**, 703 (1990); Geolip2 - Shimada, Y., *et al., J. Biochem.* **106**, 383 (1989); Cclipase - Kawaguchi, K., *et al., Nature* **341**, 164 (1989); Hv_Jhe - Hanzlik, T.N., *et al., J. Biol. Chem.* **264**. 12419 (1989); Culexestbl - Mouches, C., *et al., Proc. Natl. Acad. Sci. U.S.A.* **87**, 2574 (1990); Drosestp - Collet, C., *et al., Mol. Biol. Evol.* **7**, 9 (1990); Drosest6 - Oakeshott, J.G., *et al., Proc. Natl. Acad. Sci. U.S.A.* **84**, 3359 (1987); Ratcholest - Kissel, J.A., *et al., Bioch. Biophys. Acta.* **1006**, 227 (1989); Bovcholest - Kyger, E.M., *et al., Bioch. Biophys. Res. Comm.* **164**, 1302 (1989); Balipase - Hui, D.Y. & Kissel, J.A., *FEBS Letters* **26**, 131 (1990); Rat61 - Robbi, M., *et al., Biochem. J.* **269**, 451 (1990); Human_amest - Munger, J.S., *et al., J. Biol. Chem.* **266**, 18832 (1991); Rabbitce - Korza, G. & Ozols, J., *J. Biol. Chem,* **263**, 3486 (1988); Ratce - Takagi, Y., *et al., J. Biochem.* bf 104, 801 (1988); Mousece - Ovnic, M., *et al., Genomics* **9**, 344 (1991); Rablivest2 - Ozols, J. *J. Biol. Chem.* **264**, 12533 (1989); Humanbche - Lockridge, O., *et al., J. Biol. Chem.* **262**, 549 (1987); Rabbitbche - Jbilo, O. & Chatonnet, A., *Nucl. Acid Res.* **18**, 3990 (1990); Mousebche, Mouseache - Rachinsky, T.L., *et al., Neuron* **5**, 317 (1990); Tcalifache - Schumacher, M., *et al., Nature* **319**, 407 (1986); Tmarmache - Sikorav, J.-L., *et al., EMBO J.* **6**, 1865 (1987); Fbsache - Doctor, B.P., *et al., FEBS Letters* **266**, 123 (1990); Humanache - Soreq, H., *et al., Proc. Natl. Acad. Sci. U.S.A.* **87**, 9688 (1990); Drosache - Hall, L.M.C. & Spierer, P., *EMBO J.* **5**, 2949 (1986); D2 - Rubino, S., *et al., Dev. Biol.* **131**, 27 (1989); Dcp - Bomblies, L., *et al., J. Cell Biol.* **110**, 669 (1990); Bovthyro - Mercken *et al., Nature* **316**, 647 (1985); Glutactin - Olson *et al., EMBO J.* **9**, 1219 (1990); Neurotactin - de la Escalera *et al., EMBO J.* **9**, 3593 (1990).

of the locations of the highly conserved positions within the 3-D fold of the two structures suggests possible functional bases for their conservation. As one might expect, the region of the sequence which is the most highly conserved surrounds the catalytic Ser residue (Fig. 4). This residue is located in a tight turn in a conserved strand-turn-helix motif (Schrag *et al.*, 1991; Derewenda and Derewenda, 1991; Ollis *et al.*, 1992). The catalytic Ser in all of these enzymes is contained within a G-X-S-X-G consensus sequence. The Gly residues in this consensus sequence are conserved for steric reasons. The formation of this tight turn brings these residues within about 4 Å of the neighbouring strands (Fig. 5). Side chains at these positions would sterically interfere with the formation of this strand-turn-helix motif. This tight turn forces the

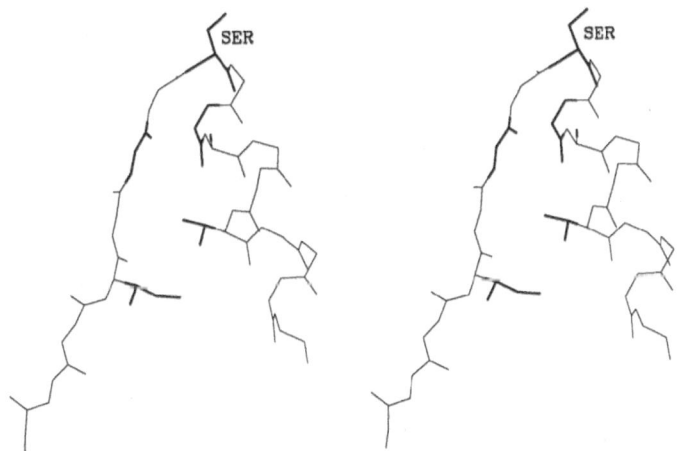

Figure 5. The stereo drawing of the strand-turn-helix supersecondary structure around the catalytic Ser. The backbone atoms of strand β_5 and helix $\alpha_{5,6}$ of GCL are shown. The conserved Gly residues of the G-X-S-X-G consensus sequence, the catalytic Ser, and residues -4, +3 and +6 (relative to Ser) important in the formation of this motif, are shown in full (bold lines).

catalytic Ser residue to adopt unfavourable ϕ,ψ angles, but projects the short side chain into a position which is accessible to the His of the triad and to the substrate. This Ser is located at the N-terminal end of an α-helix, whose dipole may help to stabilize transition states. This motif is common not only to enzymes of this family but also to other lipases (Derewenda and Derewenda, 1991; Schrag *et al.*, 1991) and to α/β hydrolase fold enzymes (Ollis *et al.*, 1992). The packing of residues at positions -4 and +6 (relative to the Ser) is also important in the stabilization of this fold. In the enzymes in which this motif has been observed, these residues are usually short or branched hydrophobic residues (Ala, Val, Ile, Leu). Comparison with the sequences of many other lipases and esterases, that show little or no sequence similarity to AChE or GCL, indicates that this supersecondary structure is likely to be present in most, if not all, of these enzymes. In α/β hydrolase fold enzymes, this motif is maintained, despite the fact that the nucleophile may be either Ser, Cys or Asp (Ollis *et al.*, 1992).

HIGHLY CONSERVED RESIDUES

Many of the invariant residues are located at the C-terminal ends of the β-strands. These residues are often at locations where the direction of the polypeptide chain changes, suggesting that the conservation of these residues maintains the proper fold of the protein. Also invariant are three salt bridges that are close to each other in the structure. These salt bridges are in the same region as the two conserved disulfide bonds. The disulfides and the salt bridges stabilize the conformations and packing of two long surface loops. These loops are located at the entrance to the active site tunnel in AChE (Sussman *et al.*, 1991), and in GCL are proposed to be involved in the conformational change which occurs at the lipid interface to allow access to the active site (Schrag *et al.*, 1991).

Most of the residues that show only low variability and are conservatively substituted are located in the central β-sheet. The β-sheet is strongly twisted. In GCL and AChE, the concave side of the β-sheet shows a predominance of long side chains

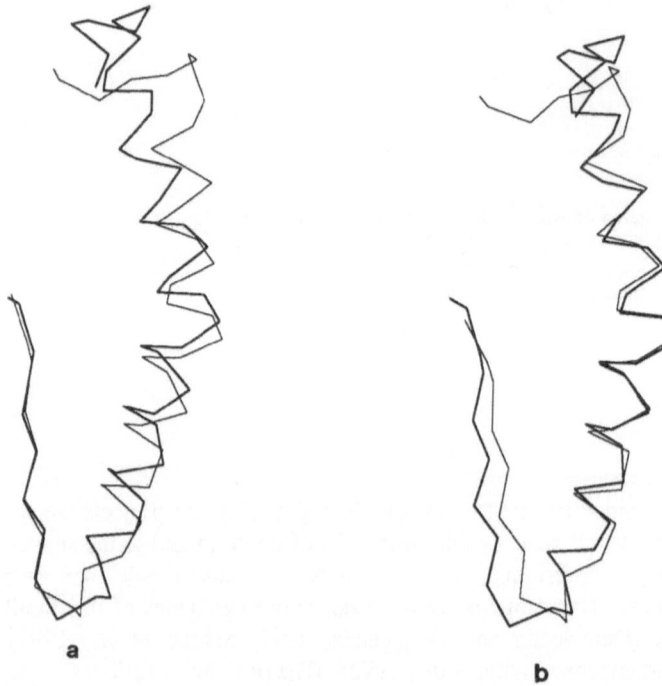

Figure 6. Superposition of the kinked helices of AChE and GCL. a) The deviations in the positions of atoms of the β-sheet were globally minimized. This results in obvious shifts in the positions of helix atoms. b) The superposition was based on the Cα atoms of this helix only and resulted in a good overlap including the kink in the middle, whereas the atoms of the adjacent β-strand are shifted.

with many aromatic rings. The majority of residues on the convex side, on the other hand, are branched hydrophobic residues (Val, Ile, Leu). The aligned sequences indicate that this pattern is maintained in all of the proteins. In general, the sequence conservation is greater in the β-sheet than in the α-helices. Although the homology in the C-terminal portion of the sequences is rather low, there are a number of highly conserved residues (mainly aromatic) in this region which are important for the packing of the secondary structural elements. In AChE and BChE there is a third conserved disulfide bridge that ties the C-terminal helix to the rest of the structure.

GLOBAL VS. LOCAL SIMILARITIES

The rms deviation of atom positions in the β-sheet are smaller than those in many of the helices and loops. The larger deviations in the loops and helices are a result of their positions nearer the surface of the protein. The β-sheets in the core of the molecules superpose very well, but slight differences in the twist of the sheet introduce mismatches which become greater as one moves from the center to the surface of the molecule. The deviations in the positions of the surface loops and helices can often be attributed to rigid body shifts of local segments of the polypeptide chain. For example, both AChE and GCL contain a long helix which has a kink in the middle (Fig.6). When one opts for minimum deviation in the superposition of the β-sheet residues, these kinked helices are shifted by about 1.5 Å relative to one another. If one considers only the helix, the superposition of the mainchain atoms, including the kink in the middle, is almost perfect. Other surface loops show similar rigid body shifts when superposition is based on minimizing the deviations in the positions of β-sheet residues.

The residues which are highly variable are primarily surface residues, some of which may be involved in quaternary associations. Other highly variable positions may be related to substrate specificity differences between the enzymes.

CONCLUSIONS

Based on the three-dimensional structures of two members of the esterase/lipase family, the sequences of 30 proteins have been aligned and analyzed. The superposition of the two structures shows clearly that the structural similarity extends throughout the entire length of the proteins despite the low degree of sequence similarity in the C-terminal portions. Highly conserved residues are mostly in the core of the protein at the edges of the strands of the β-sheet and probably are required for proper folding.

REFERENCES

Derewenda, Z.S. and Derewenda, U. (1991) Relationships among serine hydrolases: Evidence for a common structural motif in triacylglycerol lipases and esterases. *Biochem. Cell Biol.* **69**, 842-851.

DiPersio, L.P., Fontaine, R.N. and Hui, D.Y. (1990) Identification of the active site serine in pancreatic cholesterol esterase by chemical modification and site-specific mutagenesis. *J. Biol. Chem.* **265**, 16801-16806.

DiPersio, L.P., Fontaine, R.N. and Hui, D.Y. (1991) Site-specific mutagenesis of an essential histidine residue in pancreatic cholesterol esterase. *J. Biol. Chem.* **266**, 4033-4036.

Gentry, M.K. and Doctor, B.P.(1991) Alignment of amino acid sequences of acetylcholinesterases and butyrylcholinesterases. in *Cholinesterases: Structure, Function, Mechanism, Genetics and Cell*

Biology, eds. J. Massoulie, F. Bacou, E. Barnard, A. Chatonnet, B.P. Doctor and D.M. Quinn, mer. Chem. Soc.: Washington, DC. pp. 394-398.

Krejci, E., Duval, N., Chatonnet, A., Vincens, P. and Massoulié (1991) Cholinesterase-like domains in enzymes and structural proteins: functional and evolutionary relationships and identification of a catalytically essential aspartic acid. (1991) *Proc. Natl. Acad. Sci, U.S.A.* **88**, 6647-6651.

Ollis, D.L., Cheah, E., Cygler, M., Dijkstra, B., Frolow, F., Franken, S.M., Harel, M., Remington, S.J., Silman, I.,Schrag,J.D., Sussman, J.L., Verschueren, K.H.G. and Goldman, A.(1992) The α/β hydrolase fold. *Protein Eng.* **5**, 197-211.

Rossman, M., and Argos,P. (1975) A comparison of the heme binding pocket in globins and cytochrome b5. *J. Biol. Chem.* **250**, 7552-7532.

Schrag, J.D., Li, Y., Wu, S., and Cygler, M. (1991) Ser-His-Glu triad forms the catalytic site of the lipase from *Geotrichum candidum*. *Nature* **351**, 761-764.

Sussman, J.L., Harel, M., Frolow, F., Oefner, C., Goldman, A., Toker, L and Silman, I. (1991) Atomic structure of acetylcholinesterase from *Torpedo californica*: a prototypic acetylcholine--binding protein. *Science* **253**, 872-879.

STRUCTURAL ANALYSIS OF ACETYLCHOLINESTERASE AMMONIUM BINDING SITES

Isabelle Schalk,[1] Laurence Ehret-Sabatier,[1] Françoise Bouet,[2] Maurice Goeldner,[1] and Christian Hirth[1]

[1]Laboratoire de Chimie Bioorganique, Université Louis Pasteur, Faculté de Pharmacie, BP 24, F-67401 Illkirch Cedex, France
[2]Département d'Ingénierie des Protéines, Centre d'Etudes Nucléaires Saclay, 91191 Gif-sur-Yvette Cedex, France

INTRODUCTION

The identification of the amino acids residues belonging to acetylcholinesterase (AchE) binding sites has been first undertaken through physicochemical determinations or by means of chemical modification experiments. In particular, Trp, Tyr, His and Lys residues have been suggested to be involved in such interactions (Shinitzky et al., 1973; Majumdar and Balasubramanian, 1984; Fuchs et al., 1984; Page and Wilson, 1985). Later, site directed labelling experiments using photoaffinity (Kieffer et al., 1986) and affinity (Weise et al., 1990) probes, showed respectively that a tetrapeptide Gly-Ser-X-Phe (corresponding to amino acids 328 to 331 in Torpedo AchE, X is the labelled residue) and Trp_{84} belong to the quaternary ammonium binding site of the enzyme.

More recently the atomic structure of Torpedo californica AchE was determined (Sussman et al., 1991), describing the active site as a deep and narrow gorge. According to this structure the catalytic Ser_{200} lies near the bottom of this cavity probably interacting with His_{440} and Glu_{327}. The importance of His_{440} was in parallel defined by mutagenesis studies (Gibney et al., 1990). Interestingly, 40% of the surface of this active site gorge is lined by fourteen aromatic residues, suggesting an interaction between these residues and the ammonium moiety of acetylcholine.

In this context, we used an aryldiazonium salt, [^{3}H]-DDF, previously described as an efficient and specific photoaffinity label of the active site of Torpedo AchE (Ehret-Sabatier et al., 1992). With regard to the extreme reactivity of the photogenerated arylcation, [^{3}H]-DDF can tag any proximal residue, including hydrophobic or aromatic. Thus, the identification of [^{3}H]-DDF-labelled residues will allow to precise which residues are involved in the complexation of the ammonium part of acetylcholine within the binding site and thus allowing an insight in the study of this mechanism. We present here the results of the purification and Edman degradation of [^{3}H]-DDF-labelled peptides.

DDF

Multidisciplinary Approaches to Cholinesterase Functions, Edited by
A. Shafferman and B. Velan, Plenum Press, New York, 1992

RESULTS AND DISCUSSION

Fig.1 shows the extent of photoinactivation of purified Torpedo Marmorata AchE by [^{3}H]-DDF. In the absence of ligand the enzyme lost 15% of activity by irradiation, as compared to the control experiment in the dark. In the presence of [^{3}H]-DDF, 70 % of photoinactivation was observed. This inactivation was shown to be totally abolished by 2.10^{-5}M edrophonium, a selective ligand for the active site of the enzyme (Taylor and Lappi, 1975).

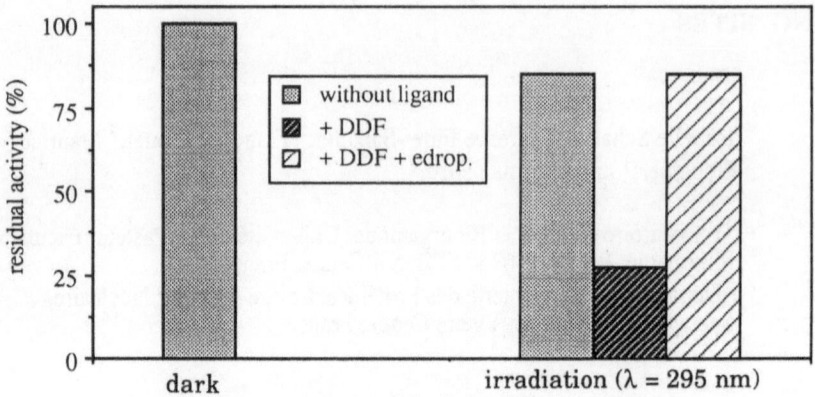

Figure 1. Irreversible inactivation of AchE by [^{3}H]-DDF

AchE was irradiated at 295 nm for 20 min with $4 \cdot 10^{-5}$M [^{3}H]-DDF (1 Ci/mmol) in the absence or in the presence of 2.10^{-5}M edrophonium. Enzymatic activities were monitored using Ellman's assay and acetylthiocholine as substrate.

After irradiation labelled AchE was subjected to proteolytic cleavage by trypsin and the digests were separated by reverse phase HPLC. The profile of radioactivity of these digests exhibited ten major radioactive peaks (fig.2), the radioactivity in dead volume belonging to free label. It is important to note that the incorporation of radioactivity in all these peaks was totally abolished by addition of a protective agent, NMe₄Br (not shown) or edrophonium, during the irradiation experiment. Thus, this profile shows that labelling with [^{3}H]-DDF

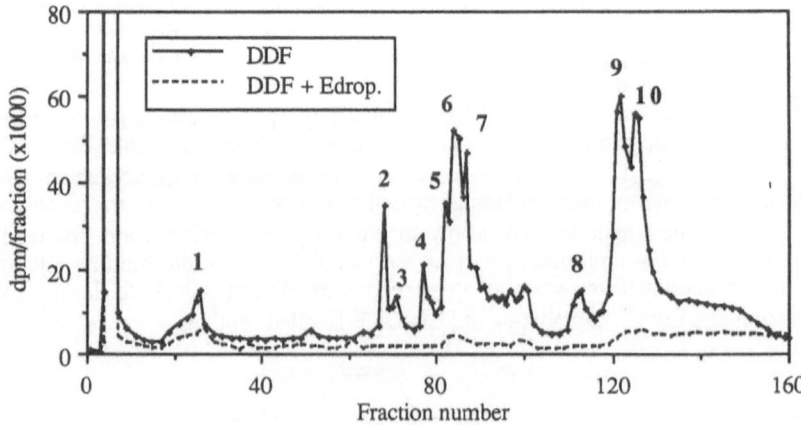

Figure 2. Reversed-HPLC profile of [^{3}H]-DDF-labelled AchE.

After irradiation AchE was digested by 5 % trypsin and the digest injected onto a Vydac C₁₈ column eluted with a linear gradient of 0.1 % TFA to 0.1 % TFA / 80 % CH₃CN in 80 min. Fractions of 0.75 ml (0.5 min) were collected. The recovery of injected radioactivity was greater than 85%.

was specifically directed at the ammonium active binding site. Moreover, knowing the specific activity of [^{3}H]-DDF (1 Ci/mmol), the total amount of radioactivity incorporated on all the labelled peptides allowed us to determine the stoechiometry of alkylation, i.e the ratio of incorporated probe per inactivated site. We found a value of 0.93 ± 0.11 (n = 7) which indicates that one mole of DDF is sufficient to block one active site.

The previous results were obtained under energy transfer conditions, i.e, using an irradiation wavelength of 295 nm in order to excite a tryptophan of the protein (Goeldner et al., 1982; Ehret-Sabatier et al., 1992). In order to check whether the labelling profile could depend of the photoactivation process, the operations were repeated under direct irradiation conditions (at 410 nm). We showed that proteolyzed labelled AchE exhibited similar HPLC profiles, indicating that the same residues were alkylated.

Peaks 2 to 7 were rechromatographed on the same HPLC column using another eluant (CH$_3$CN/nPrOH, profiles not shown). From peaks 4 and 6, we purified, in both cases, only one radioactive peptide. On the other hand, each peak 2, 3, 5 and 7 was resolved in three radioactive components (a, b, c). Thus fourteen purified peptides were subjected to Edman sequencing.

Trp279 was found to be the labelled amino acid in six fractions (2a to 3c), sequence analysis of peak 2a being shown in table 1. These six fractions contained in fact a common sequence (PQELIDVEWN), but were of different length as a result of partial proteolysis. The radioactivity incorporated on Trp279 represented 8.5% of the overall specific protein-bound radioactivity. As mentionned earlier we can assert that the labelling of this tryptophan is not correlated to the conditions of irradiation, i.e. the energy transfer between a tryptophan of the protein and the probe, since these peptides were also present in a profile of direct-irradiated labelled AchE.

The majority of radioactivity (24%) was found on **Phe330** (peptides 4 to 7b, analysis of 4 is given in table 1). Again these peptides contained a common sequence (DEGSFFLLYG), but were of different length. Finally less than 4% of total protein-bound radioactivity was attributed to a labelling of **Leu332** (TQILLGVNKDEGSFFLLY, peptide 7c, Table 1).

Table 1. Results of Edman degradation of three typical [^{3}H]-DDF-labelled peptides : 2a, 4 and 7c. The underlined values show the labelled residues, respectively Trp279, Phe330 and Leu332.

step	peak 2a			peak 4			peak 7c		
	PTH aa	pmol	cpm	PTH aa	pmol	cpm	PTH aa	pmol	cpm
1	Lys	78	382	-	-	686	Thr	47	267
2	Pro	100	465	Glu	27	1523	Gln	23	489
3	Gln	94	496	Gly	36	1395	Ile	24	792
4	Glu	50	676	Ser	8	1857	Leu	20	528
5	Leu	100	888	-	-	<u>30185</u>	Leu	25	534
6	Ile	15	672	Phe	8	4017	Gly	8	711
7	Asp	55	599	Leu	6	2163	-	-	717
8	Val	53	727	Leu	6	2190	Asn	11	900
9	Glu	35	888	Tyr	3	1289	-	-	786
10	-	-	<u>18921</u>	Gly	2	1583	Asp	3	666
11	Asn	22	7741				Glu	1	672
12							-	-	717
13							-	-	945
14							-	-	852
15							-	-	1392
16							-	-	1896
17							-	-	<u>5244</u>
18							-	-	2058
19							-	-	1356
20							-	-	1107

Interestingly the three Torpedo AchE DDF-labelled amino acids are hydrophobic residues. This is in accord with the high reactivity of the photogenerated arylcation which is able to alkylate any residue. Moreover, labelling of Phe330 confirms previous results obtained with DDF as label of Electrophorus AchE (Kieffer et al., 1986). Taken together these results point to the importance of hydrophobic aromatic residues like tryptophan and phenylalanine for the binding of ammonium moieties on AchE. This hypothesis is supported by other studies showing, on different biological systems or macrocyclic compounds, that the binding of methyl-substituted ammonium groups require not only electrostatic but also hydrophobic interactions (Dhaenens et al., 1984; Dennis et al., 1988; Dougherty and Stauffer, 1990).

In the case of Torpedo AchE the atomic structure of the enzyme allowed us to localize the DDF-labelled residues in the active site gorge (Sussman et al., 1991). In fact both Phe330 and Leu332 are positionned near the bottom of this gorge, in the same region as Trp84 previously described as being part of ammonium binding site (Weise et al., 1990). Precisely, Phe330 and Trp84 lie at equal distance (12-13 Å) from the catalytic Ser200 suggesting that both residues are at appropriate distance to complex the ammonium moiety of acetylcholine by interactions between aromatic electrons and the quaternary positive nitrogen.

On the other hand the third DDF-labelled residue, Trp279, is located close to the entrance of the active gorge, 17-18 Å above the bottom area containing the Phe330, Leu332 and Trp84.residues. Considering the size of DDF (less then 10 Å) this distance is too big to explain the labelling of Trp279 by two orientations of the probe within a single binding site. Thus we propose that DDF is recognized by two sites, located respectively one at the bottom (near Phe330) and the other at the top (near Trp279) of the active site gorge. Our experiments showed, that these two sites can be protected from DDF-inactivation either by an ammonium salt (NMe4Br) or by a specific active site ligand (edrophonium). Moreover, the fact that one mole of DDF is incorporated per inactivated site excludes the possibility of simultaneous labelling of these two sites. Labelling of one site might preclude, via dynamic interactions, the labelling of the other. This result raised the problem of the role of the region near Trp279 during the hydrolysis of natural substrate acetylcholine by AchE.

Acknowledgments : We thank Dr B. Rousseau (CEN Saclay) for preparing [^{3}H]-DDF precursor. Financial support was from the *Centre National de la Recherche Scientifique* and from the *Ministère de la Recherche et de la Technologie*.

REFERENCES

Dennis, M., Giraudat, J., Kotzyba-Hibert, F., Goeldner, M., Hirth, C., Chang, J.Y., Lazure, C., Chrétien, M. and Changeux, J.P. (1988), *Biochemistry* 27, 2346-2357.

Dhaenens, M., Lacombe, L., Lehn, J.M. and Vigneron, J.P. (1984), *J. Chem. Soc. Chem. Commun.* 1097-1099.

Dougherty, D.A. and Stauffer, D.A. (1990), *Science* 250, 1558-1560.

Ehret-Sabatier, L., Schalk, I., Goeldner, M. and Hirth, C. (1992), *Eur. J. Biochem.* 203, 475-481.

Fuchs, S., Gurari, D. and Silman, I. (1974), *Arch. Biochem. Biophys.* 165, 90-97.

Gibney, G., Camp, S., Dionne, M., MacPhee-Quigley, K. and Taylor, P. (1990), *Proc. Natl. Acad. Sci.* 87, 7546-7550.

Goeldner, M.P., Hirth, C.G., Kieffer, B. and Ourisson, G. (1982), *TIBS* 7, 310-312.

Kieffer, B., Goeldner, M., Hirth, C., Aebersold, R. and Chang, J.Y. (1986), *FEBS Letters* 202, 91-96.

Majumdar, R. and Balasubramanian, A.S. (1984), *Biochemistry* 23, 4088-4093.

Page, J.D. and Wilson, I.B. (1985), *J. Biol. Chem.* 260, 1475-1478.

Shinitzky, M., Dudai, Y. and Silman, I. (1973), *FEBS Letters* 30, 125-128.

Sussman, J.L., Harel, M., Frolow, F., Oefner, C., Goldman, A., Toker, L. and Silman, I. (1991), *Science* 253, 872-879.

Taylor, P. and Lappi, S. (1975), *Biochemistry* 14, 1989-1997.

Weise, C., Kreienkamp, H.J., Raba, R., Pedak, A., Aaviksaar, A. and Hucho, F. (1990), *Embo J.* 9, 3885-3888.

STRUCTURAL INVESTIGATIONS OF THE ACETYLCHOLINESTERASE

Ferdinand Hucho[1], Christoph Weise[1], Hans-Jürgen Kreienkamp[1],
Ute Görne-Tschelnokow[1], and Dieter Naumann[2]

[1]Institut für Biochemie, Freie Universität Berlin, Thielallee 63, 1000 Berlin
33, Germany
[2]Robert-Koch-Institut, Nordufer 100, 1000 Berlin 65, Germany

INTRODUCTION

Two of the key proteins participating in chemical transmission at nicotinic cholinergic synapses contain binding sites for acetylcholine: the nicotinic acetylcholine receptor (nAChR) and the acetylcholinesterase (AChE). Since there is no sequence homology among these proteins, the binding sites are presumably the result of convergent evolution. The elucidation and comparison of the architecture of acetylcholine binding sites from different proteins should provide interesting information concerning the structural requirements for specific binding of this cationic neurotransmitter.

Since there is no crystal structure of the nAChR in sight in the near future, its acetylcholine binding sites are being investigated using affinity labeling and protein sequencing (for a review see Changeux, 1990). But even with a three-dimensional structure at hand, the amino acid side chains interacting with a ligand can be identified most directly by affinity labeling. Cooperating with colleagues from the Estonian Academy of Sciences in Tallinn, we attempted to localize acetylcholine binding sites within the primary structure of the glycolipid-anchored homodimeric form of AChE from Torpedo californica electric tissue and of the water-soluble Cobra enzyme. This was achieved by means of the cholinium analogue ³[H] N,N-dimethyl-2-phenyl-aziridinium (DPA) which has been shown previously to react specifically with "anionic subsites" both in the active center and in the periphery of the enzyme. As a transiently reversible inhibitor it acts non-competitively with a $K_i = 6 \times 10^{-5}$ M (Purdie, 1969). Covalent reaction with the enzyme can be prevented by alkylammonium salts (Palumaa et al., 1984).

COVALENT REACTION OF DPA

DPA is formed in water spontaneously from N,N-dimethyl-2-chloro-2-phenyl-ethylamine. The kinetics of its formation and hydrolysis were investigated by Palumaa et al. (1982) and in our experiments conditions were chosen to secure maximum concentrations of the aziridinium compound: the reagent was formed immediately before use (fig. 1).

Multidisciplinary Approaches to Cholinesterase Functions, Edited by
A. Shafferman and B. Velan, Plenum Press, New York, 1992

Experiments with the Torpedo enzyme were performed with the G_2 form released from Torpedo electroplax membranes by cleavage of the phosphoglycolipid anchor with phosphatidylinositol specific phospholipase C from Bacillus thuringiensis (Kupke et al., 1989) kindly supplied by Dr. Götz, Tübingen. The AChE was subsequently purified by affinity chromatography on N-methyl-acridinium sepharose (Dudai et al., 1972); it was eluted from the affinity column with 5 mM decamethonium/0.5 M NaCl/10 mM Tris/HCl pH 7.4. This procedure afforded approximately 2 mg of homogeneous G_2 form starting from 100 g electric tissue.

The purity of the enzyme was assessed by SDS polyacrylamide gel electrophoresis. Quantitative evaluation by scanning the gel showed that the AChE band represented >97% of the total protein.

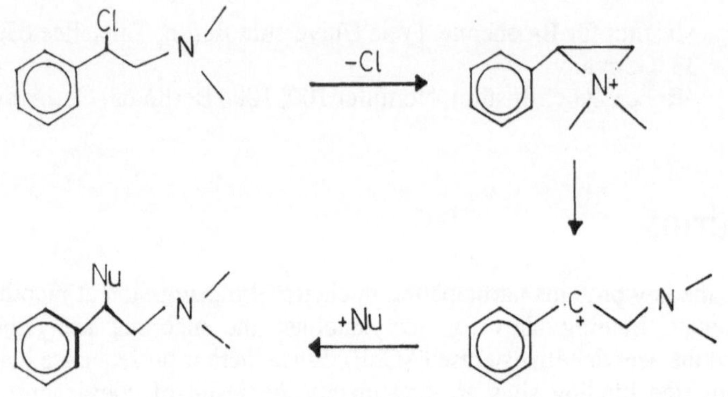

Figure 1. In aqueous solution the reactive aziridinium ion is formed from the chloro-ethyl-compound and subsequently reacts with a nucleophile

AChE from the venom of the cobra Naja naja oxiana (Tashkent Integrated Zoo Plant) was prepared by affinity chromatography as described (Raba et al., 1979).

Incubation of AChE from Torpedo californica with DPA leads to complete inactivation when 2 moles of reagent are incorporated. Decamethonium (1 mM) protects the enzyme against inactivation and partly against incorporation of the label. Propidium (O.1 mM), a ligand presumed to bind to peripheral rather than to active site anionic subsites, does protect against inactivation, but only to a small extent. The active center-specific ligand edrophonium, on the other hand, protected against inactivation much more efficiently. Quantitative experiments correlating label incorporation with inactivation (in the absence or presence of subsite-specific ligands) indicate that DPA reacts with the active center, and with peripheral sites. Furthermore, it also reacts nonspecifically to a certain degree.

LOCALIZATION OF THE LABEL WITHIN THE PRIMARY STRUCTURE

The modified enzyme was reduced and carboxymethylated. Proteolytic cleavage with a combination of trypsin and chymotrypsin and subsequent separation of the cleavage products by reverse-phase HPLC yielded the pattern shown in fig. 2 (AChE from Torpedo californica) and fig. 3 (Naja naja oxiana).

UV PROFILE OF HPLC SEPARATION

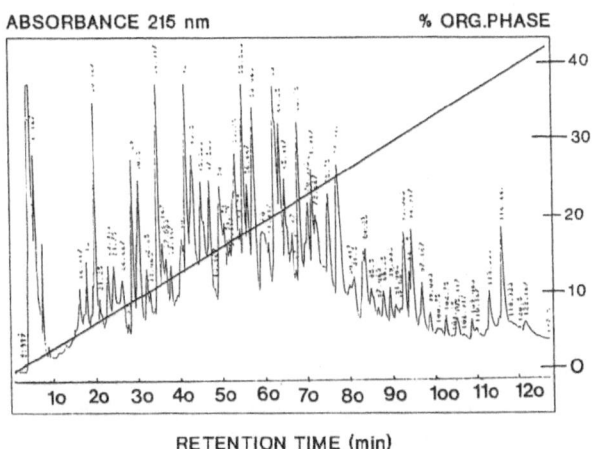

RADIOACTIVITY PROFILE OF HPLC SEPARATION

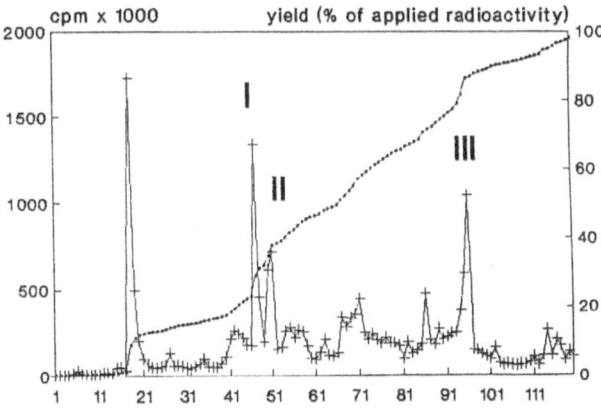

Figure 2. HPLC separation of fragments of ^{3}H-DPA-labeled AChE from Torpedo californica after tryptic/chymotryptic cleavage. Three major radioactivity peaks are seen on top of a background of unspecific labeling. (The first peak to the left is free label.)

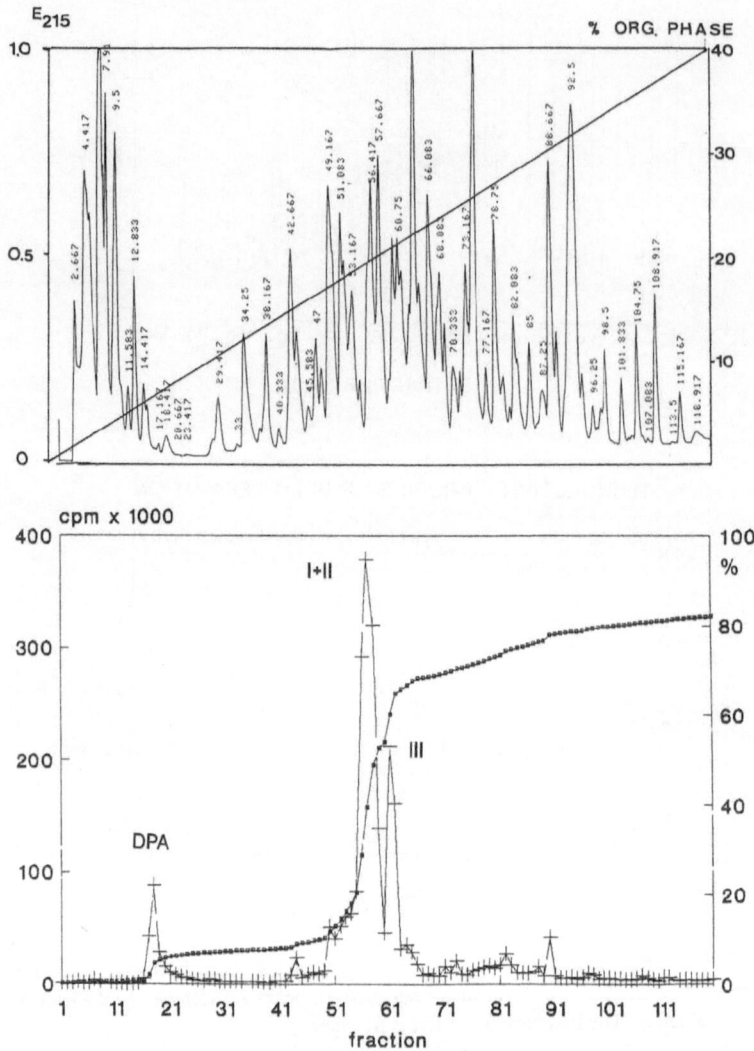

Figure 3. HPLC-separation of fragments of ^{3}H-DPA-labeled AChE from Cobra venom after tryptic/chymotryptic cleavage. In addition to a peak for the free label, two main radioactivity peaks are seen on top of a background of non-specific binding. The first of those two peaks was resolved into(two peaks by rechromatography.)

The distribution of radioactivity showed a much simpler pattern (fig.4) for the Torpedo and fig. 5 for the cobra enzyme), which is an indication of the label's specificity.

Different labeling patterns were obtained when the reaction with DPA was performed in the presence of site-specific ligands: peak III in fig. 4 disappeared when propidium competed with the affinity label (fig. 4b), while peaks I and II could be suppressed by labeling the Torpedo enzyme in presence of edrophonium (fig. 4c).

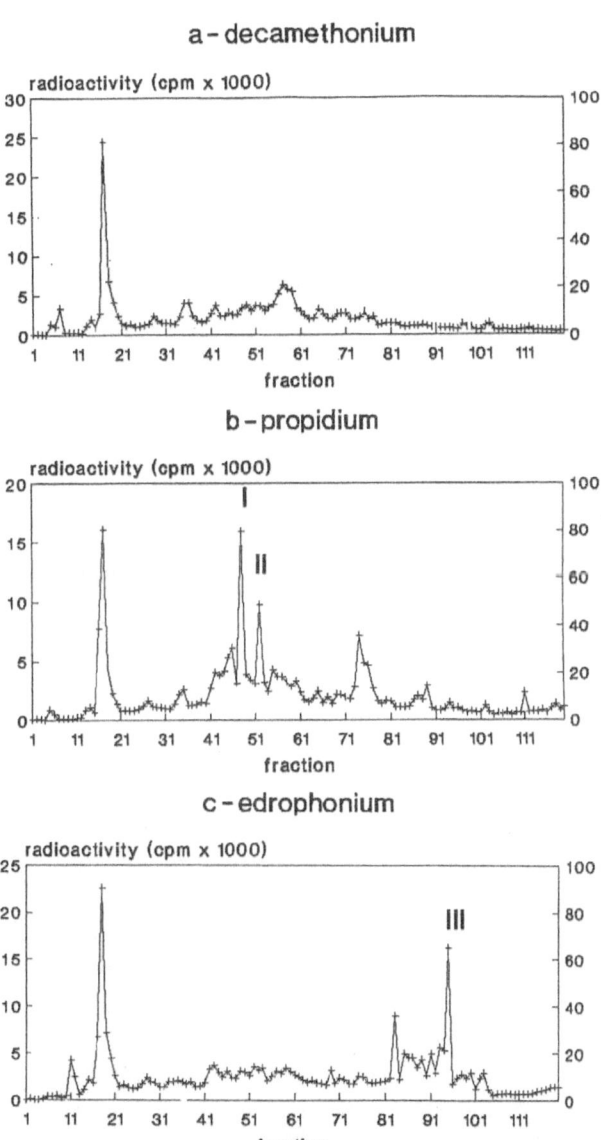

Figure 4. HPLC separation of AChE fragments in the presence of (a) 1×10^{-3}M decamethonium, (b) 1×10^{-4}M propidium and (c) 1×10^{-4}M edrophonium. Selective elimination of the radioactivity peaks I, II and III is observed with the various ligands. (The radioactivity profile of a control sample without any competing ligand is shown in fig.2)

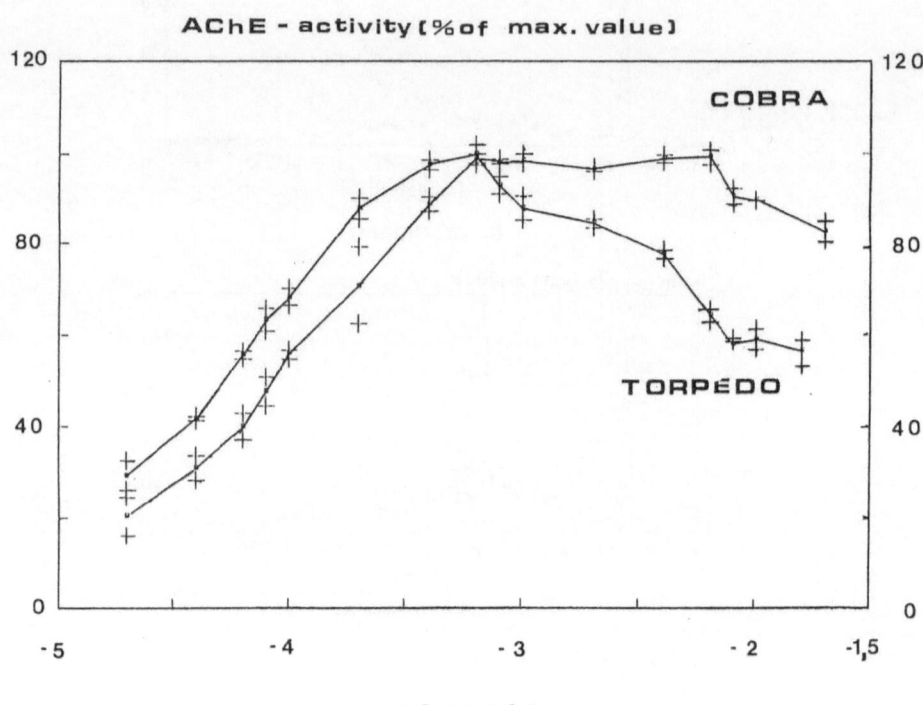

Figure 5. Protection of incorporation of DPA into AChE from Cobra venom by anionic site ligands, (a) 1×10^{-3}M decamethonium, (b) 1×10^{-4}M edrophonium and (c) 1×10^{-4}M propidium. (The control run is shown in fig.4)

126

The conclusion from these experiments was that peak I and II represent peptides from Torpedo AChE located in the active center and that peak III represents a peptide from the peripheral anionic site. No peptide corresponding to the latter was found with the cobra AChE (see below).

Table I. Automated Edman sequencing gave the following sequences

Peak	Sequence	Position	Location
I	DLFR	217-220	active center
II	SGSEMWNPN	79-87	active center
III	KPQELIDVE	270-278	peripheral site

ACHE FROM THE VENOM OF THE COBRA *NAJA NAJA OXIANA*

The active center peptide identified by affinity labeling of the cobra AChE had the sequence GAEMWNPN. Interestingly, no labeled peptide corresponding to peak III was detected in the experiments with the cobra enzyme (fig. 3). This may be due to the fact that there is no "peripheral site" in this enzyme. This interpretation is supported by kinetic investigations showing purely competitive inhibition by propidium ($K_i = 4 \times 10^{-6}$) and no substrate inhibition in a concentration range of the substrate acetylthiocholine which inhibits the Torpedo enzyme significantly (>1 mM) (fig. 6).

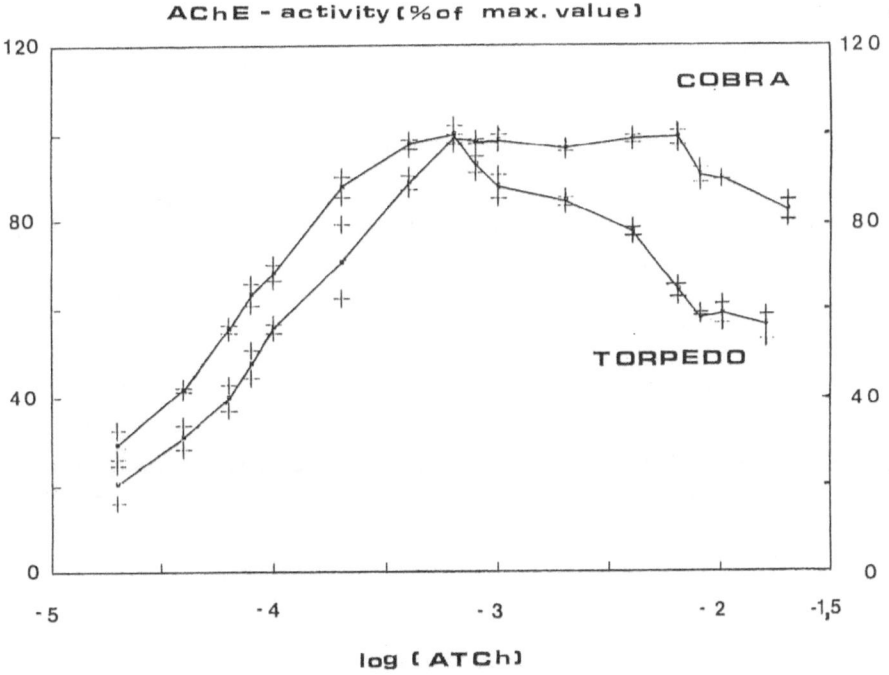

Figure 6. Dependence of enzymatic activity on substrate concentration for Cobra venom and Torpedo AChE

TRP 84 (TORPEDO ACHE) IS THE REACTION SITE IN THE ACTIVE CENTER

In both AChE from Torpedo and from cobra venom a tryptophan residue was labeled by our cationic affinity reagent. We conclude from this observation (Weise et al., 1990; Kreienkamp et al., 1991), that the substrate in the active center may be bound through an interaction of the cationic cholinium moiety with the aromatic electrons of the indolyl side chain. A similar picture emerges from affinity labeling experiments performed with the nicotinic acetylcholine receptor (Dennis et al., 1988, Galzi et al. 1990) were aromatic rather than anionic amino acid side chains are also labelled by a cationic reagent.

EVALUATION OF SECONDARY STRUCTURE BY FTIR SPECTROSCOPY

FTIR spectroscopy was used to quantitate the content of secondary structure elements in AChE from Torpedo californica. With the X-ray structure at hand (Sussman et al.,1991; we wish to thank these authors for supplying us with the coordinates) this study was undertaken to evaluate the reliability of this method for obtaining secondary structure information of other proteins, e.g. the nicotinic acetylcholine receptor.

For the FTIR experiments the protons of the enzyme had to be exchanged against deuterium. To this end, the AChE fraction was dialyzed against 15 mM NH_4HCO_3, pH 7.9, lyophylized, resuspended in 2 ml D_2O, lyophylized and resuspended again. For spectroscopy the protein concentration was finally adjusted to 22 mg/ml with a D_2O-exchanged phosphate buffer (20 mM sodium phosphate, pH 7.4, o.2 % NaN_3).

Evaluation of FTIR spectra yielded 36 % _-helical and 19 % ß-structure, compared to 32 % and 15 %, respectively, obtained from the X-ray structure. Especially for ß-sheet, these values compare favourably with values between 23 and 35% obtained in various studies with CD and Raman spectroscopy which seem to overestimate the content of ß-sheet structure (Wu et al., 1987; Aslanian et al., 1987; Manavalan et al., 1985; Aslanian et al., 1991).

CONCLUSIONS

Affinity labeling experiments are helpful for the interpretation of structures obtained by X-ray crystallography. Fig. 7 summarizes several of these studies including those from our laboratory.

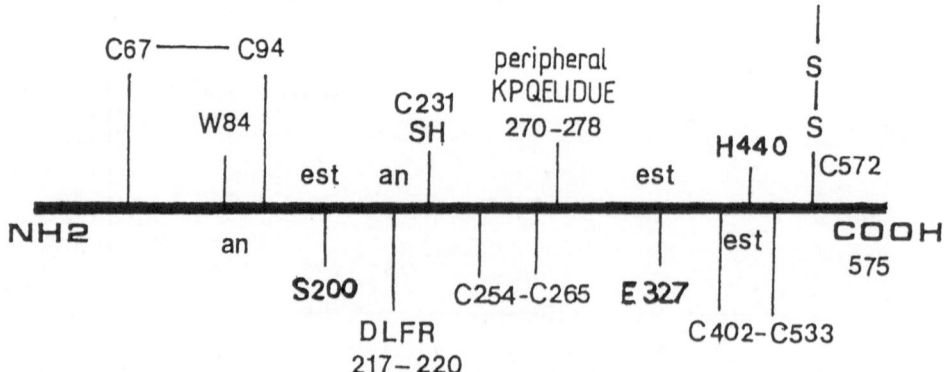

Figure 7. Molecular organization of AChE from Torpedo californica; the scheme includes the residues forming the catalytic triad and residues involved in binding at the esteratic site, at the anionic site of the catalytic center and at the peripheral anionic site. The disulfide bridges are shown according to Mac-Phee Quigley et al., 1986.

These results obtained before publication of the X-ray structure by Sussman et al. (1991) can be discussed now in the light of the known three dimensional structure. Trp 84 is located at the inner end of the deep gorge lined by many aromatic aminoacid side chains. Modeling studies (Sussman et al., 1991) have shown that this residue indeed may be in close proximity of the bound substrate.

The sequence KPQELIDVE (position 270 -278) which could be protected against affinity labeling by the peripheral-site specific ligand propidium is located at the entrance of this gorge.

Less clear is the meaning of our result that the sequence DLFR (position 217-220) is located in the active center, too. The X-ray structure poses this sequence too far away to participate in the binding of the substrate to the catalytic center. Since we have no reason to doubt the specificity of our affinity label, we have to look for further data helping to interpret this finding. At present there is no evidence that binding of the substrate induces gross conformational changes moving the sequence Asp 217 - Asp 220 closer to the active site.

ACKNOWLEDGEMENTS

The cobra enzyme used in these investigations was kindly supplied by Dr. Raivo Raba, Tallinn, Estonia. This work was supported by the Deutsche Forschungsgemeinschaft (SFB 312) and the Fonds der Chemischen Industrie.

REFERENCES

Aslanian, D., Grof, P., Balkanski, M. and Taylor, P., 1987, Raman spectroscopic study on the conformation of 11S from AChE from Torpedo californica, FEBS Lett. 219, 202-206

Aslanian, D., Grof, P., Bon, S., Masson, P., Négrerie, M., Chatel, J.M., Balkanski, M., Taylor, P. and Massoulié, J. 1991, A comparative Raman spectroscopic study of cholinesterases, Biochimie 73, 1375-1386

Changeux, J.P., 1990, Functional Architecture and Dynamics of the Nicotinic Acetylcholine Receptor: Allosteric Ligand-gated Ion Channel. Fidia Research Foundation Neuroscience Award Lecture, Vol. 4, Raven Press, Ltd., New York.

Dennis, M., Giraudat, J., Kotzyba-Hibert, F, Goeldner, M., Hirth, C. Chang, J.-Y., Lazure, C., Chrétien, M. and Changeux, J.-P., 1988, Amino acids of the Torpedo marmorata acetylcholine receptor α-subunit labeled by a photoaffinity ligand for the acetylcholine binding site, Biochemistry 27, 2346-2357

Dudai, Y., Silman, I., Shinitzky, M. and Blumberg, S., 1972, Purification by affinity chromatogrgaphy of the molecular forms of acetylcholinesterase present in fresh electric-organ tissue of electric eel, Proc.Natl.Acad.Sci.USA 69, 2400-2403

Galzi, J.-L., Revah, F., Black, D., Goeldner, M., Hirth, C. and Changeux, J.-P., 1990, Identification of a novel amino acid _- tyrosine 93 within the cholinergic ligands-binding site of the acetylcholine receptor by photoaffinity labeling, J. Biol. Chem. 265, 10430-10437

Kreienkamp, H.-J., Weise, C., Raba, R., Aaviksaar, A. and Hucho, F., 1991, Anionic subsites of the catalytic center of acetylcholinesterase from Torpedo and from cobra venom, Proc.Natl.Acad.Sci.USA 88, 6117-6121

Kupke, T., Lechner, M., Kaim, G. and Götz, F., 1989, Improved purification and biochemical properties of phosphatidylinositol- specific phospholipase C from Bacillus thuringiensis, Eur.J.Biochem. 185, 151-155

Mac-Phee-Quigley, K., Vedvick, T., Taylor, P. and Taylor, S., 1986, Profile of the disulfide bonds in acetylcholinesterase, J. Biol. Chem. 29, 13565-13570

Manavalan, P., Taylor, P. and Johnson, W.C., 1985, Circular dichroism studies of AChE conformation/ comparison of the 11S and 5.6S species and the differences induced by inhibitory ligands Biochim. Biophys. Acta 829, 365-370

Palumaa, P. Mähar A. and Järv, J., 1982, Kinetic Analysis of Butyrylcholinesterase inhibition with N,N-dimethyl-2- phenylaziridinium ion. Bioorg.Chem. 11, 394-403

Palumaa, P., Raba, R. and Järv, J., 1984, Site-specificity of butyrylcholinesterase alkylation with N,N-dimethyl-2- phenylaziridinium ion, Biochim. Biophys. Acta 791, 15-20

Purdie, J., 1969, The properties of acetylcholinesterase modified by interaction with the alkylating agent N,N-dimethyl-2- phenylaziridinium ionBiochim.Biophys.Acta 185, 122-133

Raba, R., Aaviksaar, A., Raba, M. und Siigur, J., 1979, Cobra venom acetylcholinesterase- purification and molecular properties, Eur.J.Biochem. 96, 151-158

Sussman, J.L., Harel, M., Frolow, F., Oefner, C., Goldman, A., Toker, L. and Silman, I., 1991, Atomic structure of acetylcholinesterase from Torpedo californica: A prototypic acetylcholine-binding protein, Science, 253, 872-879

Weise, C., Kreienkamp, H.-J., Raba, R., Pedak, A., Aaviksaar, A. and Hucho, F., 1990, Anionic subsites of the acetylcholinesterase of Torpedo californica, EMBO J. 9, 3885-3888

Wu, C.S., Gan, L. and Yang, J.T., 1987, Conformation similarities of the globular and tailed forms of AChE from Torpedo californica, Biochim. Biophys. Acta 911, 25-36

SUBSTRATE-SELECTIVE INHIBITION AND PERIPHERAL SITE LABELING
OF ACETYLCHOLINESTERASE BY PLATINUM(TERPYRIDINE)CHLORIDE

Robert Haas, Elizabeth W. Adams, Mark A. Rosenberry,
and Terrone L. Rosenberry

Department of Pharmacology
Case Western Reserve University
Cleveland, OH 44106 USA

INTRODUCTION

Acetylcholinesterase (AChE) catalyzes the hydrolysis of its physiological substrate acetylcholine as well as of a number of other acetic acid esters. A key feature of AChE is its speed in cleaving substrates (Rosenberry, 1975a). The second order rate constant for acetylcholine hydrolysis ($k_{cat}/K_{app} = 2 \times 10^8$ $M^{-1}s^{-1}$) approaches the value expected for a diffusion-controlled reaction. The turnover rate for acetylcholine ($k_{cat} = 2 \times 10^4$ s^{-1}) is at the upper limit of reactions catalyzed by general acid-base catalysis. Quinn (1987) has noted that an enzyme with such a high catalytic efficiency is likely to have evolved to a point where the free energies of successive transition states are nearly matched and comparable to the diffusional barrier for substrate binding.

The unique features of AChE structure that determine its catalytic power have been pursued for many years. The region of the active site has been shown to have a high net negative charge that can electrostatically attract cationic substrates (Nolte et al., 1980). Furthermore, the binding of certain cationic inhibitors to peripheral anionic sites has been shown to affect catalysis at the active site. As indicated in Fig. 1, propidium, which binds primarily to a peripheral site, and edrophonium, which binds to the active site, do not compete with each other (Taylor and Lappi 1975). In contrast, a bisquaternary ligand like ambenonium competes with both propidium and edrophonium and thus appears to bridge both sites.

In this report we introduce a new peripheral site ligand for AChE, chloro(2,2':6',2"-terpyridine)platinum(II) chloride, denoted here Pt(terpyridine)Cl. This ligand not only is cationic, but it also can form stable covalent conjugates with imidazole and thiol groups in proteins (Fig. 2; Ratilla et al., 1987). We show here that human AChE can be affinity labeled with Pt(terpyridine)Cl. Since this enzyme appears to have no free thiol groups (Rosenberry and Scoggin, 1984; Soreq et al., 1990), the enzyme residue labeled by this reagent is likely to be a histidine. This histidine residue is identified, and its location near the active site is described.

Multidisciplinary Approaches to Cholinesterase Functions, Edited by
A. Shafferman and B. Velan, Plenum Press, New York, 1992

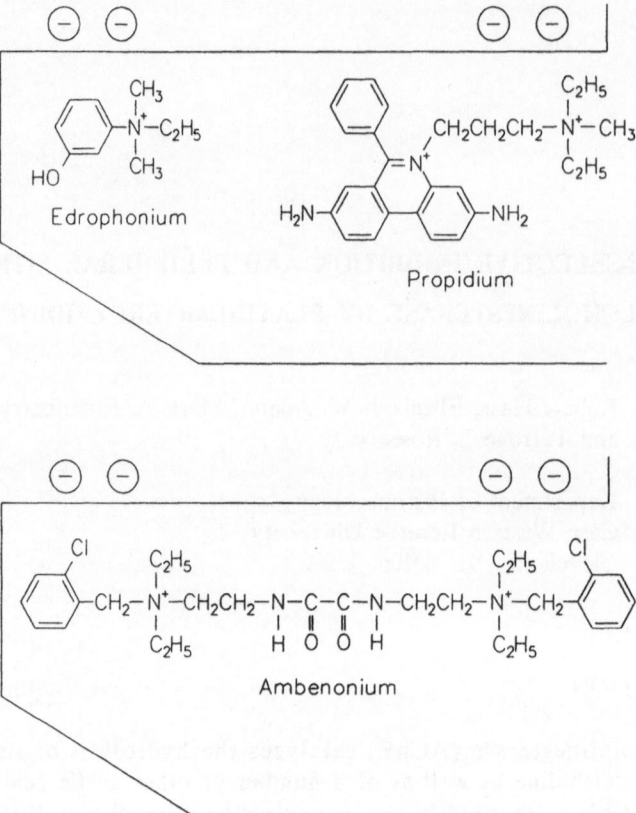

Fig. 1. **Binding of inhibitors to AChE.** The schematic diagram emphasizes the relatively large size of the active site region (solid line) and its negatively charged character (-). Edrophonium appears to bind near the catalytic triad in the active site, while propidium can bind simultaneously to a peripheral anionic site. Ambenonium occupies both sites.

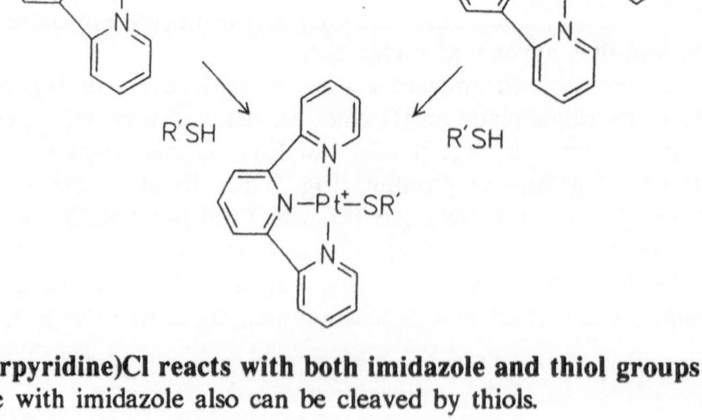

Fig. 2. **Pt(terpyridine)Cl reacts with both imidazole and thiol groups in proteins.** The conjugate with imidazole also can be cleaved by thiols.

METHODS

Materials. Histidine was coupled to Sepharose CL-4B (Pharmacia) by the procedure of March *et al.* (1974). AChE was purified by affinity chromatography on acridinium resin (Rosenberry and Scoggin, 1984). The AChE forms obtained from each species were human erythrocyte (G_2 hydrophobic), bovine erythrocyte (G_2 hydrophilic), and mouse brain (G_4 hydrophobic).

Assays. All substrate hydrolyses were measured spectrophotometrically at 25° C in solvents that contained 0.02 - 0.1% Triton X-100. Acetylthiocholine hydrolysis was measured at 412 nm (Rosenberry and Scoggin, 1984) and phenyl acetate hydrolysis at 270 nm (Rosenberry, 1975b) in 20 mM sodium phosphate (pH 7.0). Acetylcholine hydrolysis was assayed with phenol red as a coupled indicator (Sharp and Rosenberry, 1982) in 0.7 mM sodium phosphate (pH 7.0) and 40 mM NaCl.

Preparation of Pt(terpyridine)Cl derivatives. Pt(terpyridine)Cl obtained from Aldrich was used without further purification. Stock solutions of 30 mM in water were stored for up to 6 months at -20° C. Aliquots were diluted into incubation buffer (20 mM sodium phosphate, pH 7.0, 0.1% Triton X-100) for reaction with AChE as described. Model reactions of Pt(terpyridine)Cl (100 uM) with imidazole (10 mM) were monitored spectrophotometrically (Ratilla *et al.*, 1987) by the change in absorbance at 342 nm between Pt(terpyridine)Cl (ϵ_{342} = 13,300 $M^{-1}cm^{-1}$) and Pt(terpyridine)imidazole (ϵ_{342} = 16,000 $M^{-1}cm^{-1}$). Reaction of Pt(terpyridine)imidazole (100 uM) with (β-mercaptoethanol (0.3-1.0 mM) also was measured at 342 nm by formation of the Pt(terpyridine)Cl thiol adduct, which has approximately the same extinction coefficient as Pt(terpyridine)Cl.

Isolation and sequencing of tryptic fragments of Pt-AChE The AChE-Pt(terpyridine) conjugate was prepared by 1-hr incubation of a 1-ml sample of human AChE (4.2 nmol) and Pt(terpyridine)Cl (100 uM final added in two equal 10-ul amounts at 0 and 30 min incubation) in 20 mM sodium phosphate (pH 7.0), 0.1% Triton X-100. The mixture was then added to a column (0.5 x 2 cm) of histidine-linked Sepharose CL-4B, equilibrated with the resin overnight to remove excess free Pt(terpyridine)Cl, and eluted in 550-ul fractions with 20 mM sodium phospahte (pH 7.0), 0.02% Triton X-100. Since the Pt(terpyridine) adduct of AChE would not be stable to the thiol reduction step used in conventional trypsin digestion protocols, we introduced an alternative procedure involving denaturation by organic solvent (Haas and Rosenberry, in preparation). Acetonitrile was added to 30% for 10 min and then lowered to 5% by vacuum evaporation. The sample was then digested with 10 ug trypsin (Boehringer, sequencing grade) in 100 mM 2-([2-hydroxy-1,1-bis(hydroxy-methyl)ethyl]amino)ethanesulfonate (TES, pH 8), 1.5 mM calcium acetate, 5% acetonitrile overnight at 37° C (Aebersold, 1989; Fischer et al., 1991). Precipitate was removed by centrifugation at 15,000 xg for 2 min, and the sample analyzed on a Beckman 126 gradient module equipped with a a narrow bore HPLC column (Vydac C18, 2.1 x 250 mm, 300 A° pore size, 5 um particle size). Two detectors were used in tandem (a Beckman 166 UV detector at 342 nm and a Beckman 160 detector at 214 nm). The peptides were eluted at a flow rate of 0.2 ml/min with a water (0.06% TFA)-acetonitrile (0.05% TFA) gradient (5% acetonitrile for 20 min, 5 - 12% over 21 min, 12 - 38% over 260 minutes, and 38 - 65% over 27 min). Fractions were collected (1.0 min) and individual fractions selected for sequence analysis were applied to biobrene-treated filter disks for amino acid sequence analysis by a gas phase sequencer (Applied Biosystems Model 477A).

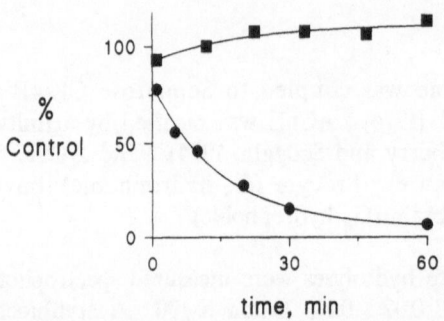

Fig. 3. **Pt(terpyridine)Cl inactivates AChE hydrolysis of acetylcholine (●) but not of phenyl acetate (■).** Reaction of Pt(terpyridine)Cl (300 nM) with AChE (10 nM) in 0.1% Triton X-100, 20 mM sodium phosphate (pH 7.0) was initiated by addition of the ligand from a 25 uM stock. At each indicated time, a 10-ul aliquot was assayed with 3.0 ml of 1.0 mM acetylcholine or 10 mM phenyl acetate as outlined in the Methods. Activities are expressed as percentages of the activity for a corresponding aliquot prior to addition of Pt(terpyridine)Cl. Lines were fitted by nonlinear regression analysis that assumed a single exponential curve. An inactivation rate constant k_{exp} = 15 $\pm$ 4 x 10^{-4} s^{-1} was calculated from 5 similar inactivation experiments assayed with acetylcholine.

RESULTS AND DISCUSSION

Pt(terpyridine)Cl selectively inactivates AChE hydrolysis of acetylcholine. AChE was incubated with Pt(terpyridine)Cl and aliquots were assayed with both acetylcholine and phenyl acetate substrates. The enzyme activity with phenyl acetate increased slightly during the incubation, but the activity with acetylcholine decreased to less than 6% of the preincubation value (Fig. 3). Since phenyl acetate hydrolysis was only slightly affected by preincubation of AChE with Pt(terpyridine)Cl, it is unlikely that the inactivation with acetylcholine involved an essential catalytic residue in the AChE active site (like [440]His in the active site catalytic triad; Sussman *et al.*, 1991). However, Pt(terpyridine)Cl did act like an affinity labeling reagent in causing the inactivation. The inactivation occurred with an apparent rate constant that was several orders of magnitude higher than that expected from the second order rate constant for the reaction of Pt(terpyridine)Cl with imidazole (0.2 M^{-1}s^{-1}; also reported by Ratilla *et al.*, 1987). The inactivation rate constant also was not proportional to the Pt(terpyridine)Cl concentration but was nearly invariant from 200 nM to 3 uM Pt(terpyridine)Cl, an observation which suggested that this ligand was binding to a peripheral enzyme site before proceeding to a covalent reaction.

Concentrations of Pt(terpyridine)Cl from 30 to 200 nM gave only 10 to 70% total inactivation of acetylcholine hydrolysis when the ligand was maintained in a 10- to 20-fold stoichiometric excess over the concentration of enzyme sites. Nevertheless, the inactivation appeared completely irreversible over at least 1-2 days in the absence of added thiol reagents.

The site at which Pt(terpyridine) is covalently linked is near the AChE active site. Kinetic competition experiments were employed as a first approach to estimate the proximity of the human AChE active site to the residue involved in covalent linkage to Pt(terpyridine). Preliminary experiments indicated that edrophonium had no effect on the rate of inactivation of acetylcholine hydrolysis, consistent with a peripheral location of the residue, but that ambenonium did slow the rate of inactivation. To provide strong evidence that the competition involved ambenonium binding to the active site, it is useful to show that the ambenonium inhibition constant for inactivation is identical to the inhibition constant for substrate hydrolysis. An indirect continuous assay was devised by monitoring AChE hydrolysis of phenyl acetate in the presence of ambenonium. When Pt(terpyridine)Cl was included in this mixture, the steady-state hydrolysis rate v increased progressively as Pt(terpyridine)Cl reacted at its peripheral site and displaced ambenonium from the enzyme (data not shown). The shift in v occurred with a single exponential rate constant k_{exp}, as predicted for a simple affinity labeling reaction. Plots of $1/k_{exp}$ vs. $1/[Pt(terpyridine)Cl]$ were linear for fixed concentrations of phenyl acetate and ambenonium, and the slopes of these reciprocal plots were linear in ambenonium concentration. These features all are predicted by a general kinetic model which assumes that ambenonium binding is mutually exclusive with both phenyl acetate and Pt(terpyridine)Cl binding but that phenyl acetate can interact with both reversible and covalent complexes of Pt(terpyridine)Cl with AChE. The apparent inhibition constant K_I for ambenonium inhibition of Pt(terpyridine)Cl affinity labeling in this system was 0.54 ± 0.22 nM, in good agreement with the expected value of 0.28 ± 0.12 nM calculated from direct ambenonium inhibition of phenyl acetate hydrolysis. This observation indicates that ambenonium binding to a single enzyme site can competitively inhibit both phenyl acetate hydrolysis and Pt(terpyridine)Cl affinity labeling. Thus the affinity labeled site appears to be relatively near the active site.

Rate constants obtained for reaction of the AChE-Pt(terpyridine) conjugate with β-mercaptoethanol indicate that a histidine residue is affinity labeled. Although only histidine residues in AChE are expected to form covalent adducts with Pt(terpyridine)Cl based on the studies of model compounds (Ratilla *et al.*, 1987) noted in the Introduction, we conducted a kinetic analysis to support this assignment. Thiols cleave Pt(terpyridine) conjugates with imidazole, and the reaction of the model compound Pt(terpyridine)imidazole with β-mercaptoethanol occurred with a second order rate constant of 21 ± 1 $M^{-1}s^{-1}$. Reaction of β-mercaptoethanol with the AChE-Pt(terpyridine) conjugate was monitored by continuous spectrophotometric assay as above, and an apparent second order rate constant of 39 ± 4 $M^{-1}s^{-1}$ was observed. The close correspondence of these two rate constants provides evidence that the AChE conjugate involves the imidazole side chain of a histidine residue.

Identification of histidine residues in AChE that are covalently labeled by Pt(terpyridine)Cl. Isolation and analysis of tryptic peptides was undertaken to identify the specific histidine residue(s) covalently labeled by Pt(terpyridine)Cl. Although the covalent conjugate appeared to be relatively stable based on the slow spontaneous reactivation of AChE activity toward acetylcholine, precautions were taken to minimize cleavage of the conjugate during its preparation for trypsin digestion. Samples were denatured with acetonitrile rather than by more conventional disulfide reduction and alkylation in guanidine hydrochloride or urea to avoid thiol cleavage of the conjugate. Trypsin digests were fractionated by reverse-phase HPLC on narrow bore columns as shown in Fig. 4. Since Pt(terpyridine) derivatives absorb at 342 nm, the HPLC effluent was monitored both at this wavelength and at 214 nm for conventional peptide detection. Fractions containing the two largest 342 nm peaks (labeled with asterisks in Fig. 4) were taken for peptide sequence analysis. The earlier peak at 145 min was sequenced for 15 cycles which gave TRPAQVLVNHE(W)(H)VL, where residues in parentheses were questionable. Based on the reported sequence of human AChE (Soreq *et al.*, 1990),

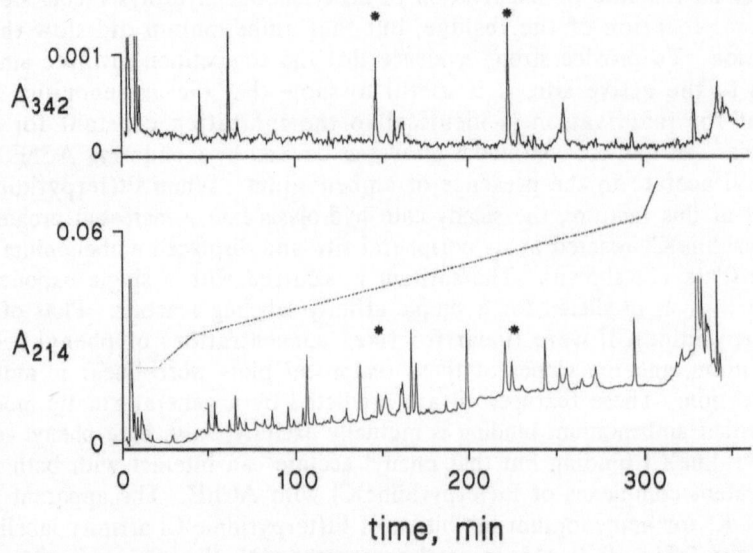

Fig. 4. Identification of tryptic fragments labeled with Pt(terpyridine) by reverse-phase HPLC. Human AChE was labeled with Pt(terpyridine)Cl, digested with trypsin, and 0.15 nmol fractionated by reverse-phase HPLC (see Methods). The effluent was monitored by tandem UV detectors at 342 nm and 214 nm, respectively. The relative percentage of solvent B in the gradient is indicated by the broken line and increased from 12% at 40 min to 38% at 300 min. Fractions of 0.2 ml (1.0 min) were collected, and those containing peaks labeled with the asterisks were taken for peptide sequencing. Peaks with 342 nm absorbance at 40-70 min were observed in sample blanks that included Pt(terpyridine)Cl but not AChE, and their relative amplitude decreased about 2-fold when the initial sample was incubated with histidine-Sepharose before digestion.

this sequence corresponded to a minimum tryptic peptide encompassing residues 268- 289 (in the torpedo AChE numbering system as shown below). The later peak at 220 min was sequenced for five cycles that gave VGVPQ. This N-terminal sequence is unique to the human AChE tryptic fragment beginning at residue 358 and extending to residue 386. Since these are the only two tryptic peptides in the human AChE sequence that have two histidine residues, these data raise questions about the selectivity of Pt(terpyridine)Cl affinity labeling of AChE. Although the apparent second order rate constant for AChE inactivation by Pt(terpyridine)Cl was several orders of magnitude higher than that for Pt(terpyridine)Cl reaction with imidazole, it is possible that the net negative charge on AChE at pH 7.0 results in an increased rate of reaction with many histidine residues on the enzyme relative to the rate with imidazole.

Identification of a histidine residue labeled by Pt(terpyridine)Cl near the human AChE active site. To determine whether histidine residues in either of the two peptides identified by sequencing are near the AChE active site, the three-dimensional structure for torpedo AChE recently reported by Sussman *et al.* (1991) was examined. We identified nine sequence segments containing 79 residues that appear to form the active site gorge described by Sussman *et al.*, and these segments are shown in Fig. 5. These workers have emphasized the abundance of hydrophobic amino acids and the high degree of sequence conservation of residues in the active

	67		114		199
Human	C Y Q Y V D T L Y P G F E G T E M W		W I Y G G G F Y S G A S S L D V Y		E S A
Bovine	C Y Q Y V D T L Y P G F E G T E M W		W I Y G G G F Y S G A S S L D V Y		E S A
Mouse	C Y Q Y V D T L Y P G F E G T E M W		W I Y G G G F Y S G A A S L D V Y		E S A
Torpedo	C Q Q Y V D E Q F P G F S G S E M W		W I Y G G G F Y S G S S T L D V Y		E S A
	* * *		*		*

225	233	273		327		432	436
Q S	W	V L V N **H** E W **H** V L P Q E S V F R F		E G S Y F L V Y G A		W	M G V P **H** G Y E I
Q S	W	D L V D **H** E W R V L P Q E **H** V F R F		E G S Y F L V Y G A		W	M G V P **H** G Y E I
Q S	W	D L V D **H** E W **H** V L P Q E S I F R F		E G S Y F L V Y G V		W	M G V P **H** G Y E I
Q S	W	E L I D V E W N V L P F D S I F R F		E G S F F L L Y G A		W	M G V I **H** G Y E I
		* * * *		*			*

Fig. 5. **Segments of the AChE amino acid sequence in the active site gorge.** The segments were identified by stereo analysis of the atomic coordinates of torpedo AChE (Sussman *et al.*, 1991) in the program FRODO. Aligned sequences of AChEs from four species are shown. The N-terminal amino acid in each segment is numbered according to the torpedo AChE sequence. Eleven sites at which at least three of the four sequences have a negatively charged residue (asterisks) and 4 sites at which a histidine residue is found in at least one sequence (bold) are noted.

site gorge among AChEs from different species, and these features are apparent from Fig. 5. Another feature that appears very significant is the distribution of negatively and positively charged residues. Eleven sites of negatively charged aspartate or glutamate residues are indicated in Fig. 5, whereas only one site is consistently a positively charged arginine or lysine residue. The net negative charge resulting from this distribution is consistent with a previous inference of an active site negative charge of -9 from kinetic data (Nolte *et al.*, 1980). This high negative charge increases the association rate for cationic ligands with the enzyme active site.

Only two segments in Fig. 5 contain histidine residues. The segment from residues 436 to 444 includes [440]His, a member of the Glu-His-Ser catalytic triad in the active site. Covalent modification of this histidine by Pt(terpyridine)Cl was unlikely because it would be expected to block hydrolysis of both phenyl acetate and acetylcholine. The segment from residues 273 to 290, on the other hand, contains the histidine residues identified as potential sites of Pt(terpyridine)Cl labeling in the earlier tryptic fragment peak sequenced in Fig. 4. This segment comprises nearly half of the rim of the active site gorge, an ideal location for a peripheral site that could influence AChE catalytic activity. The three-dimensional structure indicated that the histidines in residues 358 to 386, the sequence identified in the later tryptic fragment peak in Fig. 4, were on the enzyme surface and well separated from the active site. These histidines thus were unlikely to be involved in the selective inactivation of acetylcholine hydrolysis seen in Fig. 3.

Inspection of the segment from residues 273 to 290 in Fig. 5 indicates an interesting pattern for the three sites at which histidine residues are located. Histidine is present in all three mammalian species at residue 277, in only human and mouse AChE at residue 280, and in only bovine AChE at residue 286. This pattern suggests that a comparison of the effects of Pt(terpyridine)Cl on AChE from each species could be useful in localizing the histidine whose modification results in changes in active site properties. Purified AChE from each species was incubated with Pt(terpyridine)Cl, and aliquots were taken for phenyl acetate assay in the presence and absence of ambenonium as shown in Fig. 6. Affinity labeling of the peripheral site on AChE by Pt(terpyridine)Cl results in displacement of ambenonium

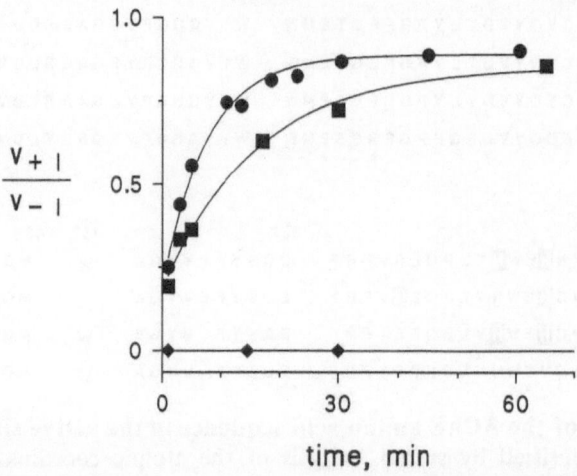

Fig. 6. Species dependence of covalent Pt(terpyridine)Cl modification of AChE.
Reaction of Pt(terpyridine)Cl (600 nM) with human (●), mouse (■) and bovine
(◆) AChE (10 nM) in 0.1% Triton X-100, 20 mM sodium phosphate (pH 7.0) was
initiated by addition of the enzyme. At each indicated time, a 20-ul aliquot was
assayed in 3.0 ml of 10 mM phenyl acetate with and without 10 nM ambenonium.
Values are expressed as ratios of the activity in the presence of ambenonium to the
activity in its absence. Lines were fitted as in Fig. 3. Inactivation rate constants
(k_{exp} = 22 ± 2 x 10^{-4} s^{-1} for human AChE and 10 ± 2 x 10^{-4} s^{-1} for mouse AChE)
were calculated from the data shown.

from the active site and in an increase in the phenyl acetate hydrolysis rate to a
value close to that of the labeled enzyme in the absence of ambenonium (as shown
in Fig. 6). However, affinity labeling is observed only with the human and mouse
enzymes and not with the bovine enzyme. Ambenonium continues to fully inhibit
bovine AChE even after 30 min of preincubation with Pt(terpyridine)Cl. This
pattern of affinity labeling is consistent with a covalent reaction of Pt(terpyridine)Cl
at [280]His, since this residue is changed to arginine in bovine AChE. In fact, survey
of the entire mammalian AChE sequences indicates that [280]His is the only histidine
residue that is conserved in human and mouse AChE but changed in bovine AChE.
Thus we conclude that Pt(terpyridine)Cl covalent modification of [280]His results in
selective inactivation of acetylcholine hydrolysis and in blockade of ambenonium
binding to the active site.

**New inferences about AChE hydrolysis of acetylcholine suggested by studies of
the AChe-Pt(terpyridine) conjugate.** Pt(terpyridine) conjugation with [280]His not only
leaves the active site relatively free to hydrolyze phenyl acetate (Fig. 3), it also
allows access of edrophonium. The K_I for edrophonium inhibition of phenyl acetate
hydrolysis by the AChe-Pt(terpyridine) conjugate was within a factor of two of that
for inhibition of control unconjugated enzyme (data not shown). Changes in catalytic
properties of the AChE-Pt(terpyridine) conjugate relative to control enzyme also are
noteworthy. While k_{cat}/K_{app} for phenyl acetate decreased to 24% of the control, it
decreased to 3% of the control for acetylcholine. Of greater significance, k_{cat} for
phenyl acetate increased about 50% over the control, but k_{cat} for acetylcholine
decreased to 9% of the control value. Such a large difference in the effects on k_{cat}
suggests that a change in the rate-limiting step for turnover of the enzyme-substrate
complex has occurred for acetycholine but not for phenyl acetate. We suggest that
the rate constant for a unimolecular transfer of acetylcholine from a peripheral

anionic site to the active site has decreased. This transfer could involve a surface diffusion pathway over the approximately 2 nm distance between the rim of the active site gorge and the catalytic residues at the base of the gorge, a process that would be equivalent to the notion of "rate-limiting induced fit" for AChE proposed several years ago (Rosenberry, 1975a). The data suggest that this transfer pathway is significantly blocked in the AChE-Pt(terpyridine) conjugate, accounting for the selective inactivation of acetylcholine relative to phenyl acetate hydrolysis.

Acknowledgements

This work was supported by grant NS16577 from the National Institutes of Health and grants from the Muscular Dystrophy Association of America. We thank Dr. Joel Sussman for providing us with the atomic coordinates of torpedo AChE.

REFERENCES

Aebersold, R., 1989, "A Practical Guide to Protein and Peptide Purification for Microsequencing" (Matsudaira, P.T., ed.), Academic Press, San Diego, CA.

Fischer, W.H., Karr, D., Jackson, B., Park, M., and Vale, W., 1991, Microsequence analysis of proteins purified by gel electrophoresis, *Methods Neurosci.* **6**: 69-84.

March, S.C., Parikh, I., and Cuatrecasas, P., 1974, A simplified method for cyanogen bromide activation of agarose for affinity chromatography, *Analyt. Biochem.* **60**: 149-152.

Nolte, H.J., Rosenberry, T.L., and Neumann, E., 1980, Effective charge on acetylcholinesterase active sites determined from the ionic strength dependence of association rate constants with cationic ligands, *Biochemistry* **19**: 3705-3711.

Quinn, D.M., 1987, Acetylcholinesterase: enzyme structure, reaction dynamics, and virtual transition states, *Chem. Rev.* **87**: 955-979.

Ratilla, E.M.A., Brothers, H.M., and Kostic, N.M., 1987, A transition-metal chromophore as a new, sensitive spectroscopic tag for proteins. Selective covalent labeling of histidine residues in cytochromes c with Chloro(2,2':6',2"-terpyridine)platinum(II) chloride, *J. Am. Chem. Soc.* **109**: 4592-4599.

Rosenberry, T.L., 1975a, Acetylcholinesterase, *Adv. Enzymol. Relat. Areas Mol. Biol.* **43**: 103-218.

Rosenberry, T.L., 1975b, Catalysis by acetylcholinesterase. Evidence that the rate-limiting step for acylation with certain substrates precedes general acid-base catalysis, *Proc. Nat. Acad. Sci. USA* **72**: 3834-3838.

Rosenberry, T.L., and Scoggin, D.M., 1984, Structure of human erythrocyte acetylcholinesterase. Characterization of intersubunit disulfide bonding and detergent interaction, *J. Biol. Chem.* **259**: 5643-5652.

Sharp, T. R. and Rosenberry, T.L., 1982, A pseudo-first order kinetic approach to measurement of acetylcholine hydrolysis by acetylcholinesterase, *J. Biochem. Biophys. Meths.* **6**: 159-172.

Soreq, H., Ben-Aziz, R., Prody, C.A., Seidman, S., Gnatt, A., Neville, L., Lieman-Hurwitz, J., Lev-Lehman, E., Ginzberg, D., Lapidot-Lifson, Y., and Zakut, H., 1990, Molecular cloning and construction of the coding region for human acetylcholinesterase reveals a G+C-rich attenuating structure, *Proc. Natl. Acad. Sci., USA* **87**: 9688-9692.

Sussman, J.L., Harel, M., Frolow, F., Oefner, C., Goldman, A., Toker, L., and Silman, I., 1991, Atomic structure of acetylcholinesterase from *Torpedo californica*: A prototypic acetylcholine-binding protein, *Science* **253**: 872-879.

Taylor, P., and Lappi, S., 1975, Interaction of fluorescence probes with acetylcholinesterase, The site and specificity of propidium binding, *Biochemistry* **14**: 1989-1997.

CRYPTIC CATALYSIS AND CHOLINESTERASE FUNCTION

Daniel M. Quinn[1], Trevor Selwood[1], Alton N. Pryor[1], Bong Ho Lee[1],
Lee-Shan Leu[1], Scott A. Acheson[1], Israel Silman[2], Bhupendra P. Doctor[3],
and Terrone L. Rosenberry[4]

[1]Department of Chemistry, The University of Iowa
Iowa City, IA 52242

[2]Department of Neurobiology, Weizmann Institute of Science
Rehovot 76100, Israel

[3]Division of Biochemistry, Walter Reed Army Institute of Research
Washington, D.C. 20307

[4]Department of Pharmacology, Case Western Reserve University
Cleveland, OH 44106

INTRODUCTION

The general features of acetylcholinesterase (AChE*) catalysis outlined in Figure 1 have
been appreciated for some time (Froede and Wilson, 1971; Quinn, 1987; Rosenberry, 1975).
The first step, nucleophilic attack by S200, is assisted by general-base catalysis by H440 and
produces a tetrahedral intermediate, which collapses to the acylenzyme by general-acid
catalyzed expulsion of choline by H440. The deacylation stage of catalysis follows a similar
series of events, with eventual expulsion from the tetrahedral intermediate of S200 of the
active site. Figure 1 also raises the possibility that E327 functions with H440 in a proton relay
network, an example of charge-relay catalysis (Blow, 1976).

---

* Abbreviations used: ACh, acetylcholine; AChE, acetylcholinesterase; ATCh, acetylcholine;
butyrylthiocholine; BzCh, benzoylcholine; DTNB, 5,5'-dithiobis (2-nitrobenzoic acid); PMPF, p-methoxy-
phenyl formate; PrTCh, propionylthiocholine; single letter amino acid codes: E, glutamate; H, histidine; S,
serine.

Multidisciplinary Approaches to Cholinesterase Functions, Edited by
A. Shafferman and B. Velan, Plenum Press, New York, 1992

141

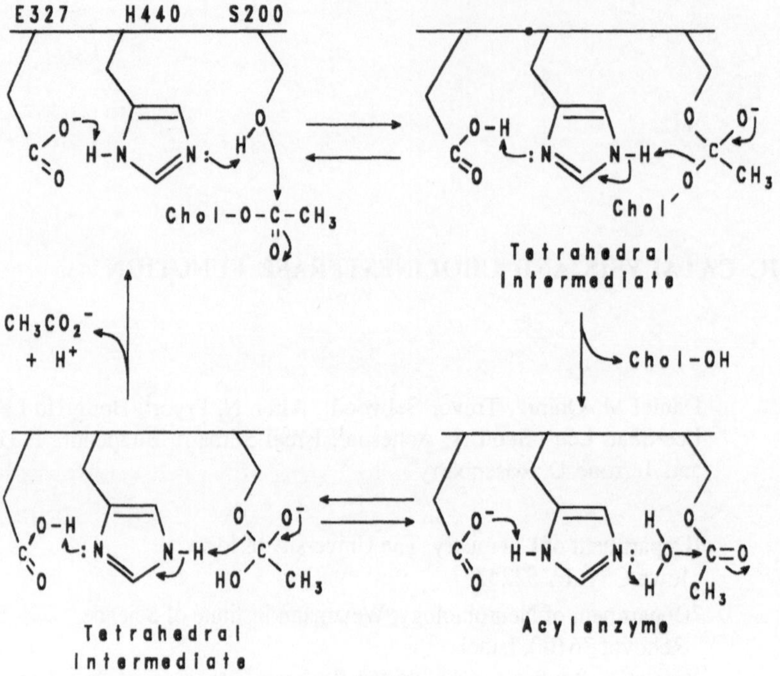

Figure 1. The Chemical Mechanism of Acetylcholinesterase Catalysis

The AChE mechanism outlined in Figure 1 bears a formal similarity to that of the serine proteases (Blow, 1976; Polgar, 1987). Despite this similarity, AChE catalysis provides the mechanistic enzymologist with a number of challenges. One is to account for the tremendous catalytic power of the enzyme. The minimal kinetic mechanism for AChE catalysis is outlined in Scheme 1, from which one derives equations 1 and 2 for the steady state rate constants. As these equations show, k_{cat} monitors events in both the acylation and deacylation stages of catalysis, while k_{cat}/K_m

$$E + A \xrightarrow[\overleftarrow{k_2}]{k_1} EA \xrightarrow{k_3} F + P \xrightarrow{k_5} E + Q$$

Scheme 1. The kinetic mechanism of AChE catalysis . A, P, and Q are ACh and the products choline and acetate, respectively. EA and F are the Michaelis complex and the acylenzyme intermediate, respectively.

$$k_{cat} = \frac{k_3 k_5}{k_3 + k_5} \tag{1}$$

$$k_{cat}/K_m = \frac{k_1 k_3}{k_2 + k_3} \tag{2}$$

monitors events up to the formation of the acylenzyme only. The k_{cat} values for hydrolyses of the physiological substrate ACh and it biomimetic surrogate ATCh are ~10^4 s^{-1}, among the highest of known turnover numbers. The corresponding k_{cat}/K_m values are $\gtrsim 10^8$ M^{-1} s^{-1}, and are rate limited by diffusion of substrate to the active site, i.e. $k_{cat}/K_m = k_1$ (Nolte et al., 1980; Bazelyansky et al., 1986). Therefore, AChE functions at the speed limit of biological catalysis (Quinn et al., 1991), a hallmark of an evolutionarily perfect enzyme.

Enzymes catalyze reactions by binding tightly to and thereby stabilizing chemical transition states (Schowen, 1978). Characterization of the structures of transition states in the AChE mechanism should therefore provide a view of the origins of the enzyme's catalytic power. However, the information that one wishes to obtain from probes of transition state structure, such as kinetic isotope effect and pH-rate measurements, is encrypted by the evolutionary perfection of AChE. For ACh hydrolysis $k_{cat}/K_m = k_1$ and therefore the isotope effect for the chemical step (k_3) is not observed. Since k_{cat} for turnover of ACh and ATCh is rate limited by both acylation and deacylation (k_3 and k_5; Froede and Wilson, 1984), isotope effects on this parameter may be difficult to interpret. A way to circumvent the cryptic catalysis of AChE function is to utilize substrates that are hydrolyzed more slowly by the enzyme than are ACh and ATCh. One expects that for slow substrates the chemical transition states will be exposed as rate determining, and hence amenable to characterization by isotope effect determinations. Accordingly, in this paper pH-rate profiles and kinetic isotope effects are compared for AChE-catalyzed hydrolyses of ATCh and of the slower substrates PrTCh, BuTCh BzCh and PMPF.

METHODOLOGICAL SECTION

Materials

Electrophorus electricus AChE was purchases from Sigma Chemical Co. Human erythrocyte AChE was purified as described by Rosenberry and Scoggin (1984). Fetal bovine serum AChE was purified as described by De La Hoz et al (1986). *Torpedo california* AChE was purified as described by Sussman et al (1988). The following reagents were used as received from the specified suppliers: ATCh chloride, BuTCh iodide, BzCh chloride, DTNB, $NaH_2PO_4 \cdot H_2O$ and $NA_2HPO_4.7H_2O$ were purchased from Sigma; NaCl was purchased from EM Science, deuterium oxide (D_2O, 99.9% D) from Isotec Ltd., and PrTCh iodide from Aldrich Chemical Co. Water for buffer preparation was distilled and deionized by passage through a Barnstead mixed-bed ion-exchange column (Sybron Corp.).

Buffers were prepared from the appropriate amounts of NaH_2PO_4 and Na_2HPO_4 (pH>5.6) and of sodium acetate and acetic acid (pH<5.6). The total buffer concentration was 0.05 M and the ionic strength was adjusted to 0.2 with NaCl. For measurements of solvent isotope effects, equivalent buffers (i.e. those in which all solute concentrations are equal) were used (Quinn and Sutton, 1991). Values of pL (L=H,D) of buffers were measured with a Corning model 125 pH meter that is equipped with a glass combination electrode; the pD of D_2O buffers was calculated by adding 0.4 to the pH meter reading (Salomaa et al., 1964).

Enzyme Kinetics and Data Treastment

AChE-catalyzed reactions were followed at 25.0 $\pm$ 0.1 ^oC by UV-visible spectrophotometry (Pryor et al., 1992). Hydrolyses of thiocholine esters were followed at 412 nm by using a coupled assay with DTNB (Ellman et al., 1961). The hydrolysis of BzCh was assayed by monitoring changes in absorbance at 240 nm. The PMPF assay was described previously (Acheson et al., 1987). The kinetic parameters V and K were determined by least-sqaueres fitting of $\{V_i, [A]\}$ data to the Michaelis-Menten equation (Wentworth., 1965):

$$V_i = \frac{V[A]}{k + [A]} \qquad (3)$$

values of k_{cat} and k_{cat}/K_m were calculated by dividing V and V/K, respectively, by the enzyme concentration. Initial velocities were determined by linear least-squares fitting of time courses for $\leq 5\%$ of total substrate turnover.

Proton Inventories

Solvent isotope effects were determined by measuring V and V/K values in equivalenty buffered H_2O and D_2O, and are expressed as $^D V = V_{H_2O}/V_{D_2O}$ and $^D V/K = (V/K)_{H_2O}/(V/K)_{D_2O}$. A proton inventory is a series of rate measurements in mixtures of buffered H_2O and D_2O of atom fraction of deuterium n, and is decribed by the Gross-Butler equation (Quinn and Sutton, 1991), which has the following form when reactant state protons do not contribute to the isotope effect:

$$k_n = k_0 \prod_1^x (1 - n + n\phi_i^T) \qquad (4)$$

Equation 4 shows that k_n, the rate constant in solvent isotopic mixtures, has a polynomial dependence on n; k_0 is the rate constant in H_2O, and the ϕ^T values are fractionation factors of each of the x transition state protons that contribute to the isotope effect. When a single proton contributes to the isotope effect, the dependence of k_n on n is linear, whereas multiple proton transfers produce nonlinear dependences on n. Pryor et al. (1992) outline statistical procedures for distinguishing between linear and nonlinear proton inventories.

RESULTS AND DISCUSSION

pH-Rate Profiles

As equation 2 illustrates, the observed rate constant k_{cat}/K_m (and hence V/K) contains microscopic rate constants that convert the free enzyme and free substrate reactant state to the acylenzyme intermediate. Therefore, the pk_a values obtained from pH-V/K profiles are ionization constants of amino acid residues in the free enzyme. Figure 2A shows a pH-V/K profile for AChE-catalyzed hydrolysis of ATCh. This profile is well-described by a single ionization of $pk_a = 6.25$, a value that is in good agreement with the long suspected intrinsic pk_a of H440 (Froede and Wilson, 1971; Quinn, 1987; Rosenberry, 1975). This result is accommodated by the mechanism shown in Figure 1. However, the pH-V/K profile changes markedly for AChE-catalyzed hydrolysis of the slow substrate BuTCh, as shown in Figure 2B. The profile for this substrate reveals two ionizations: $pK_{a1} = 4.64$ and $pK_{a2} = 6.7$. This unsual result suggests that parallel mechanisms are operating for the hydrolysis of BuTCh. For the yet slower substrate BzCh, the pH-V/K profile only depends on an ionization of $pK_a =$ 4.71; i.e., turnover of this substrate does not require that H440 be in the general-base competent neutral form. The overall picture that emerges is that AChE is a mechanistic chimera. For the physiological substrate the mechanism shown in Scheme 1 operates. However, as one increases the steric bulk in the acyl portion of the substrate , the mechanism increasingly shifts to one that depends on the basic form of a residue nearly two pK_a units more acidic than H440. It is reasonable to suggest that in this circumstance the mechanism involves general base catalysis by E199 of direct water attack on the substrate. Reactivity, isotope effect and pK_a data for choline esters are gathered in Table 1.

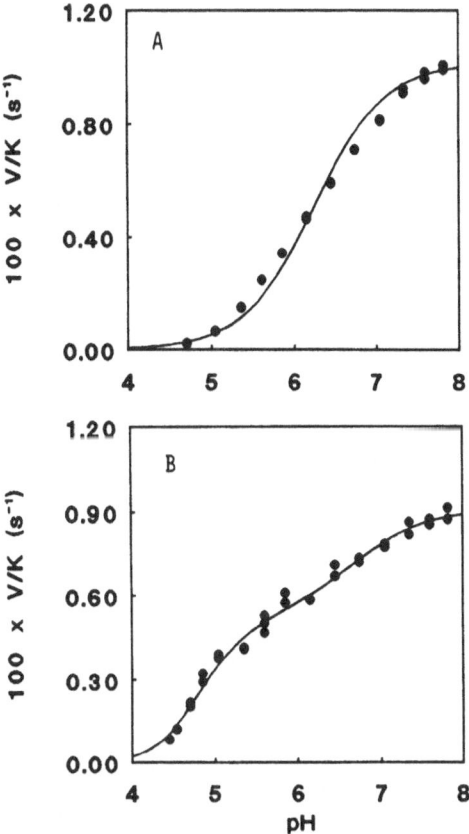

Figure 2. pH-V/K profiles for AChE-catalyzed hydrolyses of choline esters: A. ATCh hydrolysis; B. BuTCh hydrolysis.

Solvent Isotope Effects and Proton Inventories

Table 1 contains solvent isotope effects and relative reactivities for AChE-catalyzed hydrolyses of choline esters and for the aromatic ester-p-methoxyphenyl formate (PMPF). These substrates span nearly four orders of magnitude in reactivity. The data in Table 1 show that isotope effects on V are ~2, which is consistent with rate limitation by a transition state that is stabilized by proton, as outlined in Figure 1. On the other hand, isotope effects on V/K are ~1 for the most reactive substrate, ATCh, and increase to ~2 as reactivity decreases. This trend is consistent with the earlier mentioned rate limitation by diffusion of the ATCh reaction, and with increasing exposure of a chemical transition as rate limiting as substrate reactivity decreases.

Table 1. Solvent deuterium kinetic isotope effects and pH-rate effects for acetylcholinesterase-catalyzed reactions.

Substrate[1]	R[2]	$^DV/K$	^DV	pK$_a$ values [3]
ATCh	1.0	1.1±0.1	2.15±0.07	6.25±0.03
ATCh	0.9	0.9±0.1	2.15±0.05	
ATCh	0.8	1.16±0.08	2.02±0.05	
ATCh	0.4	0.9±0.1	2.30±0.08	
PrTCh	0.3	1.38±0.03	1.98±0.08	
PMPF[4]	0.04	1.09±0.01	2.6±0.1	
BuTCh[5]	0.001	1.87±0.02	1.76±0.09	4.64±0.69, 6.7±0.1
BzCh[5]	0.0002	1.9±0.2	1.7±0.3	4.71±0.03

[1]The four entries for ATCh are those for reactions catalyzed by AChEs from *E. electricus, T. california*, human erythrocytes, and fetal bovine serum, respectively. The other entries are for reactions catalyzed by *E. electricus* AChE. Except where specified, data are taken from Pryor et al (1992).

[2]Relative k_{cat}/K_m values.

[3]pK$_a$ values were calculated from dependences of V/K on pH.

[4]PMPF data are taken from Acheson et al. (1987) and Quinn (1987).

[5]pK$_a$ values for BuTCh and solvent isotope effects for BzCh are date not previously published. The pH dependence of BzCh hydrolysis was described by Quinn et al (1991).

Seven proton inventories of AChE-catalyzed hydrolyses of choline esters have been conducted (Pryor et al., 1992). These include proton inventories of V for ATcH hydrolysis catalyzed by the four AChEs listed in Table 1, proton inventories of V for *E. Electricus* AChE-catalyzed hydrolyses of PrTCh and BuTCh, and a proton inventory of V/K for *E. electricus* AChE-catalyzed hydrolysis of BuTCh. In each case the proton inventory is linear; an example inventory is shown in Figure 3A. It therefore appears that the proton transfers between H440 and E327 depicted in Figure 1 are not elements of transition state stabilization for AChE-catalyzed hydrolysis of physiological substrates. That is, AChE does not function as a charge-relay catalyst.

However, the possibility remains that manifestation of charge-relay catalysis is encrypted by the evolutionary perfection of AChE function. As mentioned in the introduction, k_{cat} (and hence V) for ATCh turnover is rate limited by both acylation and deacylation. Moreover, turnover is particularly rapid, and thus product release steps in the respective stages of catalysis may also contribute to rate limitation (Pryor et al., 1992). The AChE-catalyzed hydrolysis of PMPF is not beset with the kinetic complexity of ATCh turnover. The reaction is linearly activated by increasing concentrations of MeOH, and the corresponding Lineweaver-Burk plot shows the pattern of parallel lines that is diagnostic of ping-pong kinetics (data not shown). This result suggests that MeOH is trapping a rate-limiting formyl-AChE intermediate, and therefore the step that is monitored by k_{cat} is deacylation only. The proton inventory of V for PMPF hydrolysis is shown in Figure 3B, and is best described by a quadratic dependence of activity on the atom fraction of deuterium in the solvent. This result is consistent with the

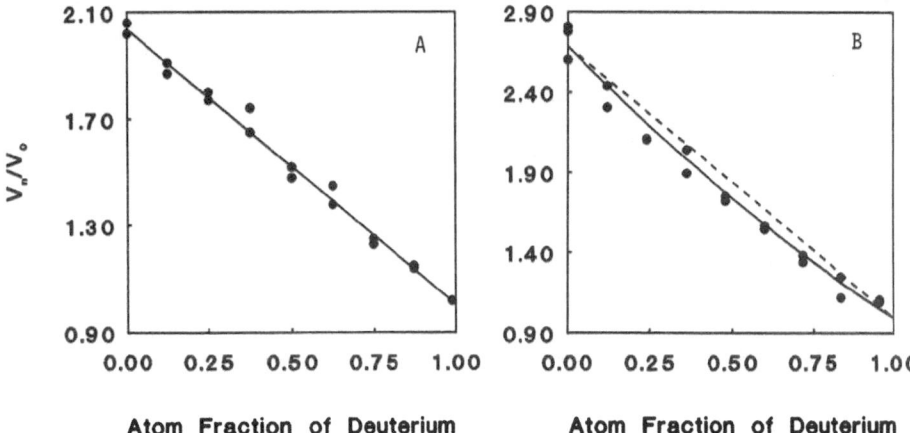

Figure 3. Proton inventorties for V of fetal bovine serum AChE-catalyzed hydrolysis of ATCh (A) and for V of *E. Electricus* AChE-catalyzed hydrolysis of PMPF (B). The dashed line in B emphasis the systematic deviation of the PMPF proton inventory from linearity.

synchronous transfer of at least two protons in the transitions state. Therefore, in a situation where a single reaction step is being monitored, charge-relay catalysis is apparently manifested.

The proton inventory and pH-rate studies described herein show that the cryptic catalyst AChE not only utilizes a variety of approaches to effect chemical catalysis, but also conceals the sources of its catalytic power from the unsuspecting view. These features of AChE should make future structure-function studies of the enzyme particularly dicey yet enticing.

Acknowledgements

Support of this work was received from NIH grants NS21334 to D.M.Q. and NS16577 to T.L.R., from a Revson Foundation grant to I.S., and from U.S. Army Medical Research and Development Command Contract No. DAMD17-89-C-9063 to I.S.

REFERENCES

Acheson, S.A., Barlow, P.N., Lee, G.C., Swanson, M.L., and Quinn, D.M., 1987, Effect of reactivity on virtual transition-state structure for the acylation stage of acetylcholinesterase-catalyzed hydrolysis of aryl ester and anilides, *J. Am. Chem. Soc.* 109:246.

Bazelyansky, M., Robey, E., and Kirsch, J.F., 1986, Fractional diffusion limited component of reactions catalyzed by acetylcholinesterase, *Biochemistry* 25:125.

Blow, D.M., 1976, Structure and mechanism of chymotrypsin, *Acc. Chem. Res.* 9:145.

De La Hoz, D., Doctor, B.P., Ralston, J.S., Riush, R.S., and Wolfe, A.D., 1986, A simplified procedure for the purification of large quantities of fetal bovine serum acetylcholinesterase, *Life Sci.* 39:195

Ellman, G. L., Courtney, K.D., Andres, V., Jr., Featherstone, R.M., 1961, A new and rapid colorimetric determination of acetylcholinesterase activity, *Biochem. Pharmacol.* 7:88.

Froede, H.C., and Wilson, I.B., 1971, Acetylcholinesterase, in: "The Enzymes," 3rd ed., Vol. 5, P.D., Boyer, ed., Academic Press, New York.

Froede, H.C.., and Wilson, I.B., 1984, Direct determination of acetylenzyme intermediate in the acetylcholinesterase-catalyzed hydrolysis of acetylcholine and acetylthiocholine, *J. Biol. Chem.* 259:11010

Nolte, H.-J., Tosenberry, T.L. and Neumann, E., 1980, Effective charge on acetylcholinesterase active sites determined from the ionic strength dependence of association rate constants with cationic ligands, *Biochemistry* 19:3705

Polgar, L., 1987, Structure and function of serine proteases, in: "Hydrolytic Enzymes," A. Neuberger and K. Brocklehurst, eds., Elsevier, Amsterdam

Pryor, A.N., Selwood, T., Leu, L.-S., Andracki, M.A., Lee, B.H., Rao, M., Rosenberry, T., Doctor, B.P., Silman, I., and Quinn, D.M., 1992, Simple general acid-base catalysis of physiological acetylcholinesterase reactions. *J. Am. Chem. Soc.* 114: in press

Quinn, D.M., 1987, Acetylcholinesterase: enzyme structure, reaction dynamics, and virtual transition states, *Chem. Rev.* 87:955.

Quinn, D.M., Pryor, A.N., Selwood, T., Lee, B.-H., Acheson, and Barlow, P.N., 1991, The chemical mechanism of acetylcholinesterase reactions: biological catalysis at the speed limit, in: "Cholinesterases: Structure, Function, Mechanism, Genetics, and Cell Biology," J. Massoulie, F. Bacou, E. Barnard, A. Chatonnet, B.P. Doctor, and D.M. Quinn, eds., American Chemical Society, Washington, D.C.

Quinn, D.M., and Sutton, L.D., 1991, Theoretical basis and mechanistic utility of solvent isotope effects, in: "Enzyme Mechanism from Isotope Effects," P.F. Cook, ed., CRC Press, Boca Raton, Florida.

Rosenberry, T.L., 1975, Acetylcholinesterase, *Adv. Enzymol. Relat. Areas Mol. Biol.* 43:103.

Rosenberry, T.L., and Scoggin, D.M., 1984, Structure of human erythrocyte acetylcholinesterase. Characterization of intersubunit disulfide bonding and detergent interaction, *J. Biol. Chem.* 259:5643.

Salomaa, P., Schaleger, L.L., and Long, F.A., 1964, Solvent deuterium isotope effects on acid-base equilibria, *J. Am. Chem. Soc.* 86:1.

Schowen, R.L., 1978, Catalytic power and transition-state stabilization, in: "Transition States of Biochemical processes," R.D. Gandour and R.L. Schowen, eds., Plenum Press, New York.

Sussman, J.L., Harel, M., Frolow, F., Varon, L., Toker, L., Futerman, A.H., and Silman, I., 1988, Purification and crystallization of a dimeric form of acetylcholinesterase from *Torpedo california* subsequent to solubilization with phosphatidyinositol-specific phospholipase C., *J. Mol. Biol.* 203:821

Wentworth, W.E., 1965, Rigourous least squares adjustment: application to some nonlinear equations, *J. Chem. Ed.* 42:96.

INFLUENCE OF IONIC COMPOSITION OF THE MEDIUM

ON ACETYLCHOLINESTERASE CONFORMATION

Harvey Alan Berman and Mark W. Nowak

Department of Biochemical Pharmacology
School of Pharmacy
State University of New York at Buffalo
Buffalo, NY 14260

INTRODUCTION

The recent crystallographic analysis of a dimeric glycophospholipid form of AchE from *Torpedo californica* (Sussman et al., 1991) provides an ideal opportunity for comparing the catalytic behaviour and recognition properties of the enzyme with its atomic structure. The globular subunit is determined to be an α/β protein containing numerous crossover motifs of the type ß-α-ß or ß-loop-ß. The catalytic residue, Ser_{200}, resides 4 A above the base of a 20 A deep cavity - or *gorge* - lined with numerous aromatic amino acid residues accounting for approximately 40 percents of the active center surface. A catalytic charge-relay triad comprising Ser_{200}-His_{440}-Glu_{327} is identified. Quite surprisingly, only few net negative amino acid residues are found within the gorge: Asp_{285} and Glu_{273} at the top, Asp_{72} about half-way into the gorge, and Glu_{199}, proximal to Ser_{200}, near the base. Overall, the atomic coordinates afford a picture of the active center as that of a deep, highly aromatic cavity, rich in *pi*-electron density. No evidence exists for an anionic choline-binding locus situated approximately 5 A from Ser_{200}.

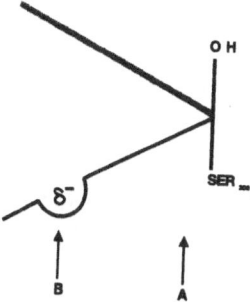

Figure 1. Topography of the active center of AchE. The shaded area designates the *n*-alkyl binding region, a region that facilitates association of uncharged agents. The negative subsite long believed to exist within the active center is indicated by δ^-. The peripheral anionic site (not shown) exists outside or at the perimeter of this operational description of the active center. A and B denote different regions within the active center at which cationic ligands associate (Berman and Leonard, 1990).

Multidisciplinary Approaches to Cholinesterase Functions, Edited by
A. Shafferman and B. Velan, Plenum Press, New York, 1992

Comparison Between Crystal Structure and Operational Behaviour of AchE

The view of AchE derived from crystallographic analysis corroborates a gratifying number of conclusions derived from kinetic, equilibrium and spectroscopic studies of a tetrameric form of AchE, as summarized in Figure 1 and enumerated below.

1. Reaction of AchE with enantiomeric methylphosphonothioates is variable and highly dependent on the nature of the groups surrounding the asymmetric phosphorus (Berman and Leonard, 1989). Based on these studies the active center was concluded to encompass a hydrophobic region, separate from the putative anionic - or choline binding - locus within the active center. The strong chiral preference (S_P > > R_P) displayed by AchE for enantiomeric methylphosphonothioates falls nicely in line with the crystallographic view placing Ser_{200} at a distance 4 A above the base of the gorge in a highly asymmetric environment containing a large proportion of aromatic amino acids.

2. Some ligands, such as decamethonium, antagonize irreversible inhibition of longer chain rather than shorter chain methylphosphonothioates, while other ligands, such as edrophonium, antagonize irreversible inhibition by methylphosphonothioates containing short- *and* long-chain leaving groups. The locus for edrophonium association is concluded to exist in proximity of Ser_{200}, within the esteratic region of the active center (Figure 1, region A). The locus for decamethonium association is concluded to be spatially removed from Ser_{200} and distal to the edrophonium binding locus (Figure 1, region B). That is to say, that the active center houses a heterogeneous population of cation binding sites (Berman and Decker, 1986a; Berman and Leonard, 1990). Crystallographic analysis of noncovalent complexes of AchE corroborate this view by indicating unique localization for structurally unrelated active center-selective ligands.

3. *Aging* - or dealkylation - of covalent organophosphonyl conjugates displays an ionic strength dependence that is *opposite* to that predicted for a reaction in which charge develops during the transition state (Berman and Decker, 1986b). This observation prompted the conclusion that the ionic strength dependence reflects the influence of ions not on the carbon-oxygen bond scission *per se* but on enzyme conformation. Transfer of AchE from a medium of high ionic strength to one of lower ionic strength might therefore be expected - through increased Coulombic repulsive interactions between net negative surface charges - to expand the protein dimensions, thereby increasing the aqueous content of the active center cavity and promoting bond breakage. The crystal structure corroborates this view by revealing a substantial distribution of net negative charge on the enzyme surface surrounding the active center gorge.

4. Finally, the high aromatic nature, that is, the high *pi*-electron density within the gorge, and in particular the presence of Trp_{84}, serves to explain the capacity for charge-transfer interactions detected through use of covalent modification of the active center with (NBD)aminoalkyl methylphosphonofluoridates (Berman, et al., 1985).

It is clear that the active center behaves as a reaction cavity which, by virtue of its depth and dimensional limitations, contains a solvent pool that differs from the bulk aqueous medium. The physical properties of this cavity govern at least some of the *covalent* reactions that occur within the cavity. The physical properties of this cavity are governed, in turn, on enzyme interactions with the aqueous medium. Any phenomenon - an increase or decrease in ionic strength, for example - that alters the charge distribution at the enzyme surface can alter characteristics of the cavity and, in turn, covalent reactivity. This relationship between ion charge at the surface and covalent reactivity holds with respect to aging; it may hold also with respect to substrate hydrolysis. It remains unknown as to what extent this relationship plays a role in governing noncovalent reactions.

Unexplained Questions

As interesting as the features the crystal structure explains are those it leaves - at least for the moment - unexplained. For example, one question of interest concerns the marked ionic strength dependence of ligand association with AchE. If no anionic site exists within the active center cavity, that is, if the choline-binding locus is not anionic in nature, then how does one explain explain the well-documented behaviour of the enzyme to associate with active center selective cations. What mechanism explains the marked dependence of cation affinity on ionic strength of the bulk medium? A related question concerns the mechanism underlying allosteric communication between the peripheral anionic site and Ser_{200}. The answers to these questions might be expected to emphasize primary sequence and positions of individual anionic amino acid residues, or the dipolar nature of enzyme secondary structure, or both. In any case, it is reasonable to expect that allosteric interactions arise from small, subtle, sub-Angstrom displacements in position of individual catalytic amino acid residues.

This paper addresses these concerns and seeks, in particular, to determine the capacity of AchE to undergo subtle shifts in atomic coordinates by examining the influence of ionic composition of the bulk medium on enzyme conformation. In so doing, we wish to highlight two salient features long associated with AchE. The first feature focuses on substrate hydrolysis and notes that the kinetics of this reaction are essentially independent of ionic composition of the surrounding medium. Such characteristics stand in contrast to the marked ionic strength dependence of ligand occupation of the peripheral anionic site and the active center. The second feature notes that the capacity of peripheral site ligands to alter substrate hydrolysis is critically dependent on the electrostatic nature of the substrate. That is, occupation of the peripheral site inhibits hydrolysis of cationic substrates in a *linear* manner while that of uncharged substrates is *nonlinear*. This behaviour reveals a pronounced dependence of covalent reactivity on the electrostatic nature of the reactants (Berman and Leonard, 1990; Berman, et al., 1991). Preferential inhibition of cationic rather than uncharged reactants appears to represent a general property of reactions within the AchE active center, since similar relationships are observed for irreversible inhibition by cationic and uncharged methylphosphonates. On this basis, we have concluded that the peripheral anionic site is linked with the active center through a principally electrostatic interaction.

This study looks at lability of AchE conformation in ionic media of different composition. Enzyme conformation is assessed through utilization of (NBD)aminoethyl methylphosphonofluoridate, one of a family of reactive probes that covalently modify the nucleophilic serine within the active center (Berman, et al., 1985). These fluorescent probes, shown in Figure 2, are of advantage in that when bound to the surface of a protein their fluorescence is governed by *static* rather than *dynamic* quenching mechanisms. As a consequence, only small changes in orientation of the probe with respect to the enzyme surface are required in order to observe measurable changes in fluorescence intensity, providing a sensitive indicator of ion-induced changes in protein conformation.

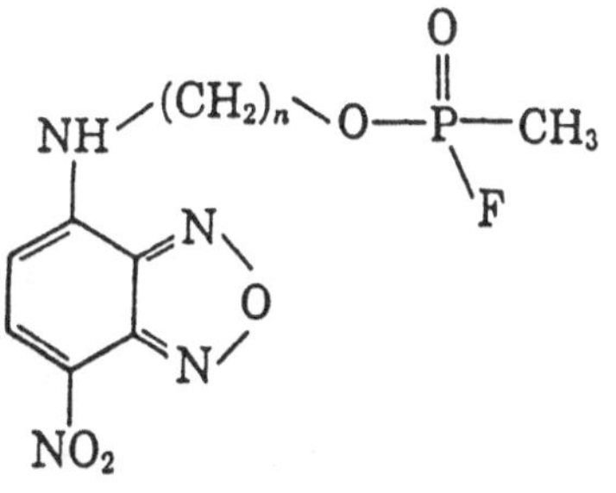

Figure 2. Structure of (NBD)aminoethyl methylphosphonofluoridate ($n=2$). Displacement of the fluoride by AchE results in formation of the covalent conjugate (NBD)aminoethyl methylphosphono-AchE, denoted (NBD)AE-MP-AchE.

RESULTS

Influence of Alkali and Alkaline Earth Ions on Fluorescence of (NBD)AE-MP-AchE

AchE from *Torpedo californica* was allowed to react with (NBD)AE-MPF to form the covalent conjugate (NBD)AE-MP-AchE. The absorption spectrum of (NBD)AE-MP-AchE appeared with a maximum at 470-475 nm and was congruent with the excitation spectrum. Upon excitation at 470 nm, the emission maximum of (NBD)AE-MP-AchE appeared at 542 nm and was characterized by a quantum yield of 0.084 (Figure 3). This value was approximately 1.4-fold greater than the value of 0.062 observed for

(NBD)aminoethanol in H₂O, and was markedly lower than observed for (NBD)aminoethanol in aprotic organic solvents. Hence, the fluorophore when covalently linked with AchE was highly quenched.

Fluorescence of (NBD)AE-MP-AchE *increased* 2-fold in the presence of Na⁺ and *decreased* 2-fold in the presence of Mg⁺⁺. Fluorescence emission spectra showing these changes are presented in Figure 3. The increase in the presence of Na⁺ was saturable, occurred without a shift in position of the excitation and emission spectra, and was compatible with an apparent dissociation constant of 12.5 ± 0.9 μM. Similar behaviour was observed upon titration with Li⁺ and K⁺. The decrease in fluorescence of (NBD)AE-MP-AchE in the presence of Mg⁺⁺ was saturable, occurred without a shift in position of the excitation and emission spectra, and was compatible with an apparent dissociation constant of 6.2 ± 0.7 μM. Similar behaviour was observed upon titration with Ca⁺⁺.

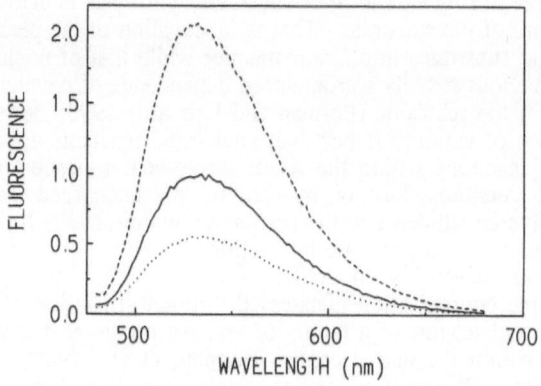

Figure 3. Influence of Na⁺ and Mg⁺⁺ on fluorescence spectrum of (NBD)AE-MP-AchE. The emission spectrum of (NBD)AE-MP-AchE, present in a 10 mM Tris-Cl buffer, pH 8.0, was measured upon excitation at 470 nm (————); the emission maximum appeared at 530 nm. Emission spectra in the presence of Na⁺ (--------) and Mg⁺⁺ (........) are shown. In all cases the excitation and emission maxima were unchanged.

Antagonism Between Alkali and Alkaline Earth Ions on the Fluorescence of (NBD)AE-MP-AchE

The effects of alkali and alkaline earth cations on fluorescence of (NBD)AE-MP-AchE were of equal magnitude but opposite in direction. It was of interest, therefore, to examine the capacity of divalent alkaline earth cations to antagonize the increase in fluorescence of (NBD)AE-MP-AchE observed in the presence of monovalent alkali ions. Solutions of (NBD)AE-MP-AchE containing different concentrations of Ca⁺⁺ were titrated with Na⁺ and changes in fluorescence monitored. In the presence of increasing concentrations of Ca⁺⁺ the resulting Na⁺ titration profiles were displaced to the right and, although they started from lower initial fluorescence values, they approached the common saturation value observed in the absence of Ca⁺⁺ (Figure 4). That is, the presence of Na⁺ was able to overcome the effect of Ca⁺⁺. Similar results were observed for antagonistic actions of Mg⁺⁺. In all cases, the shift in the titration profile was linear with Ca⁺⁺ and Mg⁺⁺ concentrations, as shown in Figure 4 for the case of Ca⁺⁺.

Fluorescence Decay Characteristics of (NBD)AE-MP-AchE in the Absence and Presence of Alkali and Alkaline Earth Ions

The physical mechanism underlying the spectral changes was examined by comparing the time-correlated decay of fluorescence intensity of (NBD)AE-MP-AchE with the steady-state fluorescence characteristics of the covalent conjugate. In the absence of alkali and alkaline earth ions the fluorescence decay of (NBD)AE-MP-AchE was complex and required analysis with an equation containing the sum of three exponential terms. Analysis with such an equation afforded lifetimes of 1.1, 5.3, and 11 nsec present in a respective ratio of 0.68, 0.18, and 0.14.

The individual fluorescence lifetimes were unchanged in the presence of Na^+. The relative proportion of the individual components, as measured by the respective pre-exponential factors, showed a distinct dependence on concentration of Na^+, where the fastest decay component decreased and the slower decay components increased with increasing Na^+ concentrations. The net effect of such alterations in the amplitudes of the individual decay components was to reduce the average lifetime, $<\tau>$, relative to the steady-state fluorescence intensity. For example, the increase in steady-state fluorescence was approximately 2-fold, whereas the net increase in $<\tau>$ was only 1.3-fold. Overall, increases in $<\tau>$ did not parallel increases in steady-state fluorescence. Such behaviour indicated that the alterations in fluorescence intensity were governed by a mechanism of *static* rather than dynamic - or collisional - quenching.

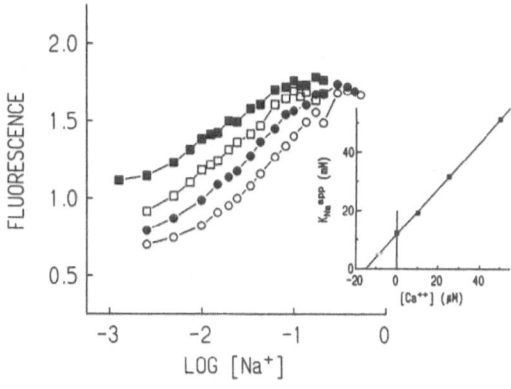

Figure 4. Antagonism by Ca^{++} of the Na^+-induced increase in fluorescence of (NBD)AE-MP-AchE. Steady-state fluorescence intensity of (NBD)AE-MP-AchE was measured upon titration with Na^+ in the presence of different concentrations of Ca^{++}: 0 (■); 10 (□); 25 (●); 50 (O) μM. The *inset* presents the relationship between the apparent dissociation constant for Na^+, determined from Scatchard plots, and the concentration of Ca^{++}. A competitive dissociation constant of 15 μM. determined from the x-intercept, is obtained for Ca^{++}, in good agreement with that obtained by direct titration.

DISCUSSION

One property of AchE that has long been known is the dramatic one-to-two order of magnitude dependence of ligand association on ionic strength of the bulk aqueous medium. This phenomenon is typified by behaviour seen for propidium, a bisquaternary cation selective for the peripheral anionic site (Taylor and Lappi, 1975), and N-methylacridinium, a monoquaternary cation selective for the active center (Nolte, et al., 1980), both of which undergo 30-100-fold increases in affinity with decreases in ionic strength of the aqueous medium. Substrate hydrolysis, in contrast, assessed with respect to k_{cat}/K_M and the individual constants k_{cat} and K_M, remains relatively constant with large changes in ionic strength (Berman and Leonard, 1990).

This study indicates that while substrate turnover is invariant with ionic composition of the medium, fluorescence intensity of a reporter ligand tethered within the active center undergoes measurable changes under similar conditions. These changes in fluorescence are seen in the presence of monovalent and divalent cations and, in both cases, the relationships between ion concentration and fluorescence intensity are saturable. The responses for alkali and alkaline earth ions differ, however, in that they are *opposite* in direction and occur over a 1000-fold different range of ion concentrations. Analysis according to a saturation binding isotherm affords apparent dissociation constants of 10 mM for Na^+ and 10 μM for Mg^{++}. Such behaviour are incompatible with a mechanism based solely on electrostatic shielding of the net surface charge of AchE or an effect proportional to the ionic strength of the bulk medium, and, by virtue of the saturable nature of the titration profile, require direct ion association with the enzyme.

Spectral Characteristics of (NBD)aminoethanol and (NBD)AE-MP-AchE

The spectral characteristics of (NBD)aminoethanol in isotropic organic media are similar to that of other (NBD)aminoalkyl adducts (Berman, et al., 1985). That is, upon transfer of (NBD)aminoethanol from H_2O to organic media of decreasing dielectric constant the absorption maxima and fluorescence excitation and emission maxima shift 15-30 nm toward shorter wavelengths and the quantum yield and lifetime increase 10-15-fold. Such changes, in that they display a near-linear dependence on the dielectric constant of the surrounding environment, indicate that the spectroscopic behavior of (NBD)aminoethanol is markedly dependent on the characteristics of the surrounding medium. If no factors other than the polarity of the immediate environment were to govern fluorescence of (NBD)aminoethanol, then it would be predicted that the spectral behavior of (NBD)aminoethyl methylphosphonates covalently associated with AchE would reflect the polarity of the active center environment. That expectation, however, is not realized.

The absorption and excitation maxima, quantum yield, and lifetime of (NBD)AE-MP-AchE closely approximate the values observed for solutions of (NBD)aminoethanol in H_2O. The emission maximum, however, approaches that of (NBD)aminoethanol in ethanol, a solvent of lower dielectric constant. Thus, the individual spectral characteristics of (NBD)aminoethanol in isotropic organic media provide no unequivocal frame of reference for deducing the physical environment within the active center. What can be concluded, based on observation of the intense optical activity exhibited by (NBD)AE-MP-AchE and broadening of the absorption spectrum (Berman, et al., 1985), is that the (NBD)aminoalkyl moiety exists in close contact with an amino acid residue in the active center. It is this interaction, the complexation of the (NBD)aminoethyl moiety with the enzyme surface, that appears to govern the fluorescence intensity of the covalent conjugate.

The ion-induced changes in (NBD)AE-MP-AchE fluorescence occur without a shift in position of the excitation and emission spectra and without corresponding increases in fluorescence lifetimes. Overall, this behaviour indicates operation of a *static* - rather than dynamic or collisional - mechanism of quenching. Static quenching suggests the proximity of a neighboring amino acid that quenches fluorescence of the (NBD)aminoethyl moiety. As a consequence, the fluorescence increases seen in the presence of Na^+ and K^+ and the decreases seen in the presence of Mg^{++} and Ca^{++} are concluded to arise from respective reductions and increases in static quenching of (NBD)aminoethyl fluorescence. Such changes in fluorescence require specific ion association with AchE and, in turn, a subsequent alteration in enzyme conformation. Since static quenching reflects a ground state complexation rather than an excited state collisional phenomenon, only small changes in distance separating the (NBD)aminoethyl moiety and the quenching residue are required to relieve quenching and therefore alter fluorescence intensity. That is, ion association with AchE results in sub-Angstrom shifts in the relationship between the (NBD)aminoethyl moiety and the active center cleft.

Interaction of the Peripheral Anionic Site and the Active Center

As mentioned in the introduction, the rates of substrate hydrolysis are essentially invariant with large changes in ionic composition of the bulk medium. This observation, coupled with findings that ligand association at the peripheral anionic site blocks completely hydrolysis of cationic rather than uncharged substrates reveals that interaction between the peripheral site and the active center is principally electrostatic in nature. This relationship appears to be general with respect to covalent reactions within the active center since peripheral site occupation blocks completely irreversible inhibition by *cationic* methylphosphonates, but blocks only incompletely inhibition by *uncharged* methylphosphonates. These observations allow the inference that while substrate hydrolysis remains constant, AchE conformation changes with changes in ionic composition of the bulk aqueous medium. This paper provides direct evidence in support of such a phenomenon and indicates, moreover, unique actions of monovalent and divalent cations on AchE conformation.

The influence of ions on AchE conformation might therefore arise from ion association at the peripheral site, for which kinetic (Changeux, 1966) and spectroscopic (Taylor and Lappi, 1975) evidence exists. One obvious mechanism for interaction of the peripheral site with the active center focuses on primary sequence of the enzyme and the relationship between individual net negative amino acid residues and the nucleophilic serine, Ser_{200}. The peripheral anionic site might be thought to interact with an *anionic subsite*, once believed to exist within the active center. However, the three-dimensional structure indicates the absence of an anionic site - or net negative residues - 5 A removed from Ser_{200}. Instead one must look to identify individual residues - some in proximity and others remote from Ser_{200} - that may play critical roles in catalysis. Interaction of the peripheral anionic site with Glu_{199}, sitting proximal to Ser_{200}, represents one logical candidate for consideration in allosteric regulation of catalysis. The importance of Glu_{199} is supported by observations that substitution of *Gln*, an uncharged amino acid, abolishes the capacity for substrate-induced inhibition (Gibney et al., 1990). Additional residues such as Glu_{443}, situated behind His_{440}, and Asp_{72}, situated between the peripheral site and active center, also merit consideration.

A second mechanism worthy of consideration focuses on the dipolar nature of enzyme secondary structure. AchE, an α/β protein, is made up of 30 percent α helix and 15 percent ß-sheet. The catalytic serine is connected to a short α helix located at the bottom of a deep hydrophobic cleft. Of significance is that α helices contain large dipole moments that point in the direction of the helix (Hol, 1985). Parallel ß sheets contain no net dipole moment; twisting of the sheet, however, confers a net dipole moment in which the positive and negative ends are localized on the respective N- and C-termini. These dipolar entities and their characteristic capacities for ion-dipole interactions with either cationic ligands or ionic electrolytes provide additional means to explain the enzyme lability in ionic aqueous media.

Functional Significance of Ion-induced Changes in AchE Conformation

The present study establishes the viewpoint that AchE conformation changes in response to subtle changes in ionic composition of the surrounding medium. Since AchE is localized within the basal lamina and along the plasmalemma circumscribing the neuromuscular junction, the enzyme can be thought to experience the ionic composition - or *changes* in the ionic composition - of the synaptic cleft during neurotransmission. In this sense, this paper supports the proposal that by virtue of the electrostatic nature of its interaction with the active center, the peripheral anionic site subserves enzyme catalysis by stabilizing the high rates of substrate hydrolysis in the presence of varying ion concentrations (Berman and Leonard, 1990; Berman, et al., 1991). Such a function to stabilize covalent reactivity and to maintain the high catalytic efficiency would seem to represent a mechanism of considerable importance in neuromuscular transmission since large excursions in catalytic efficiency would be incompatible with the requirement for rapid removal of Ach^+ from the synaptic cleft. As a consequence, it is of interest to ask to what degree the concentrations at which changes in AchE conformation are observed approximate changes in ion composition that occur within the synaptic cleft during membrane depolarization.

Jachter and Sachs (1982) describe a computer model of neuromuscular transmission that uniquely accounts for ion depletion within the synaptic cleft during the miniature endplate potential. This model describes not only the process of Ach^+ diffusion and subsequent interaction with postsynaptic receptors during the miniature endplate potential, but also incorporates contributions to the total current arising from ion conduction through channels and translational diffusion of ions within the synaptic cleft. The basis of the simulation is a *saturated disk* model in which a quantum of Ach^+ released from a presynaptic disk is considered to act on a small region of a postsynaptic disk. At the center of the saturated disk the acetylcholine receptors exist complexed with two Ach$^+$ molecules while at the fringes of the disk the receptors are either singly complexed or are free. This simple geometric model allows calculation of ion concentrations at different times during the action potential.

Under a voltage clamp of -90 mV, for example, Na$^+$ concentrations at the center of the postsynaptic disk are calculated to decrease from 140 mM to 105 mM. Under a voltage clamp of +30 mV, where the current is carried primarily by K$^+$, the synaptic concentration of K$^+$ increases from 5 mM to 15 mM. Of significance is that the reduction in Na$^+$ concentration and increase in K$^+$ concentration are comparable in magnitude to the concentrations observed to cause alterations in AchE conformation.

Although there exists no experimental evidence of ion depletion in the synaptic cleft during neurotransmission, the saturated disk model supports the viewpoint that within the restricted geometry of the cleft significant changes in ion composition may occur. Emphasizing this point in another way, the ionic environment of the neuromuscular junction, modelled in simple geometric terms, undergoes excursions in ion concentration that are comparable in magnitude to those that we have determined to be sufficient to induce changes in AchE conformation.

REFERENCES

Berman, H.A. and Decker, M.M. (1986a) Kinetic, equilibrium, and spectroscopic studies on cation association at the active center of acetylcholinesterase: Topographic distinction between trimethyl and trimethylammonium sites. *Biochim. Biophys. Acta*, 872:126.

Berman, H.A. and Decker, M.M. (1986b) Kinetic, equilibrium and spectroscopic studies on dealkylation ("Aging") of alkyl organophosphonyl-acetylcholinesterase: Electrostatic control of enzyme topography. *J. Biol. Chem.* 261:10646.

Berman, H.A. and Leonard, K.J. (1989) Chiral reactions of acetylcholinesterase probed with enantiomeric methylphosphonothioates: Noncovalent determinants of enzyme chiral preference. *J. Biol. Chem.* 264:3942.

Berman, H.A. and Leonard, K. (1990) Ligand exclusion on acetylcholinesterase. *Biochemistry* 29:10640.

Berman, H.A., Leonard, K.J., and Nowak, M.W. (1991) Function of the peripheral anionic site of acetylcholinesterase, *in* "Cholinesterases: Structure, Function, Mechanism, Genetics, and Cell Biology", J. Massoulie, F. Bacou, E. Barnard, A. Chatonnet, B.P. Doctor, and D.M. Quinn, eds. American Chemical Society Press, Washington, D.C.

Berman, H.A., Olshefski, D.F., Gilbert, M., and Decker, M.M. (1985) Fluorescent phosphonate labels for serine hydrolases. *J. Biol. Chem.* 260:3462.

Changeux, J.-P. (1966) Responses of acetylcholinesterase from *Torpedo marmorata* to salts and curarizing drugs. *Mol. Pharmacol.* 2:369.

Gibney, G., Camp, S., Dionne, M., MacPhee-Quigley, K., and Taylor, P. (1990) Mutagenesis of essential functional residues in acetylcholinesterase. *Proc. Natl. Acad. Sci.,USA* 87:7546.

Hol, W.G.J. (1985) The role of the α helix dipole in protein function and structure. *Prog. Biophys. Molec. Biol.* 45:149.

Jachter, H. and Sachs, F. (1982) Ionic depletion and voltage gradients at the endplate. *Biophysical J.* 37:311a.

Nolte, H.-J., Rosenberry, T.L., and Neumann, E. (1980) Effective charge on acetylcholinesterase active sites determined from the ionic strength dependence of association rate constants with cationic ligands. *Biochemistry* 17:3705.

Sussman, J.L., Harel, M., Frolow, F., Oefner, C., Goldman, A., Toker, L., and Silman, I. (1991) Atomic structure of acetylcholinesterase from *Torpedo californica*: a prototypic acetylcholine binding protein. *Science* 253:872.

Taylor, P. and Lappi, S. (1975) Interaction of fluorescence probes with acetylcholinesterase. The site and specificity of propidium. *Biochemistry* 14:1989.

MOLECULAR DISSECTION OF FUNCTIONAL DOMAINS IN HUMAN CHOLINESTERASES EXPRESSED IN MICROINJECTED XENOPUS OOCYTES

Averell Gnatt, Yael Loewenstein
and Hermona Soreq

Department of Biological Chemistry
Life Sciences Institute
Hebrew University of Jerusalem
Jerusalem, 91904

INTRODUCTION

The two classes of human cholinesterases (CHEs), acetylcholinesterase(acetylcholine acetyl hydrolase, ACHE, EC 3.1.1.7) and butyrylcholinesterase (acylcholine acyl hydrolase, BCHE, EC 3.1.1.8) are highly homologous proteins capable of rapidly hydrolyzing choline esters[1]. Despite their similar mechanisms of action, they differ in substrate specificity and sensitivity to various inhibitors[1,2]. Recent advances including cloning,[3-6] expression,[5,7-10] and 3 dimensional structural analysis of members of the CHE superfamily,[11,12] now enable the dissection of functional domains thereof. Within these domains, key amino acids are found which may be implicated in catalysis or in binding of various ligands. The disclosing of such key residues could lead to designing of novel therapeutic agents as well as to the unravelling of the molecular mechanisms underlying the functioning of ChEs.

A direct approach towards these goals involves the use of natural and site directed variants of human CHEs, expressed in a heterologous system, as followed by in-depth pharmacological analysis[13-15]. Analysis of the results would then be combined with structural examination of the relevant variant residues within available 3-dimensional model(s) of CHEs, to allow for an understanding of structure - function relationships.

THE EXPERIMENTAL APPROACH

Rationalized site-directed mutagenesis is dependent on a thorough understanding of the relevant structure. ChEs are members of a large superclass of proteins, all of which

Multidisciplinary Approaches to Cholinesterase Functions, Edited by
A. Shafferman and B. Velan, Plenum Press, New York, 1992

display varying levels of homology[1-5]. Biochemical studies indicated that the active site in all of these proteins should include two peptide subsites: An anionic site responsible for binding the charged choline moiety of the ACh substrate and a catalytic subsite in which the active triad is to be found[2]. Two proteins belonging to this superfamily, *Torpedo Californica* AChE[11] and *Geotrichum Candidum* lipase [12,] have been crystallysed and their 3- dimensional structure has been revealed, disclosing the peptide domains involved in these two subsites. Considering the high level of primary sequence homology and closely related biochemical traits displayed by *Torpedo* ACHE and human BCHE, it is possible to superimpose most of the structural features of human BCHE on the 3- dimensional model of the *Torpedo* enzyme[11]. Figure 1 displays a general scheme of this structure and presents the key regions which can be predicted to be involved in ligand affinity and substrate catalysis.

At the bottom of a gorge approximately 20 Angstroms deep, here denoted as the active center gorge, the catalytic triad is found. Within a presumptive strong hydrogen "sink" in this gorge, a glutamate residue in position 325 (E325) draws a hydrogen atom from a Histidine residue (H438) which, in turn, draws an hydrogen atom from the active site serine (S198). The activated S198 is then able to perform a nucleophilic attack on its substrate, exemplified in the scheme by butyrylcholine (BCh). It should be noted that H438 also comes in contact with the substrate, probably indicative of other roles it might have.

Employing BCh as a model ligand, two hydrophobic regions of binding in the active site are observed, one for the choline moiety (Choline binding site) and a second for the acid moiety (Acetyl or butyryl binding site). In contrast with the biochemical predictions, the choline binding site in CHEs was found to be hydrophobic and not anionic in nature[11], raising an intriguing discrepancy between the structural and biochemical information that we have attempted to unravel experimentally using the genetic engineering approach. In addition, both subsites in the active site center of CHEs could be assumed to interact with various parts of inhibitors such as carbamates and organophosphorus agents (OPs). Other regions of importance are the entrance into the gorge (Rim domain) and the hydrophobic lining of the gorge (Figure 1). These regions are clearly important for the penetrance of ligands into the active site center and are likely to be involved in binding inhibitors which "block" the active center gorge. In general, all regions internal to the rim and within the active site are hydrophobic, with the exclusion of several key residues such as Aspartate 70 at the top of the gorge and Glutamate 197 and Glutamate 441 within the active site center. These charged residues hence became the first candidates for experimental substitution.

In search for the actual role(s) of specific amino acid residues, *Xenopus* oocytes were microinjected with synthetic mRNA (sRNA) from 19 recombinant transcription vectors containing normal, site-directed and natural recombinant variants of human ACHE and BCHE[5,13-15] (rCHEs; TableI). The resultant proteins were assayed for activity employing the Ellman's method[16], followed by inhibition profiles performed with selected choline esters, procaine derivatives, toxic glycoalkaloids, OP agents and carbamates[13-15]. The biochemically altered features were then examined in view of the location of the altered amino acids, as it appears in the 3-dimensional model of the highly homologous *Torpedo* ACHE [11].

RESULTS

The expression system employed provided useful comparisons to natural ChEs, since both substrate specificity and selective inhibition of oocyte expressed ACHE and BCHE were nearly identical to those of the native enzymes. Biochemical alterations were displayed by

Table I. Expression of Recombinant Human CHEs in Xenopus Oocytes

Name and Alteration*	Enzyme#	(A)ctive (I)nactive	Proposed Structure - function relationship
1. A/None	ACHE(N)	A	Nearly identical to Human erythrocyte ACHE.
2. B/None	BCHE(N)	A	Nearly identical to Human serum BCHE.
3. D70G	BCHE(N)	A	Alters binding to many ligands, probably by damaging their ability to enter the gorge.
4. S198C	BCHE(S)	I	Alters active site serine. Some alterations such as S198C or T
5. S198D	BCHE(S)	I	are subtle, yet no activity was detected. This depicts
6. S198H	BCHE(S)	I	vulnerability to changes in the key triad residues.
7. S198Q	BCHE(S)	I	
8. S198T	BCHE(S)	I	
9. S425P	BCHE(N)	A	Displays normal characteristics.
10. M437D	BCHE(S)	I	Adjacent to active histidine.
11. Y440D	BCHE(S)	A	Altered choline binding site.
12. F561Y	BCHE(N)	A	Displays normal characteristics.
13. D70G Y114H	BCHE(N)	A	Y114H ameliorates effects of D70G.
14. D70G S425P	BCHE(N)	A	S425P enhances ligand resistance caused by the D70P mutation.
15. D70G F561Y	BCHE(N)	A	F561Y ameliorates effects of D70G.
16. E441G E443Q	BCHE(S)	A	Disturbs binding of a key H2O molecule in the active site; Reduces activity.
17. D70G E441G E443Q	BCHE(N+S)	A	Disturbs binding of a key H2O molecule in the active site; Alters ligand entrance; Extremely reduces activity.
18. D70G Y114H F561Y	BCHE(N)	A	Displays altered affinity to various inhibitors.
19. D70G Y114H S425P	BCHE(N)	A	Displays altered affinity to various inhibitors.

*: A = ACHE, B = BCHE   #: N = Normal variant, S = Site directed mutant Amino acid residues and their substitutions are denoted by the single letter code, whereas numbers represent the position of these residues within the primary amino acid sequence of the mature CHEs.

many of the recombinantly produced BCHEs (Table I). Together with structural analysis, these findings enabled a dissection of functional domains in the studied proteins. Mutations in BCHE were thus classified into 4 principal subgroups, according to their effects on protein function, as is detailed in the following.

I. Interference With Catalysis

5 substitutions of the active site S198 were performed. These included cysteine, aspartate, glutamine, histidine and threonine (No. 4-8, TableI). In all cases no activity was detected in the resultant rCHEs, despite the subtlety of some of the alterations, such as S198C (having a sulfur atom in place of oxygen) and S198T (containing an additional methyl group). It should be noted that the previously mutated *Torpedo* enzyme, containing an active site cysteine in place of serine, was shown to be somewhat active, (two orders of magnitude lower than control enzyme[17]).At present, we cannot exclude the possibility that similarly low levels of activity existed in the S198C mutant of BCHE, yet could not be detected in our assay system. Alternatively, S198C BCHE may be totally inactive, depicting subtle differences between the dependence of ACHE and BCHE on the presence of catalytic triad members within their active sites. Thus, altering the catalytic site triad either abolishes activity or lowers it to a point that is undetectable in our assay. Another rCHE with undetectable catalytic activity was M437D (No. 10 in table I). The substituted aspartate residue is located adjacent to the active site histidine, and it is likely that it alters the structure in this region. The vulnerability of the active site to modifications may therefore be extended to include structural or charge differences in residues adjacent to catalytic triad members.

One of the charged residues located in the active site region, E441, seems to be hydrogen bonded to E197 by one or two water molecules[12.] These water molecules may have an important function in the catalytic process. Indeed, altering E441 into glycine (No. 16 in Table I) severely lowers protein activity. Thus, the mechanism of catalysis might require finely positioned water molecules. Alternatively, or in addition, the lowered activity in the double mutant E441G, E443Q, might be the result of alterations in the structure of the substrate binding sites.

II. Disturbance of Ligand Entrance Into the Active Site Gorge

. A natural variant of serum BCHE, D70G (also termed the 'atypical' enzyme[1,18]) displayed varied affinity to numerous ligands[13-15]. IC50 values obtained from inhibition by succinylcholine (SCh), dibucaine, bambuterol, physostigmine, ecothiophate and iso-OMPA were increased by at least 10 - fold over values measured for the normal enzyme[13-15]. Although no change in Km towards BTCh was observed, the affinity towards acetylcholine was clearly reduced. Previous studies of individuals with this "atypical" enzyme have in fact displayed lowered affinity to dibucaine[19]. However, biochemical studies alone could not attribute the differences in ligand affinity to specific amino acid residues or define their position in the protein. Thus, both disturbed ligand interactions and disturbed ligand penetrance to the active site will display lowered affinity. To comprehend what determines the CHE affinity constants towards specific ligands, it is therefore crucial to assess the regions in CHEs that are involved in binding and penetrance of ligands.

In essence, the binding affinity consists of ligand penetrance in addition to its binding to the various subsites in the active site center. Thus, not only the hydrophobic binding sites but the rim, residues around the rim, and the gorge lining could contribute to the affinity constant.

Figure 1 depicts the key position, internal to the rim, of the D70 residue, one of the only charged residues in the gorge (noted as No. 3). The altered affinity displayed by D70G (which alters the charged aspartate to an uncharged residue) to some charged ligands, such as acetyl- and succinylcholine, suggests the charge as playing a role in drawing these ligands into

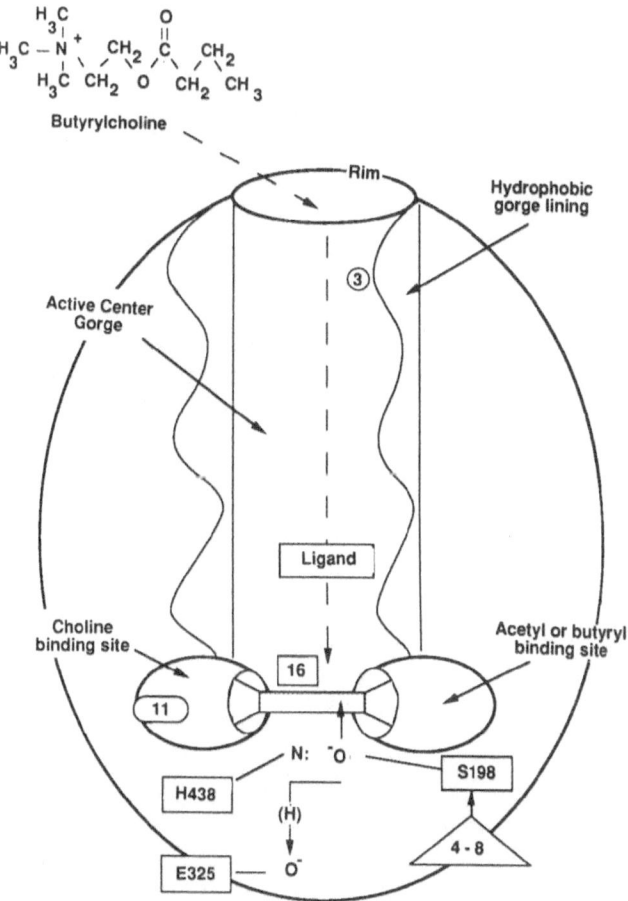

Figure 1. Structure - function BCHE model.

The BCHE model is schematically represented based on the Torpedo ACHE structure[11]. Butyrylcholine is presented as a typical BCHE ligand. Some key regions are noted: rim, active center gorge, hydrophobic gorge lining, choline and butyryl or acetyl binding sites. Circled or boxed numbers in the diagram represent the location, within this structure, of some of the BCHE variants according to their numbering in Table I. When preceded by a letter, numbers denote the location of the relevant residues along the primary amino acid sequence of the mature BCHE protein. The catalytic mechanism, including its triad and the hydrogen transfer, is schematically represented beneath the gorge, where the ligand is shown to interact with both the choline and butyryl or acetyl binding sites.

the gorge. This is supported by the unaltered affinity to this variant of cholinesters which contain bulky hydrophobic chains, such as benzoylcholine[13]. Thus, the D70 residue may constitute part of the previously predicted anionic site. This raises the possibility that the biochemically observed anionic binding site may exist, in addition to the hydrophobic choline binding site, as a composite of multiple residues to which charged ligands bind. The rim area which contains additional charged residues, might hence contain yet other sites involved in attracting charged ligands. These could function before the ligands are being pulled into the gorge by D70, and prior to their 'sliding' down the hydrophobic lining until they reach the active site.

III. Inhibition of Ligand Binding

Substitution of residues in the vicinity of the hydrophobic binding site should affect interactions with various ligands, and possibly aid in determining which site/s are involved in binding specific ligands. Such alteration in the choline binding site was found in the site-directed BCHE mutant Y440D (No. 11 in Table I). The highly hydrophobic region, which was disrupted by the introduction of an aspartate, remained fully active, although the affinity to some inhibitors was lowered by at least an order of magnitude. From dose response curves employing the BCHE specific inhibitor iso - OMPA (fig. 2) it was possible to observe a 10-fold increase in IC50 in the Y440D mutant as compared with its normal recombinant counterpart. This finding strongly suggests that the employed inhibitor interacts with the choline binding site, where this mutation has been positioned.

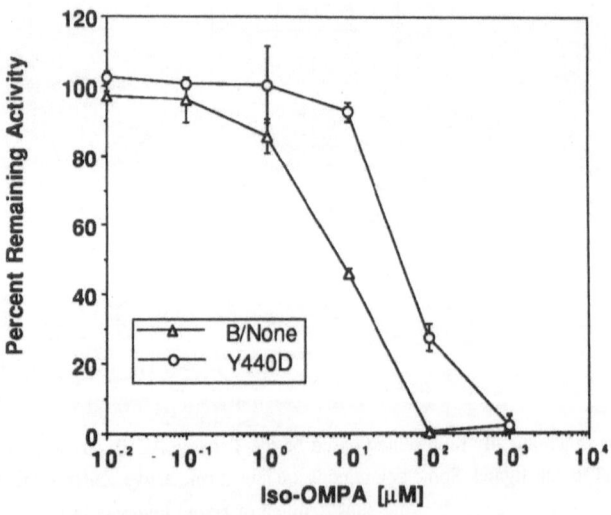

Figure 2. Iso-OMPA inhibition of Y440D

B/None: The cDNA of a normal human BCHE allele was expressed in oocytes in parallel to the site directed mutant Y440D. Error bars represent the standard deviation of the data points. Note the 10- fold shift in IC50 values towards iso-OMPA as observed from the point of 50% inhibition.

IV. Ameriorating Substitutions

The addition of other mutations to the D70G variant enables either an alleviation or increase in the affinity to various inhibitors. For example, S425P alone has no effect on any of the measured IC50 values[13-14]. Yet, when linked with D70G, it sharply decreases the affinity towards succinylcholine and dibucaine. In contrast, S425P displays an alleviating effect on the interaction of BCHE with bambuterol[15]. Other mutations, such as Y114H and F561Y, also correct for decreased levels of activities caused by the D70G alteration, in spite of the large distances between these ameliorating residues and D70 within the protein. One possibility is that the double mutants have slightly different protein backbone structures than that of D70G alone. The altered protein structure could then allow for an alleviation of the primary effects of the D70G mutant.

DISCUSSION

The implementation of several disciplines, including cDNA cloning of normal human CHEs[4-6] and natural variants[20], site - directed mutagenesis[15], expression of synthetic mRNA in *Xenopus* oocytes[5,13-15], pharmacokinetic analysis of recombinant CHE proteins[13-15,17] and further structural analysis (this manuscript) has been combined to approach functional aspects of CHEs. Two hydrophobic regions in the active site emerge from these studies as participating in the binding of ligands. One of these sites, the choline binding site, was shown to interact with the OP inhibitor iso-OMPA. However, these sites are not solely responsible for the affinity constants characteristic of CHEs, since mutating D70, located at the rim of the gorge, also causes drastic changes in the affinity to ligands.

The previously termed 'anionic' site[2] may in fact exist as a composite of variously positioned charged residues, which are together responsible for attracting and drawing charged ligands into the active site gorge. Once in the gorge, hydrophobic interactions would be involved in the positioning of such ligands within the active site center. Catalysis by CHEs then involves a triad which includes serine, histidine and glutamate[11,12]. Altering these amino acids, or ones adjacent to their positions, would either destroy activity or cause severe impairments in it. In addition, the E441G, E443Q mutant is likely to have a distorted ability to position a water molecule, resulting in extremely low activity. This implies that key water molecules, including those hydrogen bonded to the original residue in this domain, play an important role in catalysis.

ACKNOWLEDGEMENTS

We are grateful to Dr. J. Sussman, Rehovot, for providing us with the coordinates of the crystal structure of *Torpedo* ACHE. This work was supported by the Medical Research and Development Command, the U.S. Army (Grant No. DAMD 17-90-Z-0038) and by the Association Francaise Contre les Myopathies (AFM) to H.S.

REFERENCES

1. H. Soreq, and H. Zakut. Cholinesterase genes: Multileveled regulation. In: Sparkes, R.S. (Ed). "Monographs in human genetics" Vol 13, Karger, Basel (1990).
2. D.M. Quinn. Acetylcholinesterase. Enzyme structure, reaction dynamics and virtual transition states. Chem. Rev. 87:9556(1987).

3. M. Schumacher, S. Camp, Y. Maulet, M. Newton, K. Macphee-Quigley, S.S. Taylor, T. Friedmann, and P. Taylor. Primary structure of *Torpedo californica* acetylcholinesterase deduced from its cDNA sequence. Nature 319:407(1986).

4. C.A. Prody, D. Zevin-Sonkin, A. Gnatt, O. Goldberg, and H. Soreq, H. Isolation and characterization of full-length cDNA clones coding for cholinesterase from fetal human tissues. Proc. Natl. Acad. Sci. USA 84:3555(1987).

5. H. Soreq, R. Ben-Aziz, S. Seidman, A. Gnatt, L. Neville, J. Lieman-Hurvitz, Y. Lapidot-Lifson, and H. Zakut. Molecular cloning and construction of the coding region for human acetylcholinesterase reveals a G,C rich attenuating structure. Proc. Natl. Acad. Sci. USA 87:9688(1990).

6. C. McTiernan, S. Adkins, A. Chatonnet, T.A. Vaughan, C.F. Bartels, M. Kott, T.L. Rosenberry, B.N. La Du, and O. Lockridge. Brain cDNA clone for human cholinesterase. Proc. Natl. Acad. Sci. USA 84:6682(1987).

7. H. Soreq, S. Seidman, P.A. Dreyfus, D. Zevin-Sonkin, and H. Zakut. Expression and tissue specific assembly of cloned human butyrylcholinesterase in microinjected *Xenopus laevis* oocytes. J. Biol. Chem. 266:4025(1989).

8. E. Krejci, F. Coussens, N. Duval, J.M. Chatel, C. Legacy, M. Puype, J. Vandekerchkhove, J. Cartaud, S. Bon, and J. Massoulie. Primary structure of a collagenic tail subunit of *Torpedo* acetylcholinesterase: co-expression with catalytic subunit induces the production of collagen- tailed forms in transfected cells. EMBO J. 10: 1285(1991).

9. G. Gibney and P. Taylor. Biosynthesis of *Torpedo* acetylcholinesterase in mammalian cells: Functional expression and mutagenesis of the glycophospholipid-anchored form. J. Biol. Chem. 266:4025(1991).

10. B. Velan, C. Kronman, H. Grosfeld, M. Leitner, Y. Gozes, Y. Flashner, T. Sery, S. Cohen, R. Ben- Aziz, S. Seidman, A. Shafferman and H. Soreq. Recombinant human acetylcholinesterase is secreted from transiently transfected 293 cells as a soluble globular enzyme. Cell. Molec. Neurobiol. 11:143(1991).

11. J.L. Sussman, M. Harel, F. Frolow, C. Oefner, A. Goldman, L. Toker and I. Silman. Atomic structure of acetylcholinesterase from *Torpedo* californica: A prototypic acetylcholine-binding protein. Science, 23:872-879(1991).

12. J.D. Schrag, Y. Li, S. Wu, and M. Cygler. Ser-His-Glu triad forms the catalytic site of the lipase from Geotrichum candidum. Nature, 351:761-764(1991).

13. L.F. Neville, A. Gnatt, Y. Loewenstein, S. Seidman, G. Ehrlich and H. Soreq. Intramolecular relationships in cholinesterases revealed by oocyte expression of site- directed and natural variants of human BCHE. EMBO J. In press(1992).

14. L.F. Neville, A. Gnatt, Y. Loewenstein and H. Soreq. Aspartate 70 to glycine substitution confers resistance to naturally occurring and synthetic anionic-site ligands on *in vivo* produced human butyrylcholinesterase. J. Neurosci. Res. 27:452(1990).

15. L.F. Neville, A. Gnatt, S. Seidman, R. Padan and H. Soreq. Anionic site interactions in human butyrylcholinesterase disrupted by two single point mutations. J. Biol. Chem. 265:20735(1990).

16. G.L. Ellman, D.K. Courtney, V. Andres, and R.M. Featherstone. A new and rapid colorimetric determination of acetylcholinesterase activity. Biochem. Pharmacol. 7:88(1961).

17. G. Gibney, S. Camp, M. Dionne, K. MacPhee-Quigley, and P. Taylor. Mutagenesis of essential functional residues in acetylcholinesterase Proc. Natl. Acad. Sci. USA 87:7546(1990).

18. O. Lockridge and B.N. La Du. Comparison of atypical and usual human serum cholinesterase: Purification, number of active sites, substrate affinity, and turnover number. J. Biol. Chem. 253:361(1978).

19. M. Whittaker. Cholinesterases. In: Beckmann, A.D. (Ed). "Monographs in Human Genetics" Vol. 11. Karger, Basel (1986).

20. A. Gnatt, C.A. Prody, R. Zamir, J. Lieman-Hurwitz, H. Zakut and H. Soreq. Expression of alternatively terminated unusual CHE mRNA transcripts mapping to chromosome 3q26-ter in nervous system tumors. Cancer Res. 50:1983(1990)

ACETYLCHOLINESTERASE CATALYSIS -
PROTEIN ENGINEERING STUDIES

Avigdor Shafferman[1], Baruch Velan[1], Arie Ordentlich[1], Chanoch Kronman[1], Haim Grosfeld[1], Moshe Leitner[1], Yehuda Flashner[1], Sara Cohen[1], Dov Barak[2], and Naomi Ariel[1]

[1]Departments of Biochemistry
[2]Organic Chemistry
Israel Institute for Biological Research
Ness-Ziona, 70450, Israel

INTRODUCTION

Sequence conservation analysis relates the cholinesterases to a superfamily of polypeptides (Myers *et al.*, 1988; Krejci *et al.*, 1991), including enzymes such as microsomal carboxyesterase, cholesterol esterase, lysophospholipase , *Geotrichum* lipase and *Drosophila* esterase-6, as well as several noncatalytic polypeptides. AChE is the best characterized enzyme in this superfamily. Kinetic studies have indicated that the active site of AChE consists of two subsites: an anionic subsite to which the trimethylammonium group of acetylcholine binds and an esteratic subsite which interacts with the ester-bond region and mediates catalysis. Evidence also exists for an allosteric regulation of AChE activity by ligand binding to an anionic site(s) physically remote from the active site (Changeux, 1966).

The esteratic subsite of AChE resembles active sites of serine hydrolases of other enzyme families. The active-site serine (MacPhee-Quigley *et al.*, 1985) was shown by sequence comparison to reside within an amino acid stretch conserved in many hydrolases including serine peptidases (Gibney *et al.*, 1990). A histidine residue appears to be part of the active site as well (Krupka, 1966; Roskoski, 1974). The active site serine and histidine residues were identified in *Torpedo californica* AChE (TcAChE) as Ser200 and His440 by site directed mutagenesis (Gibney *et al.*, 1990). The existence of a Ser-His-Acid catalytic triad in AChE has been subjected to controversy on mechanistic grounds (Quinn, 1987). Various Asp residues have been suggested as putative triad-related carboxylic residues on the basis of evolutionary conservation (Gentry and Doctor 1991; Krejci *et al.*, 1991). The recently resolved X-ray structure of TcAChE (Sussman *et al.*, 1991) suggests however,

Multidisciplinary Approaches to Cholinesterase Functions, Edited by
A. Shafferman and B. Velan, Plenum Press, New York, 1992

165

that the active site of AChE contains Glu rather than Asp in its catalytic triad. X-ray crystallography also revealed a 20Å deep narrow "gorge", that penetrates halfway into the enzyme molecule and contains the catalytic triad 4Å away from its bottom (Sussman *et al.* 1991). This "active site gorge" is lined by 14 aromatic amino acids, which may explain results of kinetic studies (Quinn, 1987) and site directed labeling studies which have identified tryptophan (Kreienkamp *et al.*, 1991), and phenylalanine (Kieffer *et al.*, 1986) residues at or near the active center.

Recently we developed an efficient expression system for Human AChE cDNA in the 293 human cell line (Velan *et al.*, 1991a; Kronman *et al.*, 1992) which is easily amenable to site directed mutagenesis. Here we describe the use of this system to confirm the participation of specific Ser and His residues in the active site and to examine several conserved carboxylic moieties for their possible involvement in the enzymatic activity . We demonstrate that presence of Glu at position 334 (analogous to Glu327 of TcAChE) is essential for hydrolysis, while many other conserved carboxylic residues are most likely required for maintaining structural integrity. We also identify aromatic and charged amino acids in and around the "active site gorge" that are involved in allosteric regulation and modulation of AChE catalytic activity.

METHODS

Construction of Expression Vectors for AChE and its Mutants: Plasmids construction, isolation of DNA fragments, cloning and bacterial transformation were performed essentially as described in Current Protocols in Molecular Biology (Ausubel, 1987). Bipartite vectors derived from pEwCAT (Shafferman *et al.*, 1992a,b), expressing both human *ache* (Soreq *et al.*, 1990) and the reporter *cat* genes were used to express the various AChE mutants in 293 human kidney cells. Mutagenesis was performed by DNA cassette replacement (Shafferman *et al.*, 1987) into a series of HuAChE sequence variants (designated Ewl Ew4, Ew5, and Ew7) which conserve the wild type coding specificity, but carry new unique restriction sites (Shafferman *et al.*, 1992a,b).

Transient Transfection and Quantitation of AChE: At least two different clone isolates were tested for each plasmid construct. Transfection was carried on as described previously (Velan *et al.*, 1991a,b). Medium was collected, and assayed for AChE activity (Ellman *et al.*, 1961); medium of mock-transfected 293 cells served as control. Cell lysates were assayed for intracellular AChE and CAT activity (Gorman *et al.* 1982). AChE-protein mass was determined by a specific ELISA developed previously (Velan *et al.*, 1991b, Shafferman *et al.*, 1992a). AChE-protein production levels of individual mutants (average of four transfections) were calculated by normalization to CAT activity.

Structure Analysis and Molecular Graphics: Analysis of the three dimensional model of HuAChE was performed on Silicon Graphics workstation IRIS 70/GT, using the SYBYL modelling software (Trypos Inc.) The HuAChE model was constructed by methods of comparative modelling (Barak *et al.*, 1992), based on the known X-ray structure of TcAChE (Sussman *et al.*, 1991). Structural refinement by molecular mechanics was done using the MAXMIN force field and zone refinement procedure ANNEAL, both included in SYBYL. Optimization of ACh-HuAChE complex and the structures with rotated aromatic side chain of Y337, included 23 amino acids.

RESULTS AND DISCUSSION

The analysis of the mutant AChE polypeptides relied on quantitative immunological assays based on polyclonal antibodies to native HuAChE, monoclonal antibodies to conformation-dependent epitopes and antibodies to denatured HuAChE epitopes. All values obtained were then normalized by quantitation of the coexpressed *cat* reporter gene (average of four transfections) to correct for inherent variabilities in different transfection experiments. This type of assessment of productivity and structural integrity together with analysis of enzymatic activity of the various mutants provided the basis for differentiation between activity-related and production/folding-related effects and their classification into four major groups (Table 1).

Table 1. Classification of selected HuAChE mutants.

Class	HuAChE mutant	TcAChE position	Production[a] (%)	Activity[b] (%)	Comments
A	E84Q	82	100 ± 14	70 ± 18	AChE & BChE conserved
A	D131N	128	100 ± 15	70 ± 17	AChE conserved
A	D333N	326	80 ± 16	75 ± 22	AChE & BChE conserved
A	D349N	342	100 ± 12	100 ± 22	AChE & BChE conserved
B1	D95N	93	11 ± 3	13 ± 2	family conserved[c]
B1	H432A	425	20 ± 6	15 ± 4	AChE & BChE conserved
B2	D175N	172	<1	<0.1	family conserved; salt bridge
B2	D404N	397	<1	<0.1	family conserved; salt bridge
C	S203A	200	100 ± 15	<0.1	catalytic triad; family conserved
C	H447A	440	100 ± 19	<0.1	catalytic triad; family conserved
C	E334A	327	100 ± 12	<0.1	catalytic triad; E/D family conserved
C	W86A	84	40 ± 8	<0.1	catalytic center / "anionic subsite"
D	D74N	72	65 ± 5	8 ± 1	AChE & BChE conserved
D	E202Q	199	130 ± 30	30 ± 10	family conserved
D	Y337A	330	100 ± 23	25 ± 6	aromatic residue conserved in AChEs
D	F338A	331	100 ± 26	50 ± 12	AChE & BChE conserved

[a] CAT normalized ELISA values relative to wild type. [b] CAT normalized activity values relative to wild type (at 0.5 mM ATCh). [c] refers to conservation in the ChEs superfamily (cf. Krejci *et al.*, 1991).

Conserved Amino Acids Contributing to Processing/Folding of the AChE Polypeptide

Replacement of E84, D333, and D349, which are conserved in all sequenced ChEs (Gentry and Doctor, 1991), and of D131,which is conserved in all AChEs, with their corresponding amides (E->Q or D->N substitution) affected neither production levels nor enzymatic activity (Class A, Table 1). The HuAChE 3-D model based on the TcAChE structure indicates that residues E84, D131 and D349 are positioned on the outer surface of the TcAChE molecule. It is not surprising, therefore, that mutation of these residues do not affect catalytic activity. Modification of charge on the side chain of D333 , an amino acid adjacent to the active site Glu334 (see below), would be expected to affect the catalytic profile. The failure to do so is consistent, nevertheless, with HuAChE 3-D model which indicates that the charged moiety of D333 is pointed away from the catalytic triad. Thus, the role of the charged residues at the positions 84, 333 and 349 which are conserved in cholinesterases, but are not influenced by modification, remains to be determined. A different class of mutations (class B) is represented by D95N and H432A substitutions, that cause significant reduction in production without affecting activity of residual polypeptide,or by the D175N and D404N substitutions which cause complete loss of AChE protein production. Amino acids D95, D175 and D404 are strictly conserved in the cholinesterase superfamily polypeptides. Interestingly, Asp95 substitution affected the ratio of secreted/intracellular enzyme (Shafferman et al. 1992a), suggesting that the highly conserved D95 residue within the conserved small cystein loop of the ChE superfamily is related to processing or protein traffic functions. The most pronounced indication to the role of conserved charged amino acids in production/folding is provided by the mutants D175N and D404N of the rHuAChE polypeptide. The 3-D structure (Sussman et al. 1991) of AChE provides some insight into the molecular events that could lead to the destabilizing effects of mutations in these locations. D404 can be implicated in a salt bridge with R525 and in hydrogen bonding with Y382 (interestingly, both R525 and Y382, are completely conserved among the cholinesterases). These interactions may be crucial for stabilizing the interaction between three helices and can play a role in bringing together Cys409 and Cys529 which form an intramolecular disulfide bridge. A similar type of structural requirement may also explain the conservation of D175 and the effect of its substitution on production. D175 can form a salt bridge with R152 bringing together alpha-helix C and beta-strand 4 (Sussman et al., 1991). Both D175 and R152 are conserved not only among cholinesterases but also in every sequenced member of the cholinesterase superfamily (Krejci et al., 1991). Salt bridges analogous to those involving D175 and D404 in HuAChE were also revealed by X-ray structure of the distantly related enzyme Geotrichum lipase (M. Cygler, personal communication).

Ser203-His447-Glu334 Constitute the Catalytic Triad of HuAChE

The "C class" (Table 1) phenotype - effective production of a folded polypeptide devoid of catalytic activity - is represented by Ser203, Glu334 and His447 mutant polypeptides. The S203A and the H447A mutations did not alter secretion levels of AChE-protein, yet, in both mutants no enzymatic activity above background levels (less than 10^{-3} of wild type) was detectable in the cell culture medium. The efficient interaction of the nonactive mutant polypeptides with polyclonal antibodies (raised against native HuAChE) in the antigen capture ELISA (Shafferman et al., 1992b) suggests that loss of activity is not the consequence of a major conformational change. This assumption was further substantiated by examining the interaction of the mutated polypeptides with four monoclonal antibodies (AE-1 and AE-2, Fambrough et al., 1982; HR-5 and A123, Brimijoin et al. 1983) directed towards conformational epitopes. Indeed, both S203A and H447A AChE polypeptides interacted with all four monoclonal antibodies as efficiently as the wild type enzyme . It is reasonable to assume that Ser203 and His447 of human AChE, in analogy to Ser200 and His440 of

TcAChE (Gibney *et al.*, 1990) are essential for catalysis and can not be functionally substituted by alanine residues. Mutations of Glu334 provide the first biochemical confirmation of the X-ray structural model, that indicate a Glu residue in the catalytic triad of hydrolases (Sussman *et al.*, 1991). It is of interest that alanine substitutions of any of the HuAChE-triad residues are tolerated by the molecule with no major effect on folding or secretion. This seems to be in accordance with the overall fold of the hydrolase enzyme group to which the cholinesterases belong (Ollis *et al.* 1992). The catalytic residues of these enzymes are protruding from turns located on three different loops brought together to form the catalytic triad. These observations, which suggest that triad positions may eventually accomodate different amino acids, led us to generate various catalytic triad permutations in the HuAChE mold. Engineering of triads such as a "papain like": Cys-His-Glu or "trypsin like": Ser-His-Asp resulted in reduced productivity and in non detectable (less than 1% of wild type) cholinesterase activity.

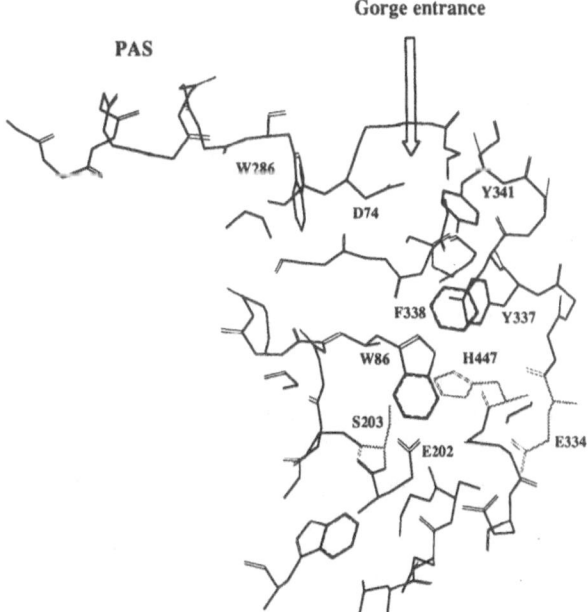

Figure 1. Longtitudal view of the "active site gorge" of HuAChE depicting residues mutated in this study. The catalytic triad (S203-H447-E334) is shown in shaded lines. Alpha- carbon trace of residues forming the rim and the walls of the "gorge" as well as the putative peripheral anionic site (PAS) is depicted .

Trp86 of HuAChE is an Essential Element of the Catalytic Center

The AChE active center has been postulated to contain a catalytic subsite as well as an anionic subsite responsible for the binding of the positive charge of the choline moiety. Tryptophan 84 of TcAChE (W86 in HuAChE) was identified by photoaffinity labeling studies to be part of the anionic site in the active center (Kreienkamp *et al.*, 1991) and according to the X-ray structure (Sussman *et al.*, 1991) is positioned in the active site gorge, 10Å away from Oγ of the catalytic serine (Fig. 1). The replacement of W86 in HuAChE by either an anionic residue, glutamate, or an alanine residue (Shafferman *et al.*, 1992b) yielded molecules not impaired in the secretion but devoid of activity (C class Table 1). The conformational integrity of the W86 mutants was assessed by quantitative ELISAs using four different monoclonal antibodies (AE-1,AE-2,HR-5 and A123). The W86 mutants and the wild type preparations were indistinguishable in these assays, suggesting that the loss in catalytic activity in W86A and W86E polypeptides is probably not a consequence of distorted folding. It is striking that out of nineteen conserved AChE positions mutated to date, the C class phenotype was manifested only by catalytic triad mutations: S203, H447 and E334 (Shafferman *et al.*, 1992a,b) and the W86 mutations. We conclude therefore that W86, is indeed a critical element in the active center. Sussman *et al.*, (1991) suggested, on the basis of modelling which involved ACh docking, that the choline moiety makes close contact with the W84 of TcAChE. In our hands, docking of ACh onto a model of the HuAChE shows that the methyl groups of the quaternary ammonium are within 3.6-4.0Å away from the indole plane of W86 in the ACh-HuAChE complex. Recent crystallographic data for an edrophonium-TcAChE complex show that the aliphatic substituents of the quaternary nitrogen of edrophonium are in a plane parallel to and within 4Å of the W84 indole ring of TcAChE (Harel *et. al.*, 1992).

Asp74 Mutations Affect Both the Active and the Peripheral Anionic Sites of HuAChE

The most informative mutations with respect to catalysis are probably those which lead to an altered AChE activity but do not impair productivity (D class in Table 1). The D74 mutants belong to this class of mutations . A natural mutation of the analogous aspartate D70 in human BChE (D70G) was implicated in the 'atypical' phenotype (McGuire *et al.*, 1989; Neville *et al.*, 1990). The BChE atypical variant is characterized by reduced affinity to charged ligands which led to the assumption that this aspartate resides within the anionic subsite. We generated therefore four mutants : D74G, D74N, D74E,and D74K and examined them in representative 'atypical' assays: resistance to succinylcholine and resistance to dibucaine (La Du *et al.*, 1990). Except for D74E all other mutations clearly conferred loss of inhibition by succinylcholine and a marked decrease in inhibition by dibucaine (Shafferman *et al.*, 1992a). It appears therefore that the atypical phenotype can be 'recreated' in an AChE, suggesting that, in both HuBChE and HuAChE, Asp70/74 assumes an analogous role in the configuration of the enzyme. Yet, the distance between this residue and the active-site serine near the bottom of the gorge is ca.16Å, far greater than the expected 4.7Å distance between the esteratic and anionic subsites (Rosenberry, 1975; Berman and Decker, 1986). This difference in distances would argue against the direct involvement of the Asp70/74 residue with the anionic subsite. Nevertheless replacements of Asp at position 74 affected the Km more than replacement in any other position, including substitution of amino acids vicinal to the active center (E202, F338, Y337; see Table 2). Furthermore, all four D74 mutants are less susceptible than the wild type to inhibition by edrophonium, a selective active center ligand, and yet none of the mutations had a significant effect on the first order rate of catalysis (Table 2). We conclude therefore that D74 in HuAChE can influence the conformation of the catalytic center through an allosteric mechanism. Surprisingly mutations at position 74 had also a significant effect on the affinity towards propidium (Table 2), the selective peripheral anionic site ligand (Taylor and Lappi, 1975) . It also appears that charge at position 74 is a dominant factor affecting the interaction of propidium with AChE (compare D74E to D74K Table 2). Furthermore, bisquaternary ligands, which presumably associate with both the active center and the peripheral anionic site, exhibit a 100 to 1000 fold decrease in affinity to the D74G, D74N or D74K mutants (Table 2). All these results, taken together, suggest that D74 is a key residue involved in some allosteric interactions between the exterior and the active center deep in the catalytic "gorge".

Peripheral Anionic and Substrate Inhibition Sites Probably Overlap

Substrate inhibition of AChE at high ACh concentrations is well documented (Quinn, 1987) and was proposed to occur directly at the active center (Froede *et al.* 1986) or indirectly through a peripheral site which changes the active center allosterically (Changeux, 1966; Rosenberry, 1975). A variety of uncompetitive AChE inhibitors of which propidium is typical, are assumed to associate specifically with this binding site (Taylor and Lappi, 1975). Replacement of D74 or of W286, the latter is immediately adjacent to a peptide previously identified as part of a peripheral anionic site (Weise *et al.*, 1990), produced AChE molecules with lower affinity for propidium (Table 2). We found also, that mutations in these two positions generated AChE molecules in which inhibition by high substrate concentration (Fig. 2 and Table 2) is partially (W286A) or completely (D74N, D74G, D74K) eliminated. These results support the kinetic studies which conclude that the peripheral and the substrate inhibition sites could overlap (Radic *et al.*, 1991).

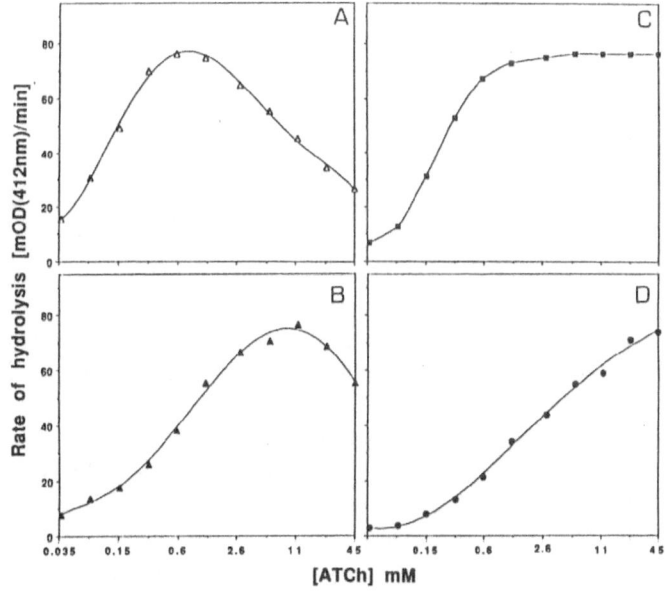

Figure 2. Substrate concentration dependence of the rate of hydrolysis of ATCh by HuAChE and selected mutants. Panel A - wild type HuAChE - normal bell shape. Panel B - Y341A mutant - shifted bell shape. Panel C - Y337A mutant - no inhibition (type I). Panel D - D74N mutant - no inhibition (type II).

Table 2. Effect of mutations in and around the active site "gorge" on kinetic parameters of HuAChE.

AChE Type[a]	$\frac{Km(mut)[b]}{Km(WT)}$	$\frac{kcat(mut)[c]}{kcat(WT)}$	IC$_{50}$ (mut)/IC$_{50}$ (WT)[d]			Substrate Inhibition Profile[e]
			EDP	PROP	DMT	
WT	1.0	1.00	1	1	1	Normal bell shape
D74E (72)	3.6	1.20	2	1.5	8	Shifted bell shape
D74N	4.1	0.70	7	4.0	100	None - type II
D74G	4.5	1.00	5	6.0	100	None - type II
D74K	32.0	1.00	4	36.0	830	None - type II
E202Q (199)	2.5	0.15	4	1.0	10	Shifted bell shape
E202D	1.8	0.04	22	1.5	14	None - type II
E202A	4.7	0.03	80	1.2	77	None - type II
W286A (279)	2.0	1.00	1.5	12.0	9	Shifted bell shape
Y337F (330)	1.4	1.00	1	0.8	1.0	Normal bell shape
Y337A	1.0	0.30	15	1.0	0.8	None - type I
F338A (331)	1.9	0.46	2	1.0	7	Normal bell shape
Y341A (334)	2.8	0.90	2	1.7	10	Shifted bell shape

[a] TcAChE amino acids numbering is provided in parentheses. [b] Wild type (WT) HuAChE Km value is 0.14 mM. [c] Wild type HuAChE kcat is $3.7 \times 10^5 \times min^{-1}$. [d] Wild type HuAChE IC$_{50}$ values for edrophonium (EDP), propidium (PROP), and decamethonium (DMT) are: 1.5, 1.5, and 8.5 µM, respectively. ATCh at 0.5mM was used in inhibition studies. [e] Representatives of the different profiles are depicted in Fig. 2.

Glu202 Modulates the Catalytic Activity of HuAChE

The analysis of mutations involving HuAChE E202, which is adjacent to the catalytic serine-203, supports some of the conclusions from an earlier site directed mutagenesis study (Gibney *et al.*, 1990) performed on the analogous residue (E199) in TcAChE. In both enzymes these mutants exhibit reduced affinity for the ACTh substrate (higher Km values), although the extent of the change is by far less pronounced in the HuAChE mutants (Table 2). The ability to quantitate the amount of enzyme involved in catalysis allowed us also to determine that E202 mutants are much less effective in the acylation and deacylation steps. This is reflected by the 6 to 30 fold decrease in their apparent first order catalytic rate constant (kcat; Table 2), and 17 to 140 fold decrease in the calculated (kcat/Km) apparent bimolecular rate constant, as compared to the wild type enzyme. The E202 mutants exhibit lower affinity to edrophonium or decamethonium but are only marginally affected in their affinity to propidium in comparison to wild type. These observations can be explained by the spatial orientation of E202 relative to the catalytic triad. The carboxyl group of E202 is within interaction distance (3.6Å) from N$^\varepsilon$ of the imidazole ring of H447, as well as from the methylene adjacent to the trimethylammonium portion of ACh-HuAChE complex. This arrangement may contribute to the stabilization of the protonated imidazolium facilitating proton transfer from Ser-203 and thus formation of the acyl-enzyme intermediate. It is possible that replacement of E202, which results in such significant changes in both Km and catalytic rates, actually disrupts the spatial arrangement in the active center. Such conformational perturbation of structure can account for the refractiveness of E202 HuAChE (Table 2) and analogous TcAChE mutants (Gibney *et. al.*, 1990) to substrate inhibition .

Aromatic Amino Acids in the Catalytic "Gorge" are Involved in the Relay of Inhibition Signals Induced by Substrate Binding at the Periphery

The D74 residue is located at the entrance to the "gorge" (Fig. 1). This fact and the pleiotropic effects of D74 mutations: lower affinities towards selective ligands of the active and the anionic peripheral sites as well as altered substrate inhibition profiles, led us to propose that D74 is an essential element in the relay of the conformational changes induced by ligand binding at the peripheral anionic site. According to our model, binding of substrate to the peripheral anionic site induces transmission of a signal from this site through tryptophan-286 (and perhaps also through Y124 and Y72) to D74 and from there down to the active center. Some clues on how this signal may be transduced from D74 down to the "gorge" is provided by mutations at positions Y341 and Y337. The Y341 residue can form a hydrogen bond with D74 (Sussman *et al.*, 1991). Replacement of Y341 by alanine generates an enzyme with a somewhat lower affinity to any of the tested ligands. The Y341A mutant is not inhibited effectively by high concentrations of acetylthiocholine (at least 20 fold higher levels than those required for inhibition of wild type HuAChE; Table 2 and Fig. 2). Proceeding down into the "gorge" from Y341 we find that F338 is a less significant element in the proposed relay while Y337 appears to be a key element. Unlike F338A, the Y337A mutant is more resistant to inhibition by edrophonium. More significantly Y337A is totally refractory to inhibition by high concentration of substrate (Fig. 2), while F338A behaves like wild type HuAChE.

Based on molecular modelling we suggest that the allosteric changes initiated at the exterior can be translated into an inhibitory effect by a physical blockage of the active center through a conformational change in Y337. Such a change can prevent or interfere with the entry of substrate to an unoccupied active site and/or reduce the efficiency of the deacylation of the acyl-enzyme complex. Molecular mechanics calculation shows that the aryl moiety of Y337 can rotate as much as 62° into the "gorge" cavity without substantially altering the energy of the system (ca. 2.0 Kcal/mole). A similar arrangement was in fact observed in the X-ray structure of TcAChE-tacrine complex (Harel *et al.*, 1992). This rotation would place the aromatic side chain of Y337 in close contact with ACh docked at the active center and thus may block the formation of ACh-HuAChE complex. According to the proposed model, alanine at position 337 will not be able to physically block the entrance to the catalytic site, while an aromatic phenylalanine at this position (Y337F) could still exert the regulatory effect at high substrate concentrations, as is indeed the case (Table 2). The crystal structure of the complex of TcAChE with edrophonium (Harel *et al.*, 1992) shows that the aromatic ring of F330 (analogous to Y337 of HuAChE) moves to make a better contact with the aromatic ring of the ligand. The indole ring of residue W279 in TcAChE (analogous to W286 in HuAChE) displays a substantial movement upon binding of edrophonium or tacrine (Harel *et al.*, 1992). This experimental observation clearly demonstrates the linkage between a conformational change in the active center and that of a distally located residue near the entrance of the "gorge". Interestingly, in human BChE, an enzyme devoid of substrate inhibition behavior, the aromatic residues W286 and Y337 of HuAChE are replaced (Prody *et al.*, 1987, McTiernan *et al.*, 1987) by alanine residues. While the relay of conformational changes via aromatic residues provides an attractive model which is consistent with the kinetic chrachteristics of the mutants we are still a long way from unveiling the complete sequence of molecular events and the actual mechanism of the allosteric changes involved in catalysis and substrate inhibition.

ACKNOWLEDGEMENTS

We would like to express our deep appreciations to Drs. Harel, Sussman and Silman for sharing data prior to their publication. We thank Tamar Sery, Dana Stein, Gila Friedman and Nechama Zeliger for their excellent technical assistance and Dr.S. Brimijoin for providing some of the monoclonal antibodies. This work was supported by the U.S. Army Research and Development Command, Contract DAMD17-89-C-9117 to A.S.

REFERENCES

Ausubel, M.F., Brent, R., Kingston, R.E., Moore, D.D., Smith, J.A., Seidman, J.G., and Struhl, K. (eds).,
   1987, *Current Protocols in Molecular Biology* (Wiley Interscience) New York.

Barak, D., Ariel, N., Velan, B.,and Shafferman, A., This volume.

Berman, H.A. and Decker, M.M., 1986, *Biochim. Biophys. Acta.* **872**,125-133.

Brimijoin, S., Mintz K. P. and Alley, M., 1983, *Mol. Pharmacol.* **24,** 513-520.

Changeux J.P., 1966, *Mol. Pharmacol.* **2**, 369-392.

Ellman, G.L., Courtney, K.D., Andres, V., and Featherstone, R.M., 1961, *Biochem. Pharmacol.* **7**, 88-95.

Fambrough, D.M., Engel, A.G. and Rosenberry, T.L., 1982, *Proc. Natl. Acad. Sci. USA.* **79,** 1078-1083.

Froede, H.C., Wilson, I.B., and Kaufman, H., 1986, *Arch. Biochem. Biophys.* **247**, 420-423.

   Gentry, M.K. and Doctor, B.P., 1991, in *Cholinesterases: Structure, Function, Mechanism, Genetics, and
   Cell Biology.* eds Massoulie, J., Bacou, F., Barnard, E.A. Doctor, B.P. and Quinn, D.M. (Am. Chem.
   Soc. Washington) pp. 394-398.

Gibney, G., Camp, S., Dionne, M., MacPhee-Quigley, K. and Taylor, P., 1990, *Proc. Natl. Acad. Sci. USA*
   87,7546-7550.

Gorman, C.M., Moffat, L.E. and Howard, B.H., 1982, *Mol. Cell. Biol.* **2**,1044-1051.

Harel, M., Silman, I., and Sussman, J., 1992, Submitted.

Kieffer, B., Goeldner, M., Hirth C., Aebersold, R., and Chang, J.-Y., 1986, *FEBS Lett.* **202**, 91-96.

Kreienkamp, H.J., Weise, C., Raba, R., Aaviksaar, A., Hucho, F., 1991, *Proc. Natl. Acad. Sci. USA.* **88**,
   6117-6121.

Krejci, E., Duval, N., Chatonnet, A., Vincens, P. and Massoulie J., 1991, *Proc. Natl. Acad. Sci. USA*, **88**,
   6647-6651.

Kronman, C.,Velan, B., Gozes, Y., Leitner, M., Flashner, Y., Lazar, A.,Marcus, D., Sery, T., Papier, A.,
   Grosfeld, H., Cohen, S. and Shafferman, A., 1992, *Gene*, in press.

Krupka, R.M., 1966, *Biochemistry* **5**,1988-1998.

La Du, B.N, Bartels, C.F., Nogueira, C.P., Hajara, A., Lightstone, H.,Van Der Spek, A. and Lockridge, O.,
   1990, *Clin Biochem.* **23**, 423-431.

MacPhee-Quigley, K., Taylor, P. and Taylor, S., 1985, *J.Biol. Chem.***260**,12185-12189.

McGuire, M.C., Nogueira, C.P., Bartels, C.F., Leightstone, H., Hajara, A., Van der Spek, A.F.L., Lockridge, O., and La Du, B.N., 1989, *Proc. Natl. Acad. Sci. USA*. **86**,953-957.

McTiernan, C., Adkins, S., Chatonnet, A., Vaughan, T.A., Bartels, C.F., Knott, M., Rosenberry, T.L., La Du, B.N. and Lockridge, O., 1987, *Proc. Natl. Acad. Sci. USA* **84**, 6682-6686.

Myers, M., Richmond, R., and Oakeshott, J., 1988, *Mol. Biol. Evol.***5**,113-119.

Neville, L.F. ,Gnatt, A., Padan, R., Seidman, S. and Soreq, H., 1990, *J. Biol. Chem.* **265**,20735-20738.

Ollis, D.L., Cheah,E., Cygler, M., Dijkstra, B., Frolow, F., Franken, M.S., Harel, M., Remington, S.J., Silman, I., Schrag, J., Sussman, J.L. Verschueren, K.H.G. and Goldman A., 1992, *Protein Engineering*, in press.

Prody, C.A., Zevin-Sonkin, D., Gnatt, A., Goldberg, O. and Soreq, H., 1987, *Proc. Natl. Acad. Sci. USA*. **84**, 3555-3559.

Quinn, D., 1987, *Chem. Rev.* **87**,955-979.

Radic, Z. , Reiner, E., and Taylor, P., 1991, *Molec. Pharmacol.* **39**, 98-104.

Rosenberry, T.L., 1975, *Adv. Enzymol. Relat. Areas Mol. Biol.* **43**,103-218.

Roskoski, R., 1974, *Biochemistry* **13**, 5141-5144 .

Shafferman, A., Velan, B., Cohen, S., Leitner, S. and Grosfeld, H., 1987, *J. Biol. Chem.* **262**, 6227-6237.

Shafferman, A. Kronman, C.,Flashner, Y., Leitner, M.,Grosfeld, H.,Ordentlich, A., Gozes, Y., Cohen,S.,Ariel, N., Barak, D., Harel, M., Silman, I., Sussman J.L., and Velan, B. 1992a, *J.Biol.Chem.* in press.

Shafferman, A., Velan, B., Ordentlich, A., Kronman, C., Grosfeld, H., Leitner, M., Flashner, Y., Cohen, S., Barak, D. and Ariel N. , 1992b, Submitted.

Soreq, H., Ben-Aziz, R., Prody, C.A., Seidman, S., Gnatt, A., Neville, A., Lieman-Hurwitz, J., Lev-Lehman, E., Ginzberg, D., Lapidot- Lifson, Y., and Zakut, H., 1990, *Proc. Natl. Acad. Sci. USA*. **87**, 9688-9692.

Sussman, J.L., Harel, M., Frolow, F., Oefner, C , Goldman, A., Toker, L. and Silman, I., 1991, *Science* **253**, 872-879.

Taylor, P. and Lappi, S., 1975, *Biochemistry*, **14**, 1989-1992.

Velan , B., Kronman, C., Grosfeld, H., Leitner, M., Gozes, Y., Flashner, Y., Sery, T., Cohen, S., Ben-Aziz, R., Seidman, S., Shafferman, A. and Soreq, H., 1991a, *Cell Mol. Neurobiol.* **11**, 143-156.

Velan, B., Grosfeld, H., Kronman, C., Leitner, M., Gozes, Y., Lazar, A., Flashner, Y., Marcus, D., Cohen, S. and Shafferman, A., 1991b, *J. Biol. Chem.* 266:23977-23984.

Weise C., Kreienkamp, H.J., Raba, R., Pedak, A., Aaviksaar, A. and Hucho, F., 1990, *EMBO J.* **9**, 3385-3388.

SITE-DIRECTED MUTAGENESIS OF FUNCTIONAL RESIDUES IN *TORPEDO* ACETYLCHOLINESTERASE

Israel Silman[1,3,] Eric Krejci[1], Nathalie Duval[2], Suzanne Bon[1], Philippe Chanal[1], Michal Harel[4,] Joel Sussman[4] and Jean Massoulié[1]

[1]Laboratoire de Neurobiologie, URA CNRS 295, Ecole Normale Supérieure, 46 rue d'Ulm, 75005 Paris, France

[2]Laboratoire des Venins, Institut Pasteur, 25 rue du Dr. Roux, 75015 Paris, France

[3]Department of Neurobiology and [4]Department of Structural Biology, Weizmann Institute of Science, Rehovot 76100, Israel

INTRODUCTION

Acetylcholinesterase (AChE) is remarkable for its very high catalytic activity relative to other serine hydrolases (Quinn, 1987). It displays a turnover number of the order of 100 μseconds (Vigny et al., 1978), and thus approaches a rate of activity at which substrate diffusion may become rate-limiting (Bazelyansky et al., 1986). The recent solution of the three-dimensional structure of *Torpedo californica* AChE has revealed that AChE displays some rather unusual, perhaps unique, structural features compared to other hydrolases in general and to serine hydrolases in particular (Sussman et al., 1991, 1992). Thus a principal structural (and presumably functional) feature of this enzyme is a deep and narrow cavity, penetrating 20 Å into it, with the catalytic triad characteristic of serine hydrolases being located close to its bottom. The function of this cavity is, as yet, obscure, but it has been named the 'aromatic gorge', since a substantial part of its surface is provided by the rings of fourteen highly conserved aromatic amino acids. Some of these aromatic rings, located near the catalytic triad, may be involved directly in substrate-binding, by dipole-dipole interaction with the quaternary group of the substrate, acetylcholine (ACh). However, the role of more distal aromatic rings, further up the gorge, remains to be clarified. The catalytic triad of AChE is unusual inasmuch as

Multidisciplinary Approaches to Cholinesterase Functions, Edited by
A. Shafferman and B. Velan, Plenum Press, New York, 1992

it contains a glutamic acid, E327 in the *Torpedo* numbering, rather than the aspartic acid residues found previously in serine hydrolases, although a conserved aspartic acid residue, D326, occurs adjacent to it. A feature which is restricted to *Torpedo* AChE is the presence of a non-conserved cysteine residue, C231, bearing a free sulfhydryl group, which is situated deep within the core of the protein, ca 8 Å from Oγ of the active-site serine. This cysteine residue is of interest because its covalent modification by several sulfhydryl reagents abolishes AChE activity (Steinberg et al., 1990), even though it is not involved in catalytic activity.

Site-directed mutagenesis is, by now, a well-established technique for elucidating the structural and/or functional role of a given amino acid (or acids) in a biologically active protein. Its use is especially productive when applied to a protein of known three-dimensional structure, since the structural information allows both rational planning of a coherent repertoire of mutations, as well as meaningful interpretation of the results thus obtained. In the following, we will describe some initial studies which employed site-directed mutagenesis to explore the role of a number of potentially interesting amino acid residues in *Torpedo* AChE.

MATERIALS AND METHODS

Mutagenesis and Expression

Site-directed mutagenesis was carried out by conventional techniques, using the gene coding for the H catalytic subunit of *Torpedo marmorata* AChE, and the mutated DNA was transfected into COS-7 cells (Duval et al., 1992). It should be noted that the amino acid sequences of *T. californica* and *T. marmorata* AChE differ at only ten positions, and none of these differences appear to be of structural or functional significance. It thus seems fully justified to use the three-dimensional structure of the *T. californica* enzyme in conjunction with site-directed-mutagenesis data generated with *T. marmorata* AChE.

Cholinesterase Assays

Cholinesterase activity was determined either radiometrically, using ^{3}H-acetylcholine, according to Johnson & Russell (1975), or spectrophotometrically, using acetylthiocholine, according to Ellman et al. (1961).

RESULTS AND DISCUSSION

Mutation of Acidic Amino Acid Residues

It was earlier shown that mutation of D397 to Asn produced a mutant which did not display enzymic activity (Krejci et al., 1991). The reason for this mutant being inactive is probably due to its playing a structural role, since inspection of the three-dimensional structure revealed that it was remote from the active site. Such a role might involve stabilization of the molecule by formation of a salt-bridge with R517 as was suggested for the equivalent residue in human AChE (Shafferman et al., 1992). Solution of the three-dimensional structure, in fact, assigned E327 as being the acidic member of the catalytic triad. The adjacent acidic residue, D326, does not, apparently, play a catalytic role, since its carboxyl group points away from the active site. It was, nevertheless, of interest to examine the effect of mutations of these two adjacent residues on the activity of the *Torpedo* enzyme.

Table 1 shows the levels of AChE activity obtained in COS cells upon expression of several such mutants. As might be predicted from the apparent involvement of E327 in the charge relay system, the E327Q mutation resulted in complete loss of enzymic activity as has also been shown to be the case for mutagenesis of the homologous residue, E334, in human AChE (Shafferman et al., 1992). The mutation, E327D, had a similar effect. Thus, the additional methylene group of glutamic acid appears to be essential to correct functioning of the catalytic triad. The mutation, D326N, of the adjacent aspartic acid residue, produces only a slight reduction in enzymic activity. This is not surprising since, as already mentioned, the carboxylate group in question points away from the active site.

Table 1. Effect of site-directed mutagenesis of Asp-326 and Gly-327 on enzymic activity of *Torpedo* AChE.

	Residue		Activity in COS
	326	327	cell extracts
Untransfected control cells	-	-	0.1
Wild type AChE	Asp	Glu	2.4
D326N	Asn	Glu	1.5
E327Q	Asp	Gln	0.15
E327D	Asp	Asp	0.15
D326N/E327Q	Asn	Gln	0.1

Activity in COS cell extracts was measured on acetylthiocholine. Activity is expressed as $\Delta OD_{412nm}/\mu l/min$.

Mutation of Cys231

Although the chemical modification data presented earlier (Steinberg et al., 1990) provided strong evidence for C231 being the site of inhibition of *Torpedo* AChE by sulfhydryl agents, it seemed, nevertheless, to be of interest to examine the effect of substituting this residue upon the susceptibility of the enzyme to such compounds. The mutant, C231S, was prepared accordingly. Under conditions where enzymic activity of the wild-type (WT) enzyme towards ACh was inhibited ca. 50% by the organomercurial, p-chloromercurisulfonic acid, no detectable inhibition of C231S could be observed. This provides direct evidence for C231S being the site of inhibition by sulfhydryl reagents. C231 is deeply buried within the interior of the enzyme, being located ca 8 Å from the active-site serine. Thus, direct steric blockage of the active-site by its chemical modification does not seem to be involved in inhibition of enzymic activity; we are currently investigating the physicochemical basis for this phenomenon (Dolginova et al., in preparation).

Mutation of Aromatic Amino Acids

As already mentioned, the cavity leading to the active site, the aromatic gorge, is lined with a large number of aromatic rings of conserved aromatic amino acid residues. AChE displays a high degree of sequence homology ($ca.$ 50%) with butyrylcholinesterase (BChE). Inspection of the residues in BChE which are homologous with those lining the aromatic gorge in AChE shows that, apart from trivial substitutions (e.g. Ser for Thr, Ile for Val), there are six major differences: All six involve substitution of an aromatic residue in AChE by a non-aromatic residue in BChE (Harel et al., 1992). *Torpedo* AChE and human serum BChE display especially high sequence homology and a favourable alignment of their sequences. This permitted accurate modelling of the BChE molecule on the basis of the three-dimensional structure of the AChE molecule (Harel et al.,1992). Comparison of the two structures suggested a structural basis for the known difference in substrate specificity between the two enzymes, which is manifested in the fact that butyrylcholine is hydrolysed very well by BChE, whereas it is a poor substrate for AChE. Thus, the rings of two aromatic residues, F288 and F290, which appeared to prevent docking of butyrylcholine in the active site of AChE, were replaced by Leu and Val, respectively, in BChE. Accordingly, the double mutant, F288L/F290V, was generated and expressed. This double mutant displayed about 10% of the activity of the wild-type towards ACh. Table 2 shows a comparison of the activity of the two mutants on acetylthiocholine and butyrylthiocholine. It can be seen that, whereas the wild type displays little or no activity on butyrylthiocholine, the double mutant displays substantial activity; in fact, at high substrate concentrations it hydrolyses the two substrates at similar rates. *iso*OMPA is an organophosphorus anticholinesterase agent with a high selectivity for BChE (Austin & Berry, 1953). This selectivity may be due to its bulky character. Table 3 shows that *iso*OMPA serves as an efficient inhibitor of the F288L/F290V double mutant under conditions where it has hardly any effect on the wild-type enzyme.

Table 2. Activity of wild type AChE and of the F288 L - F290 V double mutant on acetylthiocholine and butyrylthiocholine

	Acetylthiocholine		Butyrylthiocholine		
	1.2 mM	25 mM	1.2 mM	25 mM	
W T	100	142	n.d*	n.d*	
F288L;F290V⟩		100	120	40	105

Relative activities on acetylthiocholine and butyrylthiocholine were determined by the Ellman procedure. For both wild type and double mutant, activities are expressed relative to activity on 1.2 mM acetylthiocholine, taken as 100.

* not detectable.

Table 3. Inhibition of wild type AChE and of the F288L - F290V double mutant by *iso*OMPA

	ACTIVITY	
	110 min	200 min
WT	91	73
F288L;F290V⟩	4	2

Inhibition was carried out by incubation with 1 mM *iso*OMPA, for the times shown, in 0.1 M phosphate buffer, pH 7.0, at room temperature. The numbers represent percentage residual activity determined on ^{3}H-ACh.

Another aromatic residue which is conserved in AChE, but lacking in BChE, is W279, located close to the top of the gorge. On ligand-binding at the 'anionic' subsite of the catalytic site, W279 undergoes a conformational change even though it is *ca.* 15 Å remote from the ligand binding site. W279 thus appears to be an attractive candidate for participating in the 'peripheral' anionic site. Various bisquaternary ligands are believed to serve as powerful inhibitors of AChE because they span the two 'anionic' sites (Mooser and Sigman, 1974). Such inhibitors are known to be poor inhibitors of BChE (Main, 1976). It was, therefore, of interest to generate an AChE molecule lacking this Try residue, and the mutant, W279A, was produced and expressed accordingly. It was found to retain >80% of the activity of the wild-type. Table 4 shows IC$_{50}$ values for the inhibition of wild-type AChE and of W279A by three different AChE inhibitors, under the conditions of our radiometric assay using ^{3}H-acetylcholine. The mutation has little effect on the inhibition of AChE by edrophonium, which is a characteristic competitive inhibitor binding at the catalytic site (Wilson and Quan, 1958). By contrast, propidium, a characteristic inhibitor acting at the 'peripheral' site (Taylor and Lappi, 1975), displays greater than 10-fold lower affinity for W279A than for the wild-type. The bisquaternary inhibitor, BW284C51, which is selective for AChE (Austin and Berry, 1953) and, most likely, spans the proximal and distal binding sites for quaternary ions, displays somewhat less affinity for the mutant than for the wild type.

Table 4. IC_{50} values for inhibition of wild type AChE and of the W279A mutant by quaternary ligands.

	IC_{50}	
	WT	W279A
Propidium	3.10^{-6}	4.10^{-5}
Edrophonium	8.10^{-5}	6.10^{-5}
BW284C51	2.10^{-7}	6.10^{-7}

The numbers show the molar concentrations at which the ligands inhibited activity 50% under the conditions of the radiometric assay.

Thus, by a limited number of mutations, generated from the knowledge gained by modelling BChE on the basis of the three-dimensional structure of AChE, we have been able to reproduce several of the features which distinguish the former enzyme from the latter.

CONCLUSIONS

The initial studies which we have performed, as outlined above, have shown the predictive value of the three-dimensional structure of *Torpedo* AChE for generating mutations which have already increased our understanding of this complex and fascinating enzyme. It may be hoped that an extended program of kinetic and physicochemical analysis of these and of other mutations will greatly enhance our understanding of how AChE achieves its remarkable catalytic efficiency.

Acknowledgments

This project was supported by the Association Franco-Israélienne pour la Recherche Scientifique et Technologique, the C.N.R.S. and the U.S. Army Medical Research and Development Command under Contract DAMD17-89-C9063.

REFERENCES

Austin, L. and Berry, W.K. (1953) *Biochem. J.* 54 695.

Bazelyansky, M., Robey, C. and Kirsch, J.F. (1986) *Biochemistry* 25:125.

Duval, N., Massoulié, J. and Bon, S. (1992) *J. Cell. Biol.*, in press.

Ellman, G.L., Courtney, K.D., Andres, V. and Featherstone, R.M. (1961) *Biochem. Pharm.* 7:88.

Harel, M., Silman, I. and Sussman, J.L. (1992), these proceedings

Johnson, C.D. and Russell, R.L. (1975) *Analyt. Biochem.* 64:229.

Krejci, E., Duval, N., Chatonnet, A., Vincens, P. and Massoulié, J. (1991) *Proc. Natl. Acad. Sci. USA* 88:6647.

Mooser, G. and Sigman, D.S. (1974) *Biochemistry* 13:2299.

Quinn, D.M. (1987) *Chem. Rev.* 87:955.

Shafferman, A., Kronman, C., Flashner, Y., Leitner, M., Grosfeld, H., Ordentlich, A., Gozes, Y., Cohen, S., Ariel, N., Barak, D., Harel, M., Silman, I., Sussman, J.L. and Velan, B. 1992, *J. Biol. Chem.* in press.

Steinberg, N., Roth, E. and Silman, I. (1990) *Biochem. Internat.* 21:1043.

Sussman, J.L, Harel, M., Frolow, F., Oefner, C., Goldman, A., Toker, L. and Silman, I. (1991) *Science* 253:872.

Sussman, J.L., Harel, M. and Silman, I. (1992), these proceedings.

Taylor , P. and Lappi, S. (1975) *Biochemistry* 14:1989.

Vigny, M., Bon, S., Massoulie, J. and Leterrier, F. (1978) *Eur. J. Biochem.* 85:317.

Wilson, I.B. and Quan, C. (1958) *Arch. Biochem. Biophys.* 73:131.

STRUCTURALLY IMPORTANT RESIDUES IN THE REGION SER91 TO ASN98

OF *TORPEDO* ACETYLCHOLINESTERASE

Göran Bucht, Elisabet Artursson, Britta Häggström, Annika Osterman, and Karin Hjalmarsson

National Defence Research Establishment
Department of NBC Defence
S-901 82, Umeå, Sweden

INTRODUCTION

In cholinesterases and cholinesterase-like proteins the region Ser91 to Asn98, numbering according to acetylcholinesterase (AChE) from *T. californica*, shows a high degree of conservation. In this region two negatively charged amino acids, Glu92 and Asp93, are found as well as a cysteine, Cys94, which is disulphide bonded to Cys67[1]. The loop created by the disulphide bond contains Trp84, which has been identified as part of the anionic subsite of the active site[2]. This suggests that the highly conserved region could have an important structural role. To gain insight into the possible role of Glu92 and Asp93 in maintaining the structure and function of AChE, we have made amino acid replacements by site-directed mutagenesis.

MATERIALS AND METHODS

Mutagenesis and Expression Vectors

A cDNA clone encoding AChE, asymmetric form, was isolated from a λgt10 cDNA library[3]. To the cDNA encoding the mature form, synthetic DNA encoding the signal peptide of AChE from *T. marmorata* was ligated. A DNA encoding the glycophospholipid-anchored form of AChE was constructed by introducing a HindIII site upstream of the region encoding the variable C-terminal part of AChE. By using the HindIII site and synthetic oligonucleotides, the sequence in the region of the cDNA corresponding to Exon3A was changed to the sequence found in Exon3H[4]. Mutant sequences were generated by oligonucleotide-directed substitutions of M13 templates (Oligonucleotide-directed in vitro Mutagenesis System, Amersham International plc). The entire coding sequence of each mutant was confirmed by dideoxy sequencing (Sequenase DNA Sequencing Kit, United States Biochemical Corp.). Wild-type and mutant AChE cDNAs were inserted into a simian virus 40-derived expression vector.

Multidisciplinary Approaches to Cholinesterase Functions, Edited by
A. Shafferman and B. Velan, Plenum Press, New York, 1992

Expression in Mammalian Cells

COS-1 cells (ATCC CRL 1650) were maintained in Dulbecco's modified Eagle's medium (DMEM) supplemented with 10% fetal calf serum, 100 u penicillin/ml, and 100 µg streptomycin/ml at 37°C in a 5% CO_2, humidified atmosphere.

Cells were plated at $1\text{-}2 \times 10^6$ per 10-cm plate 16-20 hours before transfection. Transfection of cells were performed by using TransfectACE Reagent (GIBCO) and according to the manufacturers instructions. 5 µg of plasmid DNA and 50 µl TransfectACE were used for each transfection. The cells were incubated for 40 hours at 37°C, and for an additional 24 hours at 28°C following the transfection.

Measurement of AChE Activity of Transfected Cells

Cells were harvested essentially as described by Gibney et al, 1990[5]. AChE activity was assayed by a modification of the method of Ellman et al., 1961[6], in microtiter plates at 405 nm. The kinetic constants were determined by measurements in 0.1 M $NaPO_4$ buffer as a function of substrate concentration and/or inhibitor concentration in 1 cm cuvettes at 412 nm.

SDS-Polyacrylamide Gel Electrophoresis and Western Blotting

Cells from one 10-cm plate were resuspended in 0.4 ml 0.01 M sodium phosphate pH 7.2, 0.145 M NaCl. The cells in 50 µl were pelleted by centrifugation, and resuspended in 25 µl 0.01 sodium phosphate pH 7.2, 1% β-mercaptoethanol, 1% SDS, 6 M urea, and incubated at 37°C for 30 minutes. 25 µl 0.125 M Tris-HCl pH 6.8, 4% SDS, 10% glycerol, 0.5% β-mercaptoethanol, 0.1% bromophenol blue was added and the mixture was incubated at 95°C for 3 minutes. 10 µl was loaded into a slot of a 12-17.5% gradient SDS-PAG. Western blotting was performed essentially as described by Sambrook et al, 1989[7], using Immobilon-P Transfer membrane (Millipore). The primary antibody used was a rabbit polyclonal antibody directed against AChE. The secondary reagent was an alkaline phosphatase-conjugated swine polyclonal antisera directed against rabbit Ig (Dakopatts a/s, Denmark).

RESULTS

Expression of Wild-Type and Mutant AChE in COS Cells

Three mutants of AChE were constructed: E92Q and E92L in which Glu92 was replaced by Gln and Leu, respectively, and D93N in which Asp93 was replaced by Asn. Following transfection the cells were analysed for expression of AChE by enzyme activity measurements and Western blotting. Expression of active AChE has been shown by others to be at least 10-fold greater at 28°C than at 37°C[5]. Thus, expression was maximized by transfer of the cells to 28°C 24 hours before harvest of the cells.

As can be seen in Figure 1, a protein band of the expected molecular weight could be visualised in all cell extracts, except those from untransfected cells and cells transfected with vector without AChE coding sequences. The intensities of the bands indicate that the amount of AChE protein is about the same in extracts from cells transfected with vectors encoding wild-type or mutant AChE.

Activity was abolished when the glutamic residue in position 92 was substituted with glutamine or leucine (Table 1). A reduction in activity by approximately 81% was detected with an Asp → Asn substitution at position 93.

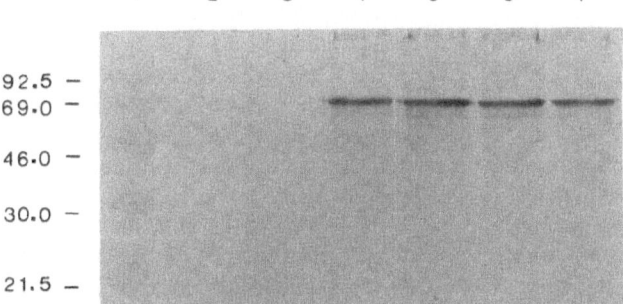

Figure 1. Western blotting following SDS-PAGE of wild-type and mutant AChE protein.
Polyacrylamide gel electrophoresis and Western blotting were performed as described in *Material and methods*. Lane 1: molecular weight markers; lane 2: untransfected cells; lane 3: vector alone; lane 4: wild-type; lane 5: E92Q; lane 6: E92L; lane 7: D93N.

Table 1. Activities and kinetic constants associated with transfection of wild-type and mutant AChE DNA.

Enzyme	Relative enzyme activity %	K_m (mM)	K_{ss} (mM)	K_i (μM) propidium	K_i (μM) edrophonium
wild-type	100	0.093	31	0.59	0.35
D93N	19.2	0.086	77	0.74	0.49
E92Q	1.0				
E92L	0.7				

Enzymatic activity and kinetic constants were measured as described in Material and Methods. The amount of AChE protein was estimated by electrophoresis followed by Western blotting. The AChE band was scanned with a video densitometer giving relative amount of AChE protein for each sample. Specific enzyme activity was calculated as ΔA/min divided by scanning units. Relative enzyme activity is expressed as % of wild-type activity.

Substitution of asparagine for the aspartic residue in position 93 has no effect on the affinity for the substrate as reflected by an unaltered K_m of the mutant (Table 1). Moreover, the K_{ss} of the mutant was only increased by a factor of about 2. This indicates that there is only a small effect on the substrate concentration dependence for catalysis of the mutant enzyme. The K_i's for edrophonium and propidium were only increased by a factor of less than 2, indicating that the mutation has no or a small effect on the binding of ligands to the peripheral or central anionic site (Table 1).

DISCUSSION

Substitution of the glutamic residue in position 92 with glutamine or leucine, results in inactive enzymes. This indicates that the glutamic residue in position 92 has an important structural role in AChE.

In the lipase from *Geotrichum candidum* a salt bridge between Arg38 and Glu103 stabilises a twist at the base of a helix-loop-helix twisted flap[8]. These residues corresponds to Arg44 and Glu92 in *Torpedo* AChE. The helix-loop-helix of the lipase corresponds to the region in AChE, residues Asp72 to Trp84, which constitutes the anionic subsite of the active site. Trp84 has by affinity labeling been identified as part of the anionic subsite[2]. Inspection of the 3-dimensional structure of AChE reveals that two hydrogen bonds stabilised by electrostatic interaction could be formed between Glu92 and Arg44[9]. By substituting the glutamic residue with glutamine or leucine this interaction is abolished and yields enzymes with no activity. Thus, it is likely that the interaction between Glu92 and Arg44 has a central role in stabilising the conformation of the active enzyme.

A substitution of Asp93 with asparagine showed a reduction of enzyme activity by approximately 81% (Table 1). Inspection of the 3-dimensional structure reveals that a hydrogen bond can be formed between Asp93 and Tyr96. This could help in stabilising the conformation of the active enzyme. However, this substitution did not effect the K_m for acetylthiocholine and had only a small effect on K_{ss} for acetylthiocholine and on K_i for edrophonium and propidium. Thus, it is likely that a general conformational change in the active site region is responsible for the reduction of enzyme activity.

ACKNOWLEDGEMENTS

The skillful technical assistance of Ulla Eriksson and Lena Lindgren is gratefully acknowledged. Prof. Palmer Taylor and Dr. Zoran Radič are thanked for valuable discussions. Part of the work were performed by Göran Bucht during a visit at Prof. Palmer Taylors laboratory, UCSD, San Diego.

REFERENCES

1. K. MacPhee-Quigley, T.S. Vedvick, P. Taylor and S.S. Taylor, Profile of the Disulfide Bonds in Acetylcholinesterase, *J. Biol. Chem.* 261:13565-13570 (1986).

2. H-J. Kreienkamp, C. Weise, R. Raba, A. Aaviksaar and F. Hucho, Anionic Subsites of the Catalytic Center of Acetylcholinesterase from *Torpedo* and from Cobra Venom, *Proc. Natl. Acad. Sci. USA* 88:6117-6121 (1991).

3. T. Claudio, M. Ballivet, J. Patric and S. Heineman, Nucleotide and Deduced Amino Acid Sequences of *Torpedo californica* Acetylcholin Receptor Gamma Subunit, *Proc. Natl. Acad. Sci. USA* 80:1111-1115 (1983).

4. Y. Maulet, S. Camp, G. Gibney, T.L. Rachinsky, T.J. Ekström and P. Taylor, Single Gene Encodes Glycophospholipid-Anchored and Asymmetric Acetylcholinesterase Forms: Alternative Coding Exons Contain Inverted Repeat Sequences, *Neuron.* 4:289-301 (1990).

5. G. Gibney, S. Camp, M. Dionne, K. MacPhee-Quigley and P. Taylor, Mutagenesis of Essential Functional Residues in Acetylcholinesterase, *Proc. Natl. Acad. Sci. USA* 87:7546-7550 (1990).

6. G.L. Ellman, D.K. Courtney, V. Andreas and R.M. Featherstone, A New and Rapid Colorimetric Determination of Acetylcholinesterase Activity, *Biochem. Pharmacol.* 7:88-95 (1961).

7. J. Sambrook, E.F. Fritsch and T. Maniatis, "Molecular Cloning. A Laboratory Manual", Cold Spring Harbor Laboratory Press, Cold Spring Harbor, New York (1989).

8. J.D. Schrag, Y. Li, S. Wu and M. Cygler, Ser-His-Glu Triad Forms the Catalytic Site of the Lipase from *Geotrichum candidum*, *Nature* 351:761-764 (1991).

9. J.L. Sussman, M. Harel, F. Frolow, C. Oefner, A. Goldman, L. Toker and I. Silman, Atomic Structure of Acetylcholinesterase from *Torpedo californica*: A Prototypic Acetylcholine-Binding Protein, *Science* 253:872-879 (1991).

A MODEL OF BUTYRYLCHOLINESTERASE BASED ON THE X-RAY STRUCTURE OF ACETYLCHOLINESTERASE INDICATES DIFFERENCES IN SPECIFICITY

Michal Harel[1], Israel Silman[2], and Joel L. Sussman[1]

[1]Dept. of Structural Biology
[2]Dept. of Neurobiology
Weizmann Institute of Science, Rehovot, Israel

INTRODUCTION

In vertebrates, two enzymes efficiently catalyze acetylcholine (ACh) hydrolysis: acetylcholinesterase (AChE) and butyrylcholinesterase (BChE)[1]. The principal role of AChE is the termination of impulse transmission at cholinergic synapses[2]. Although the second enzyme, BChE, is widely distributed, its biological role is unknown[3]. BChE derives its name from the fact that it hydrolyses butyrylcholine (BCh) at rates similar to or faster than ACh, whereas AChE hydrolyses BCh much more slowly[4]. AChE and BChE are further distinguished by their differential susceptibility to various inhibitors[5]. For example, some bisquaternary compounds, which are more potent inhibitors of AChE than their monoquaternary counterparts, bind poorly to BChE[6]. Human BChE (H-BChE) is of great interest to anaesthesiologists and geneticists, because it is responsible for the breakdown of the short-term muscle relaxant, succinylcholine[7], and because of the existence of numerous genetic variants in which the rate of succinylcholine hydrolysis is reduced[8]. Recent developments in cloning and sequencing of the cholinesterases have revealed striking sequence homology between AChE and BChE[1,8-11]. Residues 4-534 of T-AChE, which are the ones seen in the X-ray structure[12], can be aligned with residues 2-532 of H-BChE[10], with 53% identity and no deletions or additions. The residues of the catalytic triad are found in exactly the same positions (S200, E327 and H440 in T-AChE), as are the intra-chain disulfide bonds[13,14]. This marked structural similarity encouraged us to use the three-dimensional structure of T-AChE[12] to model H-BChE. We hoped, thereby, to gain an understanding of how the structural differences between the two enzymes might account for the known differences in specificity between them.

MODELLING OF H-BChE

The model building was carried out interactively, using FRODO[15,16], on an Evans & Sutherland PS390 graphics system, in order to convert the amino acid sequence of T-AChE to that of H-BChE. The H-BChE structure was energy-minimized by the simulated

Multidisciplinary Approaches to Cholinesterase Functions, Edited by
A. Shafferman and B. Velan, Plenum Press, New York, 1992

annealing program, X-PLOR[17], using the POSITIONAL refinement option. The starting model for H-BChE was the refined 2.8 Å X-ray structure of T-AChE[12] (Brookhaven access code 1ACE[18]). In order to obtain the H-BChE model, all residues in the T-AChE sequence which differ from those of H-BChE were changed accordingly. 356 side-chains of H-BChE are either identical to those of T-AChE, or have the same number of dihedral angles. The initial H-BChE model retained the experimentally determined torsion angles for these side-chains. An additional 83 residues, with fewer side-chain torsion angles in H-BChE than in T-AChE, were also allowed to retain the experimental X-ray torsion angles of T-AChE. The dihedral angles of 87 residues with longer side-chains in H-BChE than in T-AChE were fixed to their most frequently found torsion angle values[19]. In only 13 cases, where the most common rotamer of a side-chain overlapped with neighboring atoms, the second most common rotamer was taken. After 110 cycles of minimization, when the energy of the model has converged to a minimum, the C_α rms deviation between T-AChE and the energy-minimized H-BChE model is 0.28 Å. The positions of the residues of the catalytic triad, S200, E327 and H440, show only very small shifts for the H-BChE model relative to the T-AChE structure (C_α shifts of 0.1, 0.1 and 0.14 Å, respectively).

In the T-AChE structure, the catalytic triad is located close to the bottom of a ca. 20 Å deep narrow cavity, which we named the aromatic gorge, since about 40% of its surface area is lined with the rings of 14 aromatic amino acids[12]. All these residues are fully conserved in the five vertebrate AChE sequences determined so far[10,11] with the exception of a single case in which Phe is replaced by Tyr. This conservation suggests that these aromatic rings play an important role in AChE function. Indeed, various lines of evidence strongly indicate that F330 and W84, which are in close proximity to the catalytic triad, may be directly involved in binding of the quaternary group of ACh[12,20]. Although as many as 30 amino acid residues contribute, to some extent, to the lining of the gorge in the experimentally determined structure of T-AChE, comparison of the sequences of T-AChE enzyme and of H-BChE, as well as of the other known AChE and BChE sequences[21-26], shows that only 10 of these amino acids, whose side-chains face the gorge, are different in BChE relative to AChE.

Table 1. Residues with side-chains facing the active site gorge where differences are found between AChEs and BChEs.

| Residue No. | AChE | | | | BChE | | |
	Torpedo[8]	Human[23]	Mouse[24]	Bovine[25]	Human[9]	Mouse[24]	Rabbit[26]
70	Y	Y	Y	Y	N	N	N
71	V	V	V	V	I	I	I
121	Y	Y	Y	Y	Q	Q	Q
122	S	S	S	S	T	T	T
279	W	W	W	W	A	R	V
282	L	L	L	L	V	L	V
286	S	S	S	H	T	S	S
288	F	F	F	F	L	L	L
290	F	F	F	F	V	I	V
330	F	Y	Y	Y	A	A	A

Four of these changes are unlikely to be associated with the differences in enzymic properties of the two enzymes, since they involve substitution by residues with similar side-chains, i.e. V71I, S122T, L282V and S286T. The other six cases **all** involve substitution of an aromatic residue in AChE by a non-aromatic residue in BChE, viz. Y70N, Y121Q, W279A, F288L, F290V and F/Y330A.

SUBSTRATE FITTING

We previously suggested a plausible model for the docking of ACh, in an all-*trans* configuration, within the active site of T-AChE[12]. In this model, the acetyl group of ACh was positioned to make a tetrahedral bond with O^γ of S200. This resulted in the positively charged quaternary group of the choline moiety being within van der Waals distance (~ 3.5Å) of W84, whose presence within the 'anionic' site had been earlier suggested by affinity labelling experiments[20]. This assignment is confirmed by the structures of two AChE-inhibitor complexes with the anionic-site-directed competitive inhibitors, tacrine and edrophonium, which we have recently determined (Harel *et. al.*, submitted). In the H-BChE model, W84, like the catalytic triad, does not move relative to its position in T-AChE (C_α shift of 0.07 Å). Hence we can model a bound BCh molecule in the same orientation as that of ACh. When we thus try to model BCh, in the active site of AChE, it is clear that the bulkier butyryl moiety of BCh cannot fit into the 'esteratic' locus: it can be seen that two residues, F288 and F290, are near the modelled acetyl moiety of the bound ACh molecule (see Fig. 1). In the H-BChE model, however, substantial reduction in the size of the side-chains of the corresponding residues, L288 and V290, does permit the butyryl group to fit into the larger 'esteratic' pocket of the model (see Fig. 2).

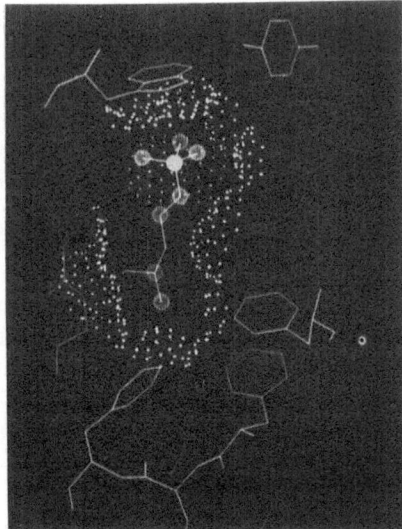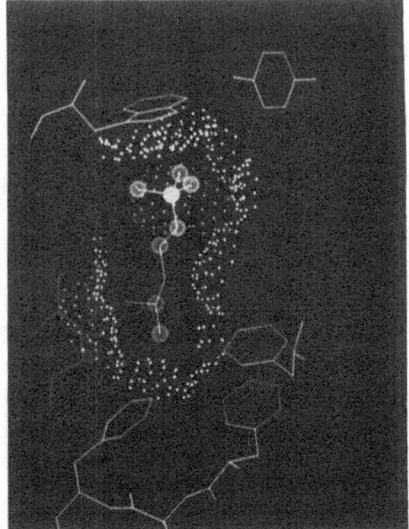

Fig. 1. Stereo view of the van der Waals surface of atoms within 3 Å of ACh in the T-AChE X-ray structure.

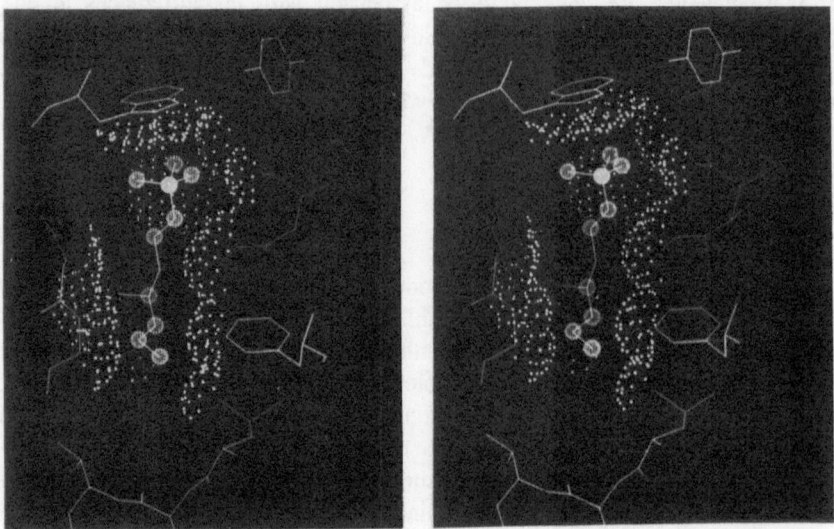

Fig. 2. Stereo view of the van der Waals surface of atoms within 3 Å of BCh in the H-BChE model.

THE PERIPHERAL ANIONIC SITE

Two other changes which involve substitution of aromatic by non-aromatic residues, are of amino acids whose aromatic side-chains undergo localized conformational changes upon binding of the competitive inhibitors, tacrine and edrophonium, as shown by our crystallographic data mentioned above (Harel *et. al.*, manuscript submitted). These changes are in W279 and F330. F330 is in close proximity to the ligand-binding site, and in both complexes, the conformational change observed involves an aromatic-aromatic interaction with the bound ligand. W279 is, however, at least 8 Å away from the 'anionic' site, and any direct contact with a small inhibitor bound there can be precluded. Is W279, which undergoes a conformational change upon binding of inhibitors to T-AChE, part of a different site? Bisquaternary ligands, such as decamethonium[27] are known to be more potent inhibitors of AChE than the corresponding monoquaternary ligands, and this has been ascribed to their binding simultaneously to the 'anionic' subsite of the catalytic site and to the 'peripheral' anionic site[6]. Binding studies of several series of *n*-alkyl *bis* ammonium ions[28,29] have shown an optimal separation of 14-15Å between the two quaternary groups, while the distance between the two indole moieties of W84 and W279 is ca. 16Å in AChE. These data can be rationalized by assuming that W279, at the mouth of the active-site gorge, is an important component of the 'peripheral' anionic site. The enhanced potency of bisquaternary compounds relative to the corresponding monoquaternary ligands does not hold for BChE[6] and the change of W279A in BChE would explain its lack of a 'peripheral' site, and thus its failure to display enhanced sensitivity to bisquaternary inhibitors.

The studies outlined above show that, for the pair of enzymes under consideration, modelling was of powerful predictive capacity. Following the predictions, it may be possible to convert AChE to a BCh-hydrolysing enzyme and to strongly reduce the affinity of a 'peripheral' ligand-binding site known to be absent in BChE.

Acknowledgements: This work was supported by U.S. Army Medical Research and Development Command Contract No. DAMD17-89-C-9063, the Association Franco-Israelienne pour la Recherche Scientifique et Technologique, the Minerva Foundation, Munich, Germany, the Kimmelman Center for Biomolecular Structure and Assembly, Rehovot, Israel.

REFERENCES

1. Chatonnet, A. & Lockridge, O. (1989) *Biochem. J.* 260:625-634.
2. Barnard, E.A. (1974) in *The Peripheral Nervous System*, J.I. Hubbard, ed. (Plenum, New York), pp. 201-224.
3. Whittaker, M. (1986) *Cholinesterase: Monographs in Human Genetics*, Vol. 2, Beckman, L., Ed. (Karger, Basel).
4. Augustinsson, K-B. (1971) in *Methods Biochem. Anal.*, Suppl. Vol. pp. 217-273.
5. Austin, L. & Berry, W.K. (1953) *Biochem. J.* 54:695-700.
6. Main, A.R. (1976) in *Biology of Cholinergic Function*, A.M. Goldberg and I. Hanin, eds. (Raven Press, New York), pp. 269-353.
7. Hobbiger, F. & Peck, A.W. (1969) *Br. J. Pharmacol.* **37**, 258-271.
8. Schumacher, M., Camp, S., Maulet, Y., Newton, M., MacPhee-Quigley, K., Taylor, S.S., Friedmann, T. & Taylor, P. (1986) *Nature*, 319:407-409.
9. Lockridge, O., Bartels, C.F., Vaughan, T.A., Wong, C.K., Norton, S.E. & Johnson, L.L. (1987) *J. Biol. Chem.* 262:549-557.
10. Gentry, M.K. & Doctor, B. P. (1991) in *Cholinesterases: Structure, Function, Mechanism, Genetics and Cell Biology*, J. Massoulié *et. al.* eds. (American Chemical Society, Washington, DC), pp. 394-398.
11. Krejci, E., Duval, N., Chatonnet, A., Vincens, P. & Massoulić, J. (1991) *Proc. Natl. Acad. Sci. USA* 88:6647-6651.
12. Sussman, J.L., Harel, M., Frolow, F., Oefner, C., Goldman, A., Toker, L. & Silman, I. (1991) *Science* 253:872-879.
13. MacPhee-Quigley, K., Vedvick, T.S., Taylor, P. & Taylor, S.S. (1986) *J. Biol. Chem.* 261:13565-13570.
14. Lockridge, O. Adkins, S. & La Du, B.N. (1987) *J. Biol. Chem.* 262:12945-12952.
15. Jones, T.A. (1987) *J. Appl. Cryst.* 11:268-272.
16. Pflugrath, W., Saper, M.A. & Quiocho, F.A. (1984) in *Methods and Applications in Crystallographic Computing*, S. Hall and T. Ashiaka, Eds. (Clarendon Press, Oxford) , pp. 404-407.
17. Brunger, A.T., Kuriyan, J. & Karplus, M. (1987) *Science* 235:458-460.
18. Bernstein, F. C., Koetzel, T. F., Williams, G. J. B., Meyer, E. F., Jr., Brice, M. D., Rodgers, J. R., Kennard, O., Schimanouchi, T. & Tasunmi, M. (1977) *J. Mol. Biol.* 112:535-542.
19. Ponder, J.W. & Richards, F.M. (1987) *J. Mol. Biol.* 193:775-791.
20. Weise, C., Kreienkamp, H.-J., Raba, R., Pedak, A., Aaviksaar, A. & Hucho, F. (1990) *EMBO J.* 9:3885-3888.
21. Sikorav, J.-L., Krejci, E. & Massoulié, J. (1987) *EMBO J.* 6:1865-1873.
22. Prody, C.A., Zevin-Sonkin, D., Gnatt, A., Goldberg, O. & Soreq, H. (1987) *Proc. Natl. Acad. Sci. USA* 84:3555-3559.
23. Soreq, H., Ben-Aziz, R., Prody, C.A., Seidman, S., Gnatt, A., Neville, L., Lieman-Hurwitz, J., Lev-Lehman, E., Ginzberg, D., Lapidot-Lifson, Y. & Zakut, H. (1990) *Proc. Natl. Acad. Sci. USA* 87:9688-9692.
24. Rachinsky, T.L., Camp, S., Li, Y., Ekstrom, T.J., Newton, M. & Taylor, P. (1990) *Neuron* 5:317-327.
25. Doctor, B.P., Chapman, T.C., Christner, C.E., De La Hoz, D.M., Gentry, M.K., Ogert, R.A., Smyth, K.K. & Wolfe, A.D. (1990) *FEBS Lett.* 266:123-127.

26. Jbilo, O. & Chatonnet, A. (1990) *Nucleic Acids Res.* 18:3990.
27. Bergmann, F., Wilson, I.B. & Nachmansohn, D. (1950) *Biochim. Biophys. Acta* 6:217-224.
28. Bergmann, F. & Segal, R. (1954) *Biochem. J.* 58:692-698.
29. Barlow, R.B. & Ing, H.R. (1948) *Nature* 161:718.

MOLECULAR MODELS FOR HUMAN AChE AND ITS

PHOSPHONYLATION PRODUCTS

Dov Barak[1], Naomi Ariel[2], Baruch Velan[2] and Avigdor Shafferman[2]

[1]Department of Organic Chemistry
[2]Biochemistry
Israel Institute for Biological Research,
P.O. Box 19, Ness - Ziona, 70450, Israel

INTRODUCTION

An important step towards a comprehensive description of the structure-function properties of AChE has been made with the elucidation of the three dimensional structure of *Torpedo californica* AChE (TcAChE)[1]. Alignment of amino acid sequences of the cholinesterase family[2] reveals over 56% sequence identity and 74% sequence homology[1] between TcAChE and human AChE (HuAChE). The highest degree of conservation is in the active site "gorge" and in the residues which are probably involved in maintaining the overall structural integrity of the protein. These observations, together with the nearly identical activity profiles of the two enzymes led us to construct a HuAChE molecular model using the TcAChE structure as a template. Such models are useful for rationalizing the structural and functional effects of site directed mutagenesis[3-6] as well as for probing the molecular basis for HuAChE reactivity. Here we provide another example for the use of this model for the analysis of the marked enantioselectivity exhibited by AChE towards organophophorous compounds[7]. In the case of isopropyl methylphosphonates (IMP) the ratio of bimolecular rate constants of inhibition (ki) for Sp-IMP vs Rp-IMP is in the range of 3 orders of magnitude[8]. This chiral selectivity is most likely a consequence of the steric requirements of the nucleophilic displacement reaction at the phosphorous atom in the enzyme environment. In order to test this assumption and also to test and refine our model, we have constructed the phosphonylation products of HuAChE, with Sp and Rp diastereoisomers of IMP.

METHODS

Molecular Modeling and Graphics: Construction of the molecular models of HuAChE and its IMP adducts, molecular graphics manipulations and amino acid replacements, were performed on a Silicon Graphics IRIS 4D/70GT workstation using SYBYL software (Trypos

Multidisciplinary Approaches to Cholinesterase Functions, Edited by
A. Shafferman and B. Velan, Plenum Press, New York, 1992

195

Inc.). Structure optimizations by molecular mechanics of either the complete structure or by zone refinements (ANNEAL), were performed using the MAXMIN force field.

Construction of the HuAChE model: The HuAChE model was constructed from the coordinates of TcAChE[1] and based on sequence alignment of cholinesterases[2]. Stretches of 20-25 aminoacids were sequentially "mutated", maintaining the backbone conformation and the general orientation of the sidechains. Following each step, the structure was zone refined with respect to the "mutated" residues and the surrounding structure. Upon completion of the replacements of the aminoacids of TcAChE by the homologous HuAChE residues, the structure was re-optimized, hydrogens were added to 34 residues comprising the active site "gorge"[1] followed by further energy optimization.

Modeling Sp and Rp-IMP adducts of HuAChE: Sp and Rp-IMP adducts of HuAChE were modeled using manual docking aided by the known X-ray structure of phosphonylated chymotrypsin[9]. Partial atomic charges reported by Kovach et. al. were used[10]. Once a reasonable structure was obtained, its energy was optimized using zone refinement which included 23 aminoacids surrounding the phosphonylated residue.

RESULTS AND DISCUSSION

Construction of the Molecular Model of HuAChE

In comparative modeling several proteins are usually employed in formulating the template for construction of a molecular model of a homologous protein, for which only the sequence is known[11]. In our case, the pronounced sequence homology (in particular in the active site "gorge") and the nearly identical enzymatic activity profiles of TcAChE and HuAChE, were assumed to be sufficient for the construction of HuAChE model based on TcAChE molecular structure, as a single template. The validity of such an assumption was tested by comparison of the resulting model, with the TcAChE X-ray structure. Cα traces of the two sets of coordinates were superimposed and the quality of fit was evaluated. The results (see Table 1) revealed only minor structural changes as reflected by the RMS (root mean square) values. Most of the observed differences are located in the periphery and not in the residues comprising the active site "gorge" or in elements responsible for maintanance of the hydrolase fold topology[12], shared by a group of hydrolases, including TcAChE. This structural similarity was not neccessarily expected from the sequence homology between the two enzymes. Therefore, at least in this case the functional analogy appears to serve as an indicator for structral similarity.

Table 1. RMS values for HuAChE model fitted onto TcAChE structure

Fitted region	RMS(Å)
Cα whole molecule	0.43
Cα active site "gorge" (34 aa)	0.39
Cα catalytic triad	0.46

The current HuAChE model includes 537 out of 583 residues since the TcAChE X-ray structure is missing the last 36 aminoacids. In addition, we need to include in our model the missing 8 residues present only in HuAChE (Glu1,Gly2,Pro113,Pro258,Pro259,Gly260, Gly261,Pro498). Further modifications will include addition of hydrogens to the whole structure followed by energy optimization. The model is constantly refined by inputs from structure-activity studies using site directed mutagenesis. Its viability can be further assessed by the ability to rationalize the interactions of the enzyme with various ligands and by correlating experimental findings with the expected structural or mechanistic effects.

Modeling of HuAChE Adducts with IMP Diastereoisomers

The marked chiral selectivity of AChE for organophosphorous inhibitors is reflected by the diastereomeric ratio (up to 4 orders of magnitude) of the bimolecular rate constants for inhibition (ki), for isopropyl methylphosphonates (IMP)[7,8]. No such selectivity was observed for BuAChE[8]. In the modeling of phosphonylation products of HuAChE by IMP diastereoisomers we assumed that for each isomer only one chiral product results[13]. Such product will be of opposite chirality to the starting IMP assuming in-line nucleophilic displacement of the leaving group by Oγ-Ser203(200). Extracts from the optimized HuAChE-IMP adducts, as viewed from the top of the "gorge", are presented in figure 1.

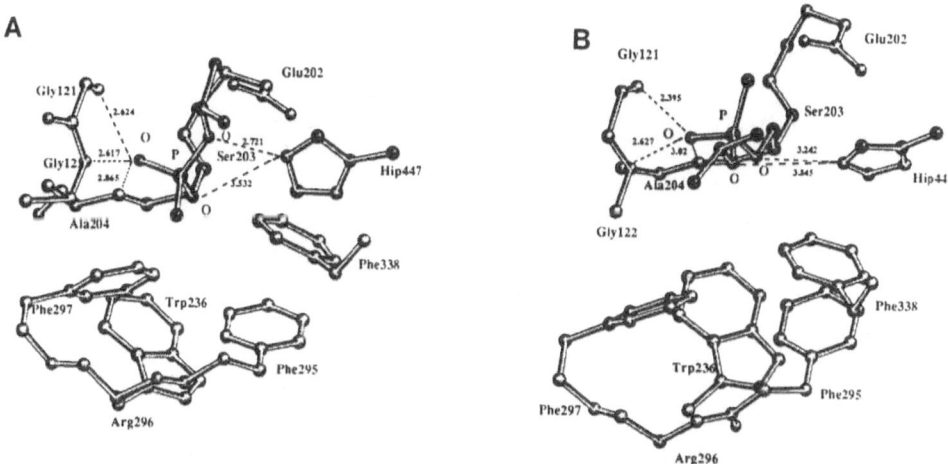

Figure 1. Model structures of the phosphonylation products of HuAChE with a) S_P-IMP (Add1); b) R_P-IMP (Add2). The extracts include part of the catalytic triad (Ser203;His447), Glu202, the putative oxyanion hole and aromatic residues (Phe295;Phe297;Trp236) which form the lower part of the lining of the "gorge". Distances between the phosphonyl oxygens, the oxyanion hole and Hip447 (protonated His447) are noted.

In both adducts the position of the phosphonyl oxygen is fixed by hydrogen bonding at the oxyanion hole (Gly121(118),Gly122(119),Ala204(201) - TcAChE numbering in parentheses). Examination of the Sp-IMP adduct (Add1, Fig.1a) shows that the isopropyl moiety projects towards the entrance of the "gorge" without any evidence for strong interaction with proximal residues. In contrast, in the Rp-IMP adduct (Add2, Fig.1b) an unfavourable conformation is forced upon the isopropyl group, as reflected by the somewhat higher energy of the optimized structure of this adduct (2.0kcal/mol). This energy difference could account in part, for the observed differences in the rate constants for inhibition by the two diastereoisomers.

Examination of the model of the Rp-IMP adduct indicates that the unfavourable interactions of the isopropyl group in this adduct could be attributed to interaction with the aryl moieties of Phe295 and Phe297. In order to test this assumption and to obtain further insight into the observed difference in stereoselectivity of AChE vs BuChE, Phe295(288) and Phe297(290) were replaced by the analogous and less bulky residues of BuChE, Leu and Val, respectively. Adducts of this "theoretical mutant" (HuAChE-Mu) were constructed and optimized (as above). Extracts of Sp-HuAChE-Mu and Rp-HuAChE-Mu are presented in figure 2. In this case, the optimized adducts did not reveal significant energy differences (<0.5 kcal/mol). The similar stability of these adducts (in contrast to Add1 vs Add2) may be understood by comparing the conformations of the isopropyl moiety in the Rp-IMP adducts (Add2 and Add4), which differ considerably. The absence of the bulky aromatic residues in positions 295 and 297 allows the isopropyl moiety in Add4 to adopt conformations which are sterically hindered in Add2.

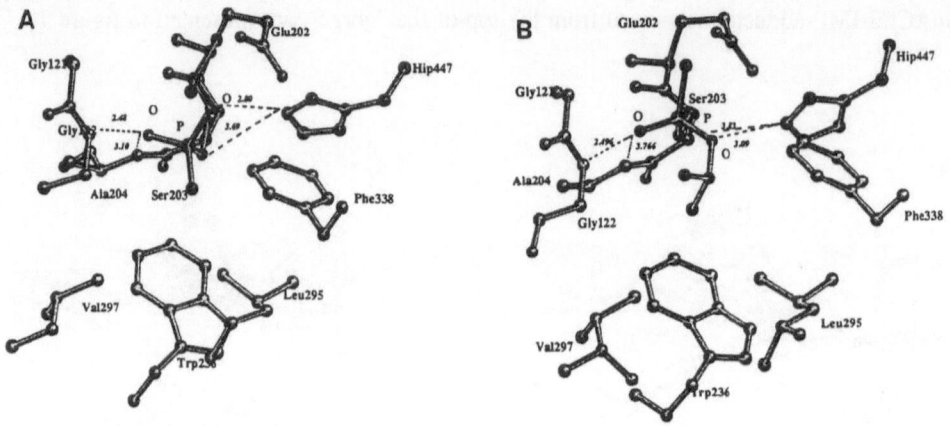

Figure 2. Model structures of the phosphonylation products of HuAChE-Mu (theoretical mutant of HuAChE, F295L, F297V) with a) S_p-IMP (Add3); b) R_p- IMP (Add4).

Thus, we can conclude that enantioselectivity of AChE towards IMP derivatives originates from differential stabilization of the O-alkyl substituent in the two diastereomers. Removal of the destabilizing elements eliminates the chiral preference in BuChE as suggested by our results with HuAChE-Mu model. Such conclusions are in agreement with SAR studies[14], which suggested that the active site of AChE should contain two hydrophobic subsites. One subsite could accomodate radicals up to n-butyl in size whereas the other was shown to be limited to acetyl groups. Jarv[14] also showed that in BuChE no size limitation for the second subsite, could be observed. Similar conclusions, pertaining the substrate specificity of BuChE vs AChE, were reached on the basis of molecular modeling of BuChE[15].

Reactivity of Chiral Phosphonylated HuAChE's in Aging and Oxime-Induced Reactivation

Model structures of Add1 and Add2 may be also used to explain the experimentally observed differences in the rate of aging for the two adducts[16]. It is probable that proton transfer to the alkoxy oxygen is the first step in the process. Since Hip447(440) is one of the possible donors, the distance between NεH447 and the alkylated oxygen of the phosphonyl moiety should influence the facility of aging. Comparison of these distances for the two adducts (Fig. 1,2) indicates that proton transfer may occur more readily in Add1 than in

Add2. Since the position of this particular oxygen atom is independent of the nature of the alkyl substituent, similar distance differences should exist in all phosphonylated AChEs rendering the Sp-adducts more reactive in the aging process than the Rp- adducts.

For chiral phosphonylation products of TcAChE no difference in rates of oxime-induced reactivation were observed[16]. Assuming that reactivation involves a specific recognition of the oxime reactivator and an in-line nucleophilic displacement at the phosphorous atom, our models indicate that Add1 should be more readily reactivated than Add2. In Add2 the isopropyl group should block the approach of a nucleophile to the phosphorpus atom from the direction opposite to OγS203 (Fig.1b). This may indicate that reactivation involves nucleophilic attact adjacent to OγS203 followed by pseudorotation of the phosphonyl moiety.

ACKNOWLEDGMENTS

We would like to express our appreciation to Drs. Sussman, Harel and Silman for providing us with the coordinates of TcAChE and sharing data, prior to publication.

REFERENCES

1. J.L. Sussman, M. Harel, F. Frolow, C. Oefner, A. Goldman, L. Toker, and I. Silman, *Science.* **253**, 872 (1991).
2. M.K. Gentry, and B.P. Doctor, in: *Cholinesterases: Structure, Function, Mechanism, Genetics, and Cell Biology.* J. Massoulie, F. Bacou, E.A. Barnard, B.P. Doctor, and D.M. Quinn, eds., Am. Chem. Soc Washington (1991), pp. 394.
3. B. Velan, H. Grosfeld, C. Kronman, M. Leitner, Y. Gozes, A. Lazar, Y. Flashner, D. Markus, S. Cohen, and A. Shafferman, *J. Biol. Chem.*, **266**, 23977 (1991).
4. A. Shafferman, C. Kronman, Y. Flashner, M. Leitner, H. Grosfeld, A. Ordentlich, Y. Gozes, S.Cohen, N. Ariel, D. Barak, M. Harel, I. Silman, J.L. Sussman, and B. Velan, *J. Biol. Chem.*, in press.
5. A. Shafferman, B. Velan, A. Ordentlich, C. Kronman, H. Grosfeld, M. Leitner, Y. Flashner, S. Cohen, D. Barak, and N. Ariel, Submitted (1992).
6. A. Shafferman, B.Velan, A. Ordentlich, C. Kronman, H. Grosfeld, M. Leitner, Y. Flashner, S. Cohen, D. Barak, and N. Ariel, This volume.
7. H.P. Benschop, and L.P.A. De Jong, *Acc. Chem. Res.* **21**, 368 (1988)
8. H.L. Boter, L.P.A. De Jong, and H. Kienhuis, in: *Interactions of Chemical Agents with Cholinergic Mechanisms,* Israel Ins. Biol. Res. 16th Annual Biology Conference (1971).
9. M. Harel, C.T. Su, F. Frolow, Y. Ashani, I. Silman, and J.L. Sussman, *J. Mol. Biol..* **221**, 909 (1991).
10. I.M. Kovach, and D. Huhta, *Teochem.*, **234**, 335 (1991).
11. J. Greer, *Proteins*, 7, 317 (1990)
12. D.L. Ollis, E. Cheah, M. Cygler, B. Dijkstra, F. Frolow, M.S. Franken, M. Harel, S.J. Remington, I. Silman, J. Schrag, J.L. Sussman, K.H.G. Verschueren, and A. Goldman, *Prot. Eng.*, in press.
13. H.A. Berman, and K. Leonard, *J. Biol. Chem.*, **264**, 3942 (1989).
14. J. Järv, *Bioorg. Chem.*, **12**, 259 (1984).
15. M. Harel, I. Silman, and J.L. Sussman, This volume.
16. H.A. Berman, and M.M. Decker, *J. Biol. Chem.*, **264**, 3951 (1989).

Keywords

Molecular models, HuChE, TcAChE, methyphosphonates, enantioselectivity

RELATIONSHIPS BETWEEN ACTIVITY OF ORGANOPHOSPHORUS INHIBITORS OF ACETYLCHOLINESTERASE AND ACCESSIBILITY OF PHOSPHORUS ATOM AS ESTIMATED BY MOLECULAR MECHANICS CALCULATIONS

Boris S. Zhorov[1], Natalya N. Shestakova[2], and
Evgeniy V. Rozengart[2]

[1]Pavlov Institute of Physiology of the Russian
Academy of Sciences, St.Petersburg, Russia
[2]Sechenov Institute of Evolutionary Physiology and
Biochemistry of the Russian Academy of Sciences
St.Petersburg, Russia

INTRODUCTION

Irreversible inhibition of acetylcholinesterase (AChE) by organophosphorus compounds is believed to be due to phosphorylation of serine hydroxyl in the enzyme active center (Cohen, Oosterbaan, 1963). The attack by the hydroxyl may occur from the different faces of phosphorus atom (Berman and Decker, 1989). The efficiency of these reactions should depend on sterical accessibility of the phosphorus atom for the attacking agent. In particular, low activity of some inhibitors may be due to low accessibility of their phosphorus atom in the reaction with the enzyme. To obtain quantitative estimates of the accessibility of different faces of the phosphorus atom, conformational flexibility of inhibitors should be taken into account. In this work, using molecular mechanics method, we have calculated all minimum-energy conformations (conformers) of 15 organophosphorus inhibitors with nitro- phenyl and S-ethylthioethyl leaving groups and different structure of phosphoryl moiety, determined the accessibility of the phosphorus atom in every conformer, and compared anti-AChE efficiency of the compounds with the total population of those conformers which have accessible phosphorus atom.

METHODS

Molecular mechanics calculations were carried out on IBM PC/AT computer and Motorolla 68020 workstation using ZMM package, the recent version of the universal

Multidisciplinary Approaches to Cholinesterase Functions, Edited by
A. Shafferman and B. Velan, Plenum Press, New York, 1992

201

conformational program (Zhorov, 1975). Equilibrium conformations were calculated by means of energy minimization with torsional and bond angles as variable geometrical parameters. Nonbonded interactions, torsional energy, bond angles deformation and loop closing potentials were taken into account. Force field of Nemethy et al. (1983) was used for nonbonded interactions, parameters for the phosphorus atom being taken from the work of Zhurkin et al. (1980). The gradient of energy was calculated by the analytical vector method (Zhorov, 1981, 1982).

The following values of bond lengths (Å) were used: S-P 2.14, S-C 1.78, P=O 1.65, P-O 1.76, P-C 1.87, O-N 1.2, O-C 1.43, N-C 1.47, C_{sp3}-C 1.54, C_{sp2}-C_{sp2} 1.39, C-H 1.07. Torsional energy was calculated as $E_{tor} = 2.0$ (1-cos 2τ) for the bond Ph-O, and as $E_{tor} = (U/2)(1+\cos 3)$ for other bonds. The following barrier heights U (kcal/mole) were used: P-S 2.0, P-O 1.0, P-C_{sp3} 3.4, P-C_{sp2} 0.0, S-C 2.0, C-C 3.0, C_{sp3}-C_{sp2} 0.0, O-C 1.0.

The energy of bond angles deformation $E_{ang} = C_a(\alpha-\alpha_0)2$ was calculated with the value of 109.5° for ideal bond angle α_0 and the following force constants C (kcal·mole^{-1}·rad^{-2}): $C_P = 100$, $C_O = 90$, $C_C = 100$. Bond angles of the methyl groups, C_{sp2}, and sulfur atoms were fixed at values of 109.5, 120, and 100°, correspondingly. Nitrophenyl group was fixed in planar conformation.

In the search for minimum-energy conformations, we used as starting points all combinations of *trans-, gauche-*, and (-)-*gauche*-conformations for ordinary bonds, all combinations of *gauche-*, and (-)-*gauche*-conformations for bonds in cyclic phosphorylic group of the compound (1), two chair conformations with axial and equatorial orientation of the bond C-P for cyclohexane fragment of the compound (14), a value of 30° for the bond O-Ph, and values of 30, 90, and 150° for the bonds C_{sp3}-C_{sp2}, and P-C_{sp2}. Other details of calculations are described elsewhere (Zhorov et al. 1991a, 1991b).

To estimate accessibility of the phosphorus atom for the attacking agent, a cylinder of radius 3 Å was drawn, its axis coinciding with the extension of the breaking bond, one of its bases touching van der Waals surface of the phosphorous atom at the face opposite the breaking bond, the other being in infinity. In more precise terms, the center of phosphorus-touching base of the cylinder was at the extension of the breaking bond X-P, at the distance 3.5 Å from the center of oxygen atom, if X = O, or 4.0 Å from the center of sulfur atom, if X = S. Conformer was considered to be accessible in the chosen direction unless center of any atom fell within the cylinder.

To evaluate possible correlation between conformational characteristics of inhibitors and their anti-AChE activities, the population of conformers was calculated:

$$p_i = 100 \cdot \exp(-E_i/RT)/[\sum_{j=1}^{N} \exp(-E_j/RT)],$$

were N is the number of conformers of the compound, E - conformational energy, i,j - indexes of conformers, R - the universal gas constant, T - temperature equal to 310 °K.

Anti-AChE activities of the compounds considered were taken from the data on inhibition of house fly head AChE (Fukuto and Metcalf, 1965) and bovine erythrocyte AChE (O'Brien, 1960; Brestkin and Godovikov 1978).

RESULTS AND DISCUSSION

First of all, we considered compounds (1)-(4) with *p*-nitrophenyl leaving group

(Fukuto and Metcalf, 1965; see Table 1). In this series, cyclic compounds (1)-(3) lack anti-AChE activity, while their acyclic analog (4) is a potent irreversible inhibitor of AChE. Cyclic (1) and acyclic (4) compounds are similar in hydrophobic properties, volume of phosphoryl groups and the rate of alkaline hydrolysis. This suggests that the rate of anti-AChE action of these compounds may be controlled by stereochemical factors.

Table 1. Antiacetylcholinesterase activity (I_{50})[a], rate of alkaline hydrolysis (k_{OH})[a], number of minimum-energy conformations (N), and population of productive conformers (p) of the compounds $R^1R^2P(O)-O-C_6H_4-4-NO_2$

Comp.	$-R^1$	R^2-	k_{OH}	I_{50}, mM	N	p,%
1	$-O-CH_2-CH_2-CH_2-O-$		1.56	> 1.3	25	0
2	$-O-CH(CH_3)-CH_2-CH_2-O-$		0.69	> 1.3	"	"
3	$-O-CH_2-C(CH_3)_2-CH_2-O-$		1.49	> 1.3	"	"
4	$-O-C_2H_5$	C_2H_5-O-	1.56	$2.6 \cdot 10^{-5}$	102	27

[a] (Fukuto and Metcalf, 1965)

Energy minimization of the compound (1) revealed 25 equilibrium conformations. Five of them have energy less than 0.5 kcal/mole, six-membered ring being in chair conformation with axial or equatorial orientation of the breaking bond. The energy of other forms is greater than 4 kcal/mole. For each equilibrium conformer of the compound (1), we have calculated accessibility of two faces of the phosphorus atom opposite to the leaving group and to the bond P=O. The results indicate that compound (1) has no conformers with the phosphorus atom accessible from both faces. Hence, low anti-AChE activity of the compound (1) may be explained by low effectiveness of the attack of nucleophilic agent.

Conformational properties of six-membered ring in compounds (2) and (3) are similar to those in compound (1). Methyl substituents in the ring reduce furthermore the accessibility of the phosphorus atom. Hence, low activity of the compounds (2) and (3) may be also explained by the absence of conformers with the accessible phosphorus atom.

Calculation of the compound (4), the active acyclic analog of the compound (1), has revealed 102 minimum-energy conformations with energy less than 4 kcal/mole. The population of conformers with the accessible face of the phosphorus atom opposite the leaving group is equal to 27 %. Thus, in the series of compounds (1)-(4), the only effective AChE inhibitor (4) has conformers with the accessible face of the phosphorus atom opposite the leaving group. Such conformers are named below as productive (see Fig. 1).

Low anti-AChE activity of the compounds (1) - (3) may be explained also by energetically unfavorable inversion of six- membered ring which have to attend inversion of configuration of the phosphorus atom in the reaction of nucleophilic substitution. But to rationalize comparable obedience of the compounds (1) and (4) for the alkaline hydrolysis, additional suggestion is necessary that alkaline hydrolysis of the compounds, unlike their reaction with AChE, occurs without inversion of the phosphorus atom.

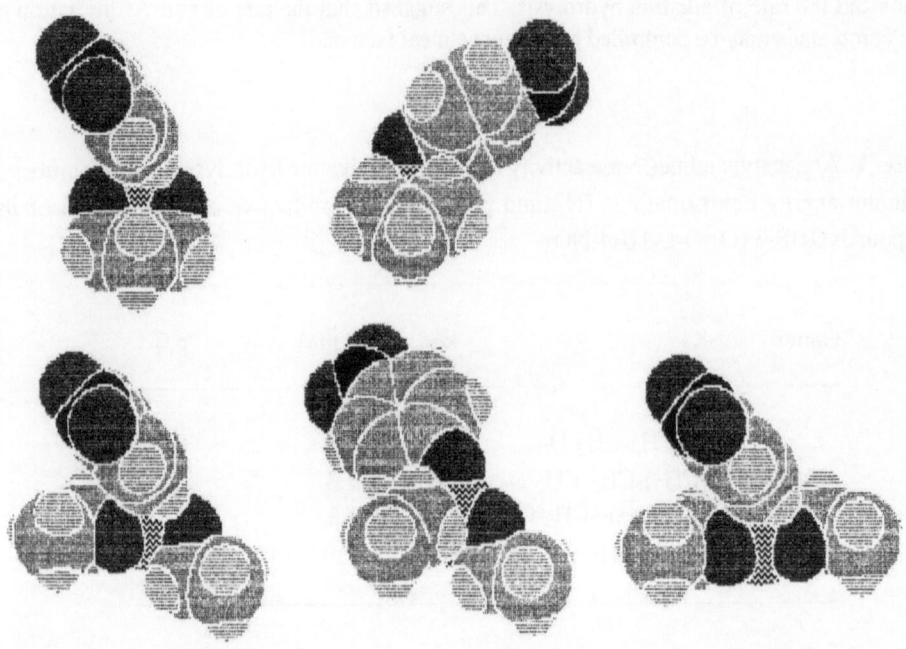

Fig. 1. Minimum-energy conformations of the compounds (1) (upper line) and (4) visualized with the help of Desk Top Molecular Modeling package (Oxford Electronic Publishing). Breaking bond P-O is pointed upward, the bond P=O is masked by the phosphorus atom. The face of the phosphorus atom opposite the breaking bond is accessible only in the third conformer of the compound (4). This conformer is productive, while other conformers shown are non-productive.

Table 2. Antiacetylcholinesterase activity (pI_{50})[a], number of minimum-energy conformations (N), and population of productive conformers (p) of the compounds $R_2P(O)\text{-}O\text{-}C_6H_4\text{-}4\text{-}NO_2$

Comp.	R	pI_{50}	N	p,%
5	$O\text{-}CH_2\text{-}CH_2\text{-}CH_3$	7.6	110	41
6	$O\text{-}CH(CH_3)_2$	6.0	110	7
7	$O\text{-}CH_2\text{-}CH_2\text{-}CH_2\text{-}CH_3$	7.8	910	40
8	$O\text{-}CH_2\text{-}CH(CH_3)_2$	6.3	105	35

[a] (O'Brien, 1960)

Considering that two alternative explanations of dramatic difference in anti-AChE activity of the compounds (1) and (4) are possible, we calculated compounds (5)-(8 with *p*-nitrophenyl leaving group and acyclic phosphoryl moieties. In this series, compounds with propoxy (5) and butoxy (7) radicals are essentially more active than their branched isomers (6) and (8) (Table 2). Taking into account that minimum-energy conformations of the compound (4) with *gauche* or (-)-*gauche* disposition of the fragment P-O-C-C were proven to have energy greater than 0.6 kcal/mole, we diminished number of energy minimizations of the compounds (5), (7), and (8), using only *trans* conformation of the fragment P-O-C-C in starting points. The results demonstrate that in each pair of isomers, activity changes in one direction with the population of productive conformers (see Table 2).

Next, we examined relationships between accessibility of the phosphorus atom and anti-AChE activity of the compounds (9)-(15) with S-ethylthioethyl leaving group and varied structure of phosphoryl group (Brestkin and Godovikov, 1978; see Table 3). Trying 243 (3[5]) starting points of the model compound $(CH_3)_2P(O)$-S-CH_2-CH_2-S-CH_2-CH_3, we found that in the set of 27 lowest energy conformers (E < 1 kcal/mole), fragments P-S-C-C and C-S-C-C turned out to be in *trans* form only. Therefore, in subsequent calculations we used only *trans* starting conformations of these fragments, i.e. 9 starting points of the chain P-S-C-C-S-C-C.

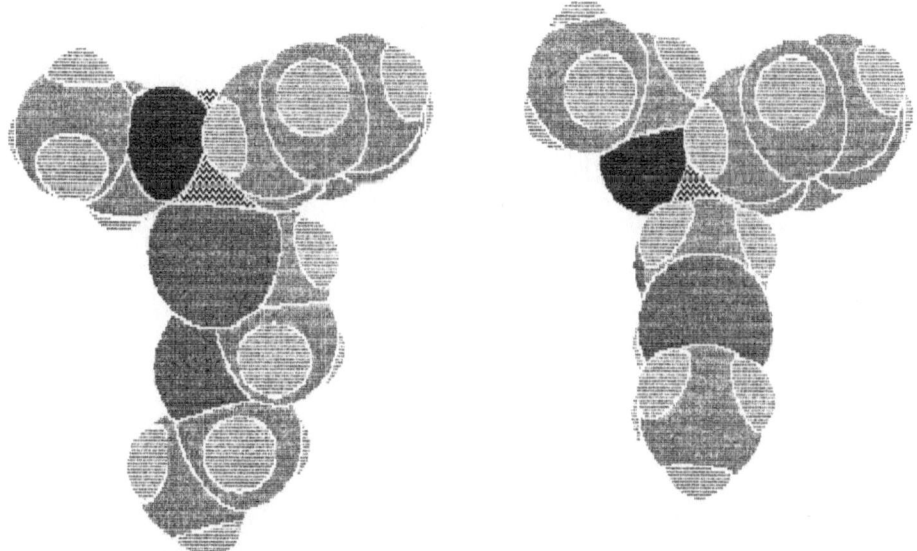

Fig. 2. Minimum-energy conformations of the compound (10) with non-accessible (left) and accessible (right) face of the phosphorus atom opposite the leaving group. Breaking bond is pointed upwards, bond P=O is hidden beyond the phosphorus atom.

The results are given in Table 3. Compound (9), the most potent inhibitor in this series, has 60% of productive conformers while non-effective cyclic derivative (15) has no productive conformers. Fig. 2 presents, for example, productive and non-productive conformers of the compound (10). Population of the productive conformers correlates with the rate of AChE inhibition (k_{II}). Coefficient of correlation is equal to 0.92.

According to calculations, population of conformers with the accessible face of the phosphorus atom opposite the bond P=O does not correlate with the activity of the inhibitors, varying from 0 to 16 % in different ligands.

Inasmuch as the correlation revealed could be sensitive to the criterion of accessibility of the phosphorus atom, we repeated calculations with other values of the radius r of the endless cylinder whose vacancy proves a conformer to have accessible phosphorus atom (see Method). Varying r from 1 to 5 Å with 0.5 Å increment, we found that the maximum of correlation coefficient corresponds to $r = 3$ Å.

Table 3. Antiacetylcholinesterase activity (k_{II})[1], number of minimum-energy conformations (N), and population of productive conformers (p) of the compounds $R^1R^2P(O)SC_2H_4SC_2H_5$

Comp.	$-R^1$	R^2-	k_{II}, $M^{-1}min^{-1}$	N	p,%
9	$-CH_3$	C_2H_5O-	$5.0 \cdot 10^4$	75	59
10	$-C_6H_5$	"	$3.1 \cdot 10^4$	121	35
11	$-CH_2C_6H_5$	"	$1.5 \cdot 10^4$	202	38
12	$-OC_2H_5$	"	$6.4 \cdot 10^3$	563	24
13	$-OCH(CH_3)_2$	$(CH_3)_2CHO-$	$2.3 \cdot 10^3$	367	12
14	$-C_6H_{11}(cyclo)$	C_2H_5O-	$1.5 \cdot 10^2$	198	8
15	$-NH-CH_2-CH_2-CH_2-O-$		$< 10^2$	17	0

[1] (Brestkin and Godovikov, 1978)

The correlation revealed indicates that active irreversible inhibitors of AChE should have well-populated conformations with the accessible face of the phosphorus atom opposite the leaving group. This deduction is consistent with the concept that reaction of nucleophilic substitution in the active center of AChE occurs through the intermediate complex in which oxygen atom of the serine hydroxyl is located opposite the breaking bond. Interestingly, the serine hydroxyl in the active center of AChE is believed to attack acetylcholine, the endogeneus ligand of AChE, in the direction normal to the plane of acetyl group. The most remarkable feature of the fully extended conformation of acetylcholine which was suggested recently to be productive (Zhorov et al., 1991b) is the complete accessibility of the faces of acetyl group normal to the vector of the attack by serine. Evidently, the reaction of AChE with both substrates and irreversible organophosphorus inhibitors is most effective if these ligands approach the serine hydroxyl in the enzyme active center with the electrophilic reactant atom positioned on plane or convex rather than on concave face of the molecules.

REFERENCES

Berman, H.A., and Decker, M.M., 1989, Chiral nature of covalent methylphosphonyl conjugates of acetylcholinesterase, *J.Biol.Chem.* 264:3951.

Brestkin, A.P., and Godovikov, N.N., 1978, Combined type of inhibition of cholinesterases by organophosphorus compounds, *Uspehi Khimii* 47:1608 (In Russian).

Cohen, J.A., Oosterbaan, R.A., 1963, The active site of acetylcholinesterase and related esterases and its reactivity towards substrates and inhibitors, Handbuch exp. *Pharmakol.* 15:299.

Fukuto, T.R., and Metcalf, R.L., 1965, Reactivity of some 2-p-nitrophenoxy-1,3,2,-dioxaphospholane 2-oxides and -dioxaphosphorinane 2-oxides, *J.Med.Chem.* 8:759.

Nemethy, G., Pottle, M.S., and Scheraga, H.A., 1983, Energy parameters in polypeptides. 9. Updating of geometrical parameters, nonbonded interactions, and hydrogen bond interactions for the naturally occurring amino acids, *J.Phys.Chem.* 87:1883.

O'Brien,R.D., 1960, "Toxic Phosphorus Esters. Chemistry, Metabolism, and Biological Effects", Academic Press, New York London.

Zhorov, B.S., 1975, Computer modeling of the three-dimensional structures of organic compounds, *Avtometriya* N1:23 (In Russian).

Zhorov, B.S., 1981, Vector method for calculating derivatives of energy of atom-atom interactions of complex molecules according to generalized coordinates, *Zh. Strukt. Khim.* 22:8 (in Russian).

Zhorov, B.S., 1982, Vector method for calculating derivatives of the energy of deformation of valence angles and torsion energy of complex molecules according to generalized coordinates, *Zh. Struct. Khim.* 23:3 (in Russian).

Zhorov, B.S., Brovtsyna, N.B., Gmiro, V.E., et al., 1991a, Dimensions of the ion channel in neuronal nicotinic acetylcholine receptor as estimated from analysis of conformation-activity relationships of open channel blocking drugs, *J.Membr.Biol.* 121:117

Zhorov, B.S., Shestakova, N.N, and Rozengart, E.V., 1991b, Determination of productive conformation of acetylcholinesterase substrates using molecular mechanics, *Quant.Struct.-Activity Relat.* 10:205.

Zhurkin, V.B., Poltev, V.I., and Florent'ev, V.L, 1980, Atom- atom potential functions for conformational calculations of nucleic acids, *Molek.Biologiya* 14:1116 (In Russian).

FACTORS INFLUENCING ACETYLCHOLINESTERASE REGULATION IN SLOW AND FAST SKELETAL MUSCLES

Janez Sketelj, Neva Črne-Finderle, and Igor Dolenc

Institute of Pathophysiology
School of Medicine
61000 Ljubljana, Slovenia

INTRODUCTION

Acetylcholinesterase (AChE) is an indispensable functional component of the neuromuscular junction. Therefore, it is highly concentrated at this small fraction of the sarcolemma (Marnay and Nachmansohn, 1938; Brzin and Zajiček, 1958). After revealing the molecular polymorphism of AChE, it was shown that it is mostly the A12 asymmetric AChE form that is concentrated in the motor endplates (Hall, 1973). There is significant activity of other AChE forms extrajunctionally although their precise function, if any, is unknown. The asymmetric AChE forms are not specific for the neuromuscular junction, but become restricted to the endplates during muscle development (Sketelj and Brzin, 1980).

Both junctional and extrajunctional AChE is under neural control (Guth et al., 1964). An interesting aspect of this nerve-muscle interaction is represented by significant differences in the patterns of AChE molecular forms in fast and slow muscles of different animals (Gisiger and Stephens, 1982; Bacou et al., 1982; Groswald and Dettbarn, 1983). Soleus (SOL) and extensor digitorum longus (EDL) muscles of the rat are fairly good representatives of slow and fast muscles, although they are not completely homogeneous (Ariano et al., 1973). Heterogeneity of contractile properties of muscle fibers, based on myosin polymorphism and metabolic enzyme patterns, is mostly determined by the pattern of electromechanical activity triggered by motor axons in individual motor units (Pette and Vrbova, 1985). The pattern of AChE molecular forms in the rat muscles can also be modified by different patterns of electrical stimulation of muscles (Lomo et al., 1985). Therefore, electromechanical activity may play an important role in AChE regulation in muscles. However, experiments on denervated fast and slow muscles in the rabbits and guinea pigs (Bacou et al., 1982; Lai et al., 1986) demonstrated that intrinsic differences between the two types of muscles may also influence AChE regulation.

We expanded these studies and, in addition, introduced the model of muscle regeneration (Sketelj et al., 1987, 1991) to further elucidate these problems. In the present work, special emphasis is laid first on regulation of the extrajunctional AChE molecular forms in the rat SOL and EDL muscles under conditions of denervation

Multidisciplinary Approaches to Cholinesterase Functions, Edited by
A. Shafferman and B. Velan, Plenum Press, New York, 1992

209

and reinnervation. Second, modifications of neural input to regenerating SOL and EDL muscles (tenotomy, cross-innervation) were used to reveal the importance of intrinsic properties of muscles, i.e. information mediated by sattelite cells of either slow or fast muscles to their descendent regenerating muscle fibers, in respect to regulation of AChE molecular forms.

METHODS

Male Wistar rats, weighing 180-200 g at the time of surgery were used in experiments. The rats were anaesthetized by ether or pentobarbitone (Vetanarcol, Werfft-Chemie, Wels, Austria, 50 mg/kg i.p.). The pups of Wistar strain of albino rats of both sexes were used in experiments dealing with postnatal muscle development.

Surgical Procedures

Denervation: The sciatic nerve was sectioned in the thigh of one leg. The proximal nerve stump was sutured to the surrounding muscles to prevent reinnervation of calf muscles.

Muscle paralysis: Five mice units of botulinum toxin, type A (Oculinum, Smith-Kettlewell, San Francisco, CA) were diluted in 0.1 ml of saline and injected subcutaneously into the lateral part of the calf. Complete paralysis of the calf muscles developed in 24 hours and lasted for at least two weeks.

Reinnervation: Sciatic nerve was crushed on one side at the midthigh level with nonserrated haemostat for 30 seconds. The nerves were left to regenerate for three weeks. By that time the toe-spreading reflex recovered indicating reinnervation of the calf muscles.

Muscle Regeneration

Nerve-intact graft (non-transplanted regenerating muscle). Care was taken to cut all blood vessels to the SOL muscle and not to damage the motor nerves in order to ensure complete reinnervation (Carlson et al., 1981). Bupivacaine was injected into the muscle together with an irreversible AChE inhibitor phospholine (Ayerst Pharmaceutical Co., New York, N.Y.). Threreafter, tenotomy of the injured SOL muscle was performed in a group of animals by excising about 5 mm of its tendon and diverting and suturing the distal part of the muscle more proximally to fasciae of the underlying muscles.

Homo- or cross-transplanted regenerating muscle. SOL muscle was excised and incubated in a solution of bupivacaine (0.5%) and phospholine (0.1 mM) for 20 min. It was then sutured either to its own tendons (homotransplanted) or to the tendons of the excised EDL muscle on the contralateral leg (cross-transplanted). The motor nerve which had innervated the muscle in place of which lay the regenerate was sutured into a small cut of the fascia of the transplanted muscle by an epineurial suture. Analogous procedure was used for EDL muscles.

Velocity Sedimentation Analysis of AChE Molecular Forms

Molecular forms of AChE were often analysed separately in endplate-rich (junctional) and endplate-free (extrajunctional) regions of muscles. Histochemical procedure for AChE described by Koenig and Rieger (1981) was used to demonstrate motor endplates. Muscles were cut longitudinally into two bands and incubated in a histochemical medium for 30 min at room temperature. Junctional and extrajunctional regions of muscles were isolated under low magnification.

Isolated muscle samples were immediately homogenized in ice cold medium (1 mg per 20 ul) containing 1M LiCl, 50mM Tris-HCl pH 7.3, 1% Triton X-100 and a set of antiproteases (10 mM EGTA, 2 mM benzamidine, 1 mg/ml bacitracin). Further procedure was essentially the same as described earlier (Sketelj et al., 1987).

RESULTS AND DISCUSSION

In the fast EDL muscle of the rat, 4S and 6S (G1 and G2) AChE forms contribute 55 ± 5% (n=6) of total AChE, whereas in the SOL only 33 ± 2% (n=6). The opposite is true about the asymmetric 16S (A12) and 13S (A8) forms: they comprise about 24 ± 6% of AChE in the EDL and 50 ± 4% in the SOL (see also Figure 1 A,D). A very conspicuous difference between the two muscles is the ratio between the 13S (A8) and 10S (G4) AChE forms. This ratio is ten times higher in the SOL (1.7), in which the 13S AChE is well expressed and the 10S AChE is low, than in the EDL (0.14) in which the 13S AChE is hardly detectable and 10S AChE is a prominent form. There is also an important qualitative difference in AChE regulation between the two muscles. The asymmetric AChE forms which are concentrated in the motor endplates in both muscles (Figure 1 B,E), are present also in the extrajunctional regions in the SOL (Figure 1 C), but are virtually absent in these regions of the EDL (Figure 1 F).

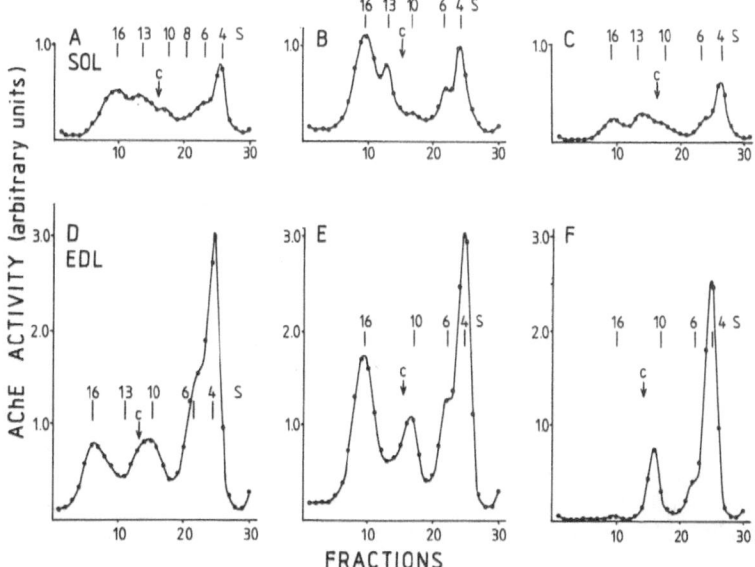

Figure 1. Velocity sedimetation analysis of AChE molecular forms in normal SOL (A- whole muscle, B- junctional region, C- extrajunctional region) and EDL muscle (D- whole muscle, E- junctional region, F- extrajunctional region) of adult rats.

AChE patterns were fairly reproducible and the patterns presented in the figure are typical representatives of three to six muscle samples. AChE activity, given as specific activity per unit of wet weight of muscle sample, is in arbitrary units comparable in all presented gradients. The top of the gradients is on the right-hand side. Peaks of AChE activity are identified by their approximate sedimentation coefficients. Arrow C shows the position of catalase (11.3 S) in the gradients.

This difference between fast and slow muscles may be quite general since it has been demonstrated also in the guinea pig (Randall et al., 1985) but there is still some doubt about the rabbit (Bacou et al.., 1982; Lai et al., 1986).

In the rat muscle, segregation of the asymmetric AChE forms to the motor endplate is a developmentally regulated process taking place during the first month after birth (Sketelj and Brzin, 1980). Accordingly, the difference in this respect between the fast and slow muscles stems from this period. At day 16 after birth, the 16S (A12) AChE form predominates in the extrajunctional region of the SOL, and is also a major molecular form extrajunctionally in the EDL (Figure 2). Downregulation of synthesis and accumulation of the A12 AChE takes place extrajunctionally during the next few weeks in both muscles. However, repression is virtually complete in the EDL but only partial in the SOL. Interestingly, but at present eluding explanation, the 13S AChE (A8) is left as the predominant asymmetric form extrajunctionally in the SOL.

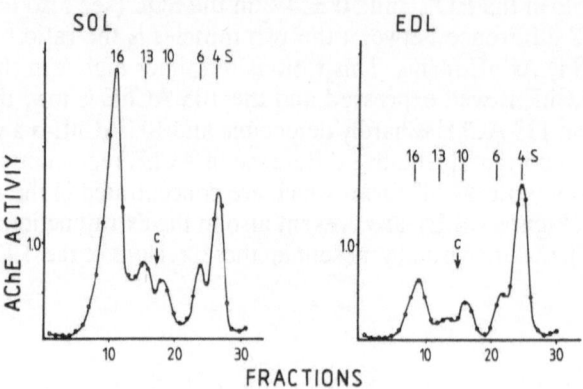

Figure 2. Velocity sedimentation analysis of AChE molecular forms in the extrajunctional regions of immature SOL and EDL muscles of 16 day-old rats. For details see the legend to Figure 1.

Phenotypic expression of many muscle properties is under neural control (Pette and Vrbova, 1985) and most of them are controlled by the pattern of muscle stimulation imposed by motoneurons (Lomo et al., 1984). The pattern of AChE molecular forms in the electrically stimulated denervated SOL also depended on the pattern of electrical stimulation of the muscle (Lomo et al., 1985). If the differences regarding AChE regulation in slow and fast muscles were maintained directly by different patterns of neural stimulation, one would expect that they would disappear after denervation. However, this is not the case. In both muscles, the junctional 16 S AChE dropped precipitously after denervation. Differences in the extrajunctional AChE regulation, though, were maintained for at least seven weeks after denervation. The 16 S AChE was still present, although reduced, in the SOL, and even increased later on. Activity of the extrajunctional A12 AChE in the denervated EDL remained more or less negligeable during the seven week period of denervation (Sketelj et al., 1992).

Surprisingly, denervation uncovered another difference in regulation of the extrajunctional AChE in SOL and EDL muscles. During the first few days after denervation, the 10 S AChE form (G4) became by far predominant AChE form extrajunctionally in the EDL (Figure 3 A). This has been observed also in other fast muscles of the rat (Gregory et al., 1989; Decker and Berman, 1990). There was not even a trace of such a response in the denervated SOL in which activity of all molecular forms extrajunctionally decreased rather uniformly (Figure 3 C). Exactly the same changes occur in muscles paralysed by botulinum toxin which blocks quantal but not non-

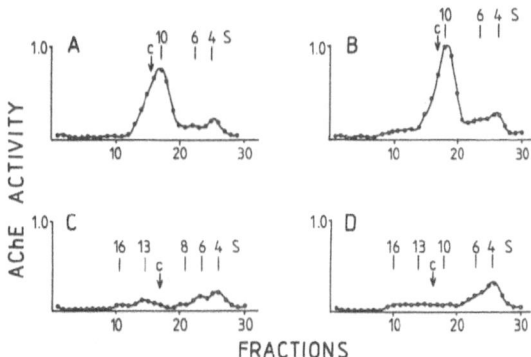

Figure 3. Velocity sedimentation analysis of AChE molecular forms in the extrajunctional regions of the inactive EDL (A- denervated, B- paralysed by botulinum toxin) or SOL muscle (C- denervated, D- paralysed by botulinum toxin). For details see the legend to Figure 1.

quantal acetylcholine release from the nerve ending (Thesleff, 1989) (Figure 3 B, D). This fact corroborates the view that electromechanical activity plays an important but, taking into account the observed differences between denervated SOL and EDL, not exclusive role in AChE regulation extrajunctionally.

AChE regulating mechanisms in both muscles also react differently to reinnervation: there is an overshoot of activity in the SOL but only slow recovery in the EDL (Groswald and Dettbarn, 1983). We demonstrate here that the reason for this difference lies in different behaviour of the extrajunctional 16 S (A12) AChE in both muscles. First signs of reinnervation of calf muscles after high sciatic nerve crush appear about two weeks after crush (Misulis and Dettbarn, 1985). There is an explosion of A12 AChE activity extrajunctionally in the SOL during the next week (Figure 4 A), but virtually no A12 AChE appeared in the extrajunctional regions of reinnervated EDL (Figure 4 B). It seems as if the extrajunctional nuclei in the SOL, although partially inhibited, remain 'embryonic-like' in respect to A12 AChE production while those in the EDL are more or less irreversibly repressed during early postnatal development.

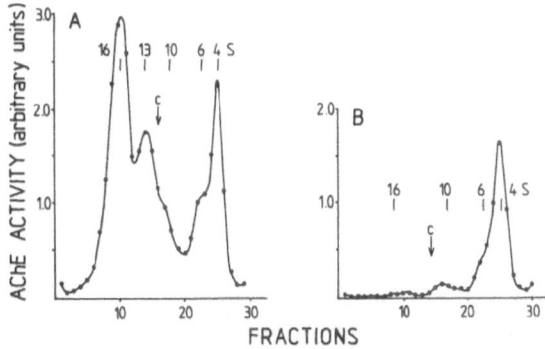

Figure 4. Velocity sedimentation analysis of AChE molecular forms in the extrajunctional regions of the reinnervated SOL and EDL muscles three weeks after crush of the sciatic nerve. For details see the legend to Figure 1.

This difference may be due either to the fact that the extrajunctional nuclei in both muscles are in an unknown way exposed to the efects of different neural stimulation patterns during the critical early postnatal period, or that the nuclei in the two

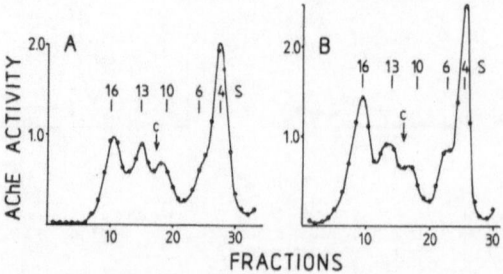

Figure 5. Velocity sedimentation analysis of AChE molecular forms in control regenerating SOL muscle (A) and tenotomized regenerating SOL muscle (B) two weeks after muscle injury. For details see the legend to Figure 1.

muscles are intrinsically differently susceptible to neural regulatory influences, possibly due to different myoblast lineages contributing to SOL or EDL muscles during their ontogenetic development (Miller and Stockdale, 1987; Grove, 1989).

There are several indirect pieces of evidence supporting the 'intrinsic differences' hypothesis. Characteristic differences between AChE patterns in SOL and EDL muscles appear already during the first ten days after birth (Dettbarn et al., 1985; Sketelj et al., 1991) when the patterns of the neural stimulation of both muscles are still quite similar (Navarette and Vrbova. 1983). Regenerating SOL and EDL muscles are produced by fusion of proliferating myoblasts, derived from 'sleeping' satellite cells in muscles (Carlson, 1978). There are some differences in AChE regulation even in non-innervated regenerating EDL and SOL muscles (Sketelj et al., 1991). Typical AChE patterns of both muscles arise immediately upon reinnervation of regenerating muscles (Črne et al., 1991) although neural stimulation is not normal because of muscle spindle destruction during muscle injury (Diwan and Milburn, 1986).

Tenotomy of SOL muscle causes significant temporary reduction of EMG activity (Vrbova, 1963; Krpati et al., 1972). However, the pattern of AChE molecular forms in innervated tenotomized regenerating SOL muscle was not significantly different from that in a control (Figure 5 A, B). Normal pattern of neural impulses is therefore not prerequisite to produce the typical SOL pattern of AChE molecular forms in the regenerating SOL.

We tested the 'intrinsic diferences' hypothesis also more directly by cross-transplantation of regenerating muscles. Typical AChE patterns of either homo- or cross-transplanted 6-week old regenerating muscles are presented in Figure 6. At such an early period of regeneration, the pattern of AChE molecular forms is primarily determined by the muscle of origin and not by the motor nerve which innervates the regenerating muscle. This visual impression is corroborated by quantitative evaluation of the ratios between the 13S and 10S AChE forms. They are 0.25 in EDL to EDL (n=6), 1.6 in SOL to EDL (n=6), 2.4 in SOL to SOL (n=6) and 0.5 in EDL to SOL (n=8) transplanted regenerating muscles. Values higher than 1 are typical for the SOL pattern, and less than 1 for the EDL pattern.

Since neither SOL nor EDL muscles of the rat are composed of homogenous slow or fast motor units (Adriano et al., 1973) it would be possible that regenerating SOL fibers originating from 'slow' satellite cells would be preferentially reinnervated by a few 'slow' motoneurons contributing to the EDL nerve, and vice versa. Fiber typing analysis that we performed, however, speaks against such possibility (results not shown). Most plausible explanation, therefore, is that, due to intrinsic differences between the satellite cells of fast and slow muscle cells, AChE regulating mechanisms in the muscle fibers arising from these cells respond differently to neural stimulation

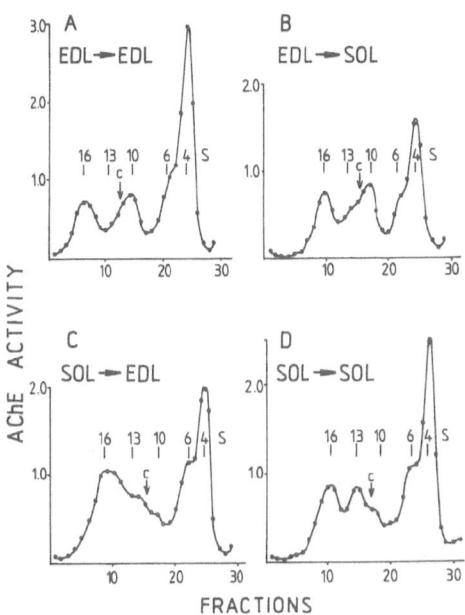

Figure 6. Velocity sedimentation analysis of AChE molecular forms in either homotransplanted (EDL->EDL, SOL->SOL) or cross-transplanted (EDL->SOL, SOL->EDL) regenerating SOL and EDL muscles six weeks after muscle injury. For details see the legend to the Figure 1.

regardless of its pattern. Preliminary results obtained in 13-week old regenerating muscles, however, indicate that later on the neural influence predominates and decisively modifies AChE regulation in mature regenerating muscles. These results, therefore, corroborate the hypothesis that during normal development of slow and fast muscles neural influences protect and amplify the AChE synthetic patterns somehow predetermined in the myoblast lineages contributing to these muscles during their ontogenetic development.

ACKNOWLEDGEMENTS

We thank Mr. Boris Pečenko for skillful technical assistance. The work was supported by a grant from Ministry of Science and Technology of Slovenia and by US-Yugoslav Joint Board DHHS grant JF 912.

REFERENCES

Ariano, M.A., Armstrong R.B., and Edgerton, V.R., 1973, Hindlimb muscle fiber populations of five mammals, J. Histochem. Cytochem. 21:51.

Bacou, F., Vigneron, P., and Massoulie, J., 1982, Acetylcholinesterase forms in fast and slow rabbit muscle, Nature 296:661.

Brzin, M., and Zajiček, J., 1958, Quantitative determination of acetylcholinesterase activity in individual endplates of normal and denervated gastrocnemius muscle, Nature 181:626.

Carlson, B.M., 1978, A review of musle transplantation in mammals, Physiol. Bohemoslov. 27:387.

Carlson, B.M., Hnik, P., Tuček, S., Vejsada, R., Bader, D.M., and Faulkner, J.A., 1981, Comparison between grafts with intact nerves and standard free grafts of the rat extensor digitorum longus muscle, Physiol. Bohemoslov. 30:505.

Črne, N., Sketelj, J., and Brzin, M., 1991, Influence of innervation on molecular forms of acetylcholinesterase in regenerating fast and slow skeletal muscles, J. Neurosci. Res. 28:315.

Decker, M.M., and Berman, H.A., 1990, Denervation-induced alterations of acetylcholinesterase in denervated and nondenervated muscle, Exp. Neurol. 109:247.

Dettbarn, W.-D., Groswald, D., Gupta, R.C., and Misulis, K.E., 1985, Use and disuse and control of acetylcholinesterase activity in fast and slow twich muscle of rat, in: "Molecular Basis of Nerve Activity," J.-P. Changeux, F. Hucho, A. Maelicke, and E. Neumann, eds., Walter de Gruyter, New York.

Diwan, F.H., and Milburn, A., 1986, The effects of temporary ischaemia on rat muscle spindles, J. Embryol. Exp. Morphol. 92:223.

Gisiger, V., and Stephens, H., 1982, Acetylcholinesterase content in both motor nerve and muscle is correlated with twich properties, Neurosci. Lett. 31:301.

Gregory, E.J., Hodges-Savola, C.A., and Fernandez, H.L., 1989, Selective increase of tetrameric (G_4) acetylcholinesterase activity in rat hindlimb skeletal muscle following short-term denervation, J. Neurochem. 53:1411.

Grove, B.K., 1989, Muscle differentiation and origin of muscle fiber diversity, CRC Crit. Rev.Neuro biol. 4:201.

Groswald, D.E., and Dettbarn, W.-D., 1983, Nerve crush induced changes in molecular forms of ace-tylcholinesterase in soleus and extensor digitorum muscles, Exp. Neurol. 79:519.

Guth, L., Albers, R.W., and Brown, W.C., 1964, Quantitative changes in cholinesterase activity of denervated muscle fibers sole plates, Exp. Neurol. 10:236.

Hall, Z.A., 1973, Multiple forms of acetylcholinesterase and their distribution in endplate and non-endplate regions of rat diaphragm muscle, J. Neurobiol. 4:343.

Karpati, G., Carpenter, S., and Eisen, A., 1972, Experimental core-like lesion and nemaline rods, Arch. Neurol. 27:237.

Koenig, J., and Rieger, F., 1981, Biochemical stability of AChE molecular forms after cytochemical staining: Postnatal focalization of the 16 S AChE in rat muscle, Dev. Neurosci. 4:249.

Lai, J., Jedrzejczyk, J., Pizzey, J.A., Green, D., and Barnard, E.A., 1986, Neural control of the forms of acetylcholinesterase in slow mammalian muscles, Nature 321:72.

Lomo, T., Gundersen, K., Hennig, R., and Westgaard, R., 1984, The role of impulse patterns in main-taining and regulating contractile properties in intact and chronically denervated and stimulated rat skeletal muscles, in: "Recent Achievement in Restorative Neurology: Upper Motor Neurone Functions and Disfunctions," J.C. Eccles, and M.R. Dimitrijevič, eds., A.G. Karger, Basel.

Lomo, T., Massoulie, J., and Vigny, M., 1985, Stimulation of denervated rat soleus muscle with fast and slow activity pattern induces different expression of acetylcholinesterase molecular forms, J. Neurosci. 5:1180.

Marnay, A., and Nachmansohn, D., 1938, Cholinesterase in voluntary muscle, J. Physiol (Lond). 92:37.

Miller, J.B., and Stockdale, F.E., 1987, What muscle cells know that nerves don't tell them, TINS 10:325.

Misulis, K.E, and Dettbarn, W.-D., 1985, Is fast fiber innervation responsible for increased acetylcho-linesterase activity in reinnervating soleus muscles?, Exp. Neurol. 89:204.

Navarette, R., and Vrbova, G., 1983, Changes of activity patterns in slow and fast muscles during postnatal development, Dev. Brain Res. 8:11.

Pette, D., and Vrbova, G., 1985, Invited review: Neural control of phenotypic expression in mammali-an muscle fibers, Muscle & Nerve 8:676.

Randall, W.R., Lai, J., and Barnard, E.A., 1985, Acetylcholinesterase of muscle and nerve, in: "Molecu-lar Basis of Nerve Activity," J.-P. Changeux, F. Hucho, A. Maelicke, and E. Neumann, eds., Walter de Gruyter, Berlin.

Sketelj, J., and Brzin, M., 1980, 16 S acetylcholinesterase in endplate-free regions of developing rat diaphragm, Neurochem. Res. 5:653.

Sketelj, J., Črne, N., and Brzin, M., 1987, Molecular forms and localization of acetylcholinesterase and nonspecific cholinesterase in regenerating skeletal muscles, Neurochem. Res. 12:159.

Sketelj, J., Črne-Finderle, N., Ribarič, S., abd Brzin, M., 1991, Interactions between intrinsic regula-tion and neural modulation of acetylcholinesterase in fast and slow skeletal muscles, Cell. Mol. Neurobiol. 11:35.

Sketelj, J., Črne-Finderle N., and Brzin, M., 1992, Influence of denervation forms of junctional and extrajunctional acetylcholinesterase in fast and slow muscles of the rat, Neurochem. Internat. (in press).

Thesleff, S., 1989, Botulinal neurotoxins as tools in studies of synaptic mechanisms, Quart. J. Exp. Physiol. 74:1003.

Vrbova, G., 1963, Changes in motor reflexes produced by tenotomy, J. Physiol. (Lond). 166:241.

COMPARTMENTALIZATION OF ACETYLCHOLINESTERASE mRNA AND PROTEIN EXPRESSION IN SKELETAL MUSCLE *IN VITRO* AND *IN VIVO*: IMPLICATIONS FOR REGULATION AT THE NEUROMUSCULAR JUNCTION

Richard L. Rotundo, Bernard J. Jasmin,
Richard K. Lee, and Susana G. Rossi

Department of Cell Biology and Anatomy
University of Miami School of Medicine
Miami, Florida 33101

INTRODUCTION

Skeletal muscle fibers are large multinucleated cells which arise from the fusion of up to several hundred mononucleated myoblasts. Because of their large size they can exhibit regionalized differences in the expression of several muscle-specific genes including those encoding synaptic components. Since normal development of cholinergically-innervated skeletal muscle fibers requires that all necessary synaptic components be expressed, assembled, and localized to the appropriate regions of the plasma membrane, this regionalization probably plays an important role in localizing molecules such as AChE to the sites of nerve-muscle contact.

Early studies on the distribution of "soluble" cytosolic proteins in skeletal muscle had suggested that their distribution was essentially random throughout the sarcoplasm and that their mRNAs were diffusible and translated at distances from the nucleus of origin (Frair and Peterson, 1983; Frair et al., 1979; Mintz and Baker, 1967). However, more recent studies from several laboratories have provided strong evidence that the expression of several muscle-specific genes is compartmentalized in multinucleated skeletal muscle fibers and that their mRNAs, as well as the translated proteins, tend to be localized around the nucleus of origin (Pavlath et al., 1989; Ralston and Hall, 1989;

Rotundo, 1990). Localization of acetylcholine receptor (AChR) α-subunit mRNA by *in situ* hybridization in tissue-cultured myotubes shows that only a subset of nuclei express this transcript and that it remains in the proximity of the nuclei expressing it (Harris et al., 1989; Horvitz et al., 1989).

The accumulation of AChR α-subunit mRNA in innervated regions of skeletal muscle fibers was first demonstrated by Merlie and Sanes (1985) in mouse diaphragm. Subsequent studies using a more precise *in situ* hybridization approach have shown that the AChR transcripts are concentrated at the vertebrate neuromuscular junction (Fontaine and Changeux, 1989; Fontaine et al., 1988; Goldman and Staple, 1989;

Multidisciplinary Approaches to Cholinesterase Functions, Edited by
A. Shafferman and B. Velan, Plenum Press, New York, 1992

Brenner et al., 1990) and that the expression of the epsilon subunit mRNA occurs exclusively in the junctional region of the fiber (Brenner, et al., 1990). Thus not only can a strong case be made for the compartmentalization of gene expression in skeletal muscle, evidence is beginning to accumulate indicating that the regulation of the genes encoding these proteins is also compartmentalized.

Expression of AChE is also compartmentalized in skeletal muscle fibers, both in tissue-cultured cells and *in vivo*. We have previously shown that skeletal muscle fibers are compartmentalized with respect to the expression and translation of AChE mRNAs, and that the polypeptide chains are assembled in the vicinity of the nucleus of origin (Rotundo, 1990). Using tissue-cultured mosaic quail-mouse skeletal muscle fibers and species-specific antibodies, we now show that AChE oligomeric forms expressed within a nuclear compartment are localized to regions of the cell surface overlying the nucleus of origin, thus giving rise to specialized cell surface nuclear domains. In vivo, AChE is highly concentrated at the neuromuscular junction and this specialized region of the plasma membrane overlies a characteristic accumulation of sub-synaptic or "fundamental" myotube nuclei. Using a quantitative mRNA PCR technique to quantitate AChE transcript levels in single cells we demonstrate that innervated regions of individual muscle fibers exhibit a large increase in the amount of AChE mRNA compared to adjacent nerve-free segments. Our results, together with those from other laboratories, suggest that regulation of AChE at the neuromuscular junction is spatially as well as temporally controlled and involves both neural input and muscle activity.

Biogenesis and Sorting of AChE In Skeletal Muscle

All oligomeric forms of AChE are exportable glycoproteins translated on the rough endoplasmic reticulum (RER). The polypeptide chains are co-translationally glycosylated and the globular forms assembled in the RER. The catalytic subunits appear to be synthesized as inactive precursors; only a subset are assembled into catalytically active oligomers and transported through the Golgi apparatus and to the cell surface. The majority of the polypeptides (70-80%) are rapidly degraded shortly after synthesis via a non-lysosomal mechanism. This non-lysosomal pathway is not well understood in any cell type, nor are the mechanisms responsible for targeting the proteins for degradation. In the case of AChE, the polypeptide chains destined for rapid degradation appear to enter the earliest (cis) elements of the Golgi apparatus, as defined by acquisition of N-acetylglucosaminyl residues on their N-linked oligosaccharides, before being sorted to the site of degradation (proteolysis).

The catalytically active AChE dimeric and tetrameric forms continue transit through the Golgi where their high mannose oligosaccharides are processed to more complex forms prior to externalization at the plasma membrane, either as cell surface-associated molecules or secreted into the extracellular milieu. During transit through the trans Golgi a subset of catalytic subunits become attached to the three-stranded collagen-like tail shortly before transport to the muscle cell surface where they become associated with the extracellular matrix. However, in skeletal muscle cultures the collagen-tailed forms do not appear as soluble molecules in the medium, unlike the dimeric and tetrameric AChE forms. This observation suggests a very efficient retention mechanism for the asymmetric form once released at cell surface. How this retention mechanism is established over a particular region of the muscle fiber is currently unknown.

Clusters of Cell Surface AChE Are Localized Primarily Over the Myotube Nuclei

Extracellular acetylcholinesterase molecules are not homogeneously distributed over the surface of skeletal muscle fibers *in vivo* nor over the myotubes in culture. *In vivo*, AChE is concentrated at the neuromuscular junction where the density of catalytic subunits is about ten-fold higher than the adjacent non-innervated regions of the fiber (Salpeter, 1967). The cell surface distribution of AChE has been studied using monoclonal α-AChE antibodies and indirect immunofluorescence. Following differentiation and onset of spontaneous contraction in culture, AChE appears as discreet patches of enzyme on the upper surface of the myotubes and has a more diffuse punctate distribution on the undersurface where the myotube is in contact with the substratum. Analysis of the distribution of AChE clusters on the upper surface of the myotubes shows that most of the clusters (>95%) are localized within a 30μm radius of a nucleus, whereas the average distance between nuclei is approximately 65 μm (Rossi and Rotundo, 1991; and 1992, in preparation). These observations suggested that each nucleus is capable of organizing the AChE molecules exclusively on the region of membrane above it.

Skeletal Muscle Fibers Are Functionally Compartementalized With Respect to the Transcription and Translation of AChE

To determine whether AChE mRNAs encoded by a given nucleus were capable of diffusing within the myofiber, or were restricted to a given region we took advantage of the fact that quails express two allelic variants of the AChE polypeptide chain, α and ß, which differ by an apparent 10 kDa on SDS gels (Rotundo, et.al., 1988). Myoblasts can be isolated from quail embryos which are heterozygous or homozygous for either of the two allelic variants. Moreover, these two different subunits can be assembled randomly in the RER and are physiologically as well as immunologically indistinguishable (Rotundo, 1990). When isotopically-labeled AChE is extracted and immunoprecipitated from heterozygous myotubes and analyzed on non-reducing SDS gels the distribution of disulfide-bonded AChE dimers was 25% α/α, 50% α/ß, and 25% ß/ß indicating random assembly in the RER. In contrast, when mosaic myotubes were made by combining equal numbers of homozygous α/α and ß/ß myoblasts the resulting dimeric AChE forms were almost entirely homodimeric. Thus translation and assembly of the AChE molecules must have occurred on the RER surrounding the nucleus encoding that particular transcript.

These experiments clearly showed that muscle cells are functionally divided into compartments, encompassing the region surrounding a given nucleus, in contrast to earlier notions that muscle mRNAs were free to diffuse and be translated throughout the cell. Furthermore, these studies provided the physical basis for postulating that locally transcribed and translated membrane proteins would be selectively transported and localized to specialized regions of the cell surface overlying the nucleus of origin, such as occurs at the neuromuscular junction.

Muscle Fiber Nuclei Expressing AChE Localize This Synaptic Component On the Overlying Region of the Membrane

To directly test the hypothesis that surface membrane proteins encoded by a given nucleus are selectively localized to the overlying regions of the plasma membrane we made mosaic muscle fibers by co-culturing quail and mouse myoblasts such that the quail nuclei comprised a minority of the myotube nuclei (Rossi and Rotundo, 1991, and in preparation). Using a species-specific antibody to localized surface AChE clusters we can now show that quail AChE clusters formed only over the quail nuclei, whereas AChE enzyme from both species was distributed throughout the muscle fiber. Thus a very selective and specific mechanism exists for localizing synaptic components to the cell surface overlying the nucleus of origin. This mechanism begins with localized transcription, translation, and assembly of the synaptic components and ends with the selective retention of these molecules on the overlying membrane.

Regulation of AChE Biogenesis in Tissue-Cultured Muscle

Although considerable information has accumulated regarding early events in the biogenesis of AChE isoforms, there is still much more that needs to be learned before we can begin to understand their regulation. What is clear from many studies both *in vivo* and in tissue-cultured cells (reviewed in Toutant and Massoulié, 1987; Rotundo, 1987) is that regulation of AChE is complex and involves a multitude of events leading up to the final pattern and relative abundance of the individual oligomeric forms expressed in a given tissue. A comprehensive discussion of these events is beyond the scope of this chapter and hence we will focus only on material presented at this meeting.

Myoblasts in tissue culture proliferate and fuse into multinucleated skeletal muscle myotubes over a period of about 3-5 days. During this period they express only the globular forms of AChE. Following differentiation between 3-5 days in culture the myotubes become spontaneously contractile at which time the collagen-tailed asymmetric form appears. The appearance of this form requires depolarization of the plasma membrane associated with spontaneous contraction and is inhibited by agents that block voltage-dependent sodium channels such as tetrodotoxin (TTX) (Rieger et al., 1980). In quail muscle cultures inhibition of spontaneous contraction by TTX results in a diminution of total AChE activity and loss of the asymmetric AChE form, yet has no effect on the rates of AChE translation, assembly, activation, or intracellular degradation (Fernandez-Valle and Rotundo, 1988; 1989). Only the rate of AChE secretion increased by 100% and could quantitatively account for the decrease in cell-associated enzyme in TTX-treated cells. These studies imply that regulation of asymmetric AChE expression can occur at the level of oligomer assembly and does not necessarily involve transcriptional or translational controls. However, much remains to be done before we understand the complex regulatory events responsible for determining the expression and levels of AChE present at the neuromuscular junction.

Organization of Skeletal Muscle Nuclei *In Vitro*

Junctional nuclei, nuclei associated with the subsynaptic sarcoplasm in the region of the neuromuscular junction, were first described last century by the histologists Khune and Ranvier and documented through their detailed camera lucida drawings. However, because of the size of the junctional region and the presence of many other cells (Schwann cells, satellite cells, and fibroblasts to name a few) it has not been possible to accurately study their distribution.

During the past year we developed a method for cleanly removing all associated cells from individual muscle fibers and staining myofiber nuclei with a fluorescent DNA intercalating dye while at the same time visualizing the neuromuscular junction with fluorescently-labeled α-bungarotoxin. Under these conditions the junctional nuclei are easily distinguished by their morphology and distribution from the extrajunctional nuclei. Quantitation of these nuclei shows that the average neuromuscular junction of Quail fast twitch skeletal muscle fibers in the 40-70 um diameter range contains from 12 to 24 junctional nuclei lying directly beneath the region of high ACh receptor density. It is this very same region which also expresses at least an order of magnitude higher levels of AChE mRNA.

Adult Skeletal Muscle Fibers Express Higher Levels of AChE mRNA at the Neuromuscular Junction: Evidence for Compartmentalization of Gene Expression *In Vivo*

To determine whether increased local expression of AChE mRNA could account at least in part for the higher levels of AChE at the neuromuscular junction we have studied individual muscle fibers dissected from the singly-innervated quail PLD muscle (Jasmin, et al., 1991; Lee et al., 1991; Jasmin et al., in preparation). Muscle fibers were first stained by the Karnovsky-Roots method to visualize AChE at the neuromuscular junction. Single muscle fibers were teased apart and micro-dissected into junctional, proximal extrajunctional, and distal extrajunctional segments and extracted individually. The mRNA was reverse transcribed into cDNA which in turn was amplified by PCR along with the appropriate standards. Our results show that innervated regions of muscle fibers have a much higher level of AChE mRNA expression, more than an order of magnitude, than the distal extrajunctional segments. Using known copy number internal standards for comparison, we can show that AChE mRNA transcripts at the neuromuscular junction constitute an intermediate abundance mRNA, compared to almost undetectable levels in extrajunctional regions. It should be emphasized that AChE is now only the second molecule for which increased levels of specific mRNA can be demonstrated at the neuromuscular junction. The first and only demonstration was of increased acetylcholine receptor mRNA at the neuromuscular junction by four laboratories (Merlie and Sanes, 1985; Fontaine et al., 1988; Fontaine et al., 1989; Goldman and Staple, 1989; Brenner, Witzmann, and Sakmann, 1990).

SUMMARY AND CONCLUSIONS

In summary, we have been able to show that mRNAs encoding cell surface proteins such as AChE in skeletal muscle fibers are preferentially translated and assembled in the organelles surrounding the nucleus of transcription, and that once assembled these molecules are transported and preferentially localized to regions of the cell surface overlying the nucleus of transcription. *In vivo*, accumulations of AChE mRNA occur in the innervated region of the muscle fiber indicating preferential expression of the AChE gene in this domain. Our working hypothesis is that specific signals from the overlying region of the membrane are responsible for maintaining active transcription from the junctional nuclei which in turn encode the very proteins responsible for maintaining the structural and functional integrity of the neuromuscular synapse. Concurrently, spontaneous contraction generates signals along the entire length of the fiber that would be responsible for the down-regulation or repression of AChE transcription in the non-innervated portions of the fiber, without which little or no translation of AChE would occur.

ACKNOWLEDGEMENTS

We would like to acknowledge the expert technical assistance of Anna Gomez and Francesca Barton in some of these studies. This research was supported by grants from the National Institutes of Health and the Muscular Dystrophy Association of America to RLR.

REFERENCES

Brenner, H.R., Witzemann, V. and Sakmann, B., (1990), Imprinting of Acetylcholine Receptor mRNA Accumulation in Mammalian Neuromuscular Synapses. Nature, Lond. 344: 544.

Fernandez-Valle, C. and Rotundo, R.L., (1989), Regulation of Acetylcholinesterase Synthesis and Assembly by Muscle Activity: Effects of Tetrodotoxin *J. Biol.Chem.* 264: 14043.

Fernandez-Valle, C. and Rotundo, R. L., (1988), Phorbol Esters Stimulate Appearance of Asymmetric Acetylcholinesterase in TTX-Treated Quail Muscle Cultures. Sc. Neurosci. Abstr. vol 14: 1163.

Fernandez-Valle, C. and Rotundo, R.L., (1992), Regulation of Asymmetric Acetylcholinesterase: Phorbol Esters Mimic the Effects of Muscle Activity in Tissue-Cultured Muscle. (in preparation).

Fontaine, B. and Changeux, J.-P., (1989), Localization of Nicotinic Acetylcholine Receptor Alpha-Subunit Transcripts during Myogenesis and Motor Endplate Development in the Chick. J.Cell Biol. 108:1025.

Fontaine, B., Sassoon, D., Buckingham, M. and Changeux, J-P. (1988), Detection of the Nicotinic Acetylcholine Receptor Alpha-subunit mRNA By In Situ Hybridization at Neuromuscular Junctions of 15-day-old Chick Striated Muscles. The EMBO Journal 7:603.

Frair, P.M. and Peterson, A.C., (1983), The Nuclear-Cytoplasmic Relationship in 'Mosaic' Skeletal Muscle Fibers from Mouse Chimaeras. Experimental Cell Research 145:167.

Frair, P.M., Strasberg, P.M., Freeman, K.B. and Peterson, A.C., (1979), Mitochondrial Malic Enzyme in Mosaic Skeletal Muscle of Mouse Chimeras. Biochemical Genetics 17:693.

Goldman, D. and Staple, J., (1989), Spatial and Temporal Expression of Acetylcholine Receptor RNAs in Innervated and Denervated Rat Soleus Muscle. Neuron 3:219.

Harris, D.A., Falls, D.L. and Fischbach, G.D., (1989), Differential Activation of Myotube Nuclei Following Exposure to an Acetylcholine Receptor-inducing factor. Nature 337:173.

Horovitz, O., Knaack, D., Podleski, T.R. and Salpeter, M.M., (1989), Acetylcholine Receptor Alpha-Subunit mRNA Is Increased by Ascorbic Acid in Cloned L5 Muscle Cells: Northern Blot Analysis and In Situ Hybridization. J.Cell Biol. 108:1823.

Jasmin, B.J., Lee, R.K., and Rotundo, R.L., (1991), Analysis of Acetylcholinesterase Gene Expression at the Avian Neuromuscular Junction by Quantitative PCR. J. Cell Biology 115:30a.

Jasmin, B.J., Lee, R.K., and Rotundo, R.L., Quantitation of Acetylcholinesterase mRNA Copy Number Along Single Muscle Fibers: Compartmentalization at the Neuromuscular Junction. (in preparation).

Lee, R.K., Jasmin, B.J., and Rotundo, R.L., (1991), Selective Acetylcholinesterase mRNA Accumulation at the Avian Neuromuscular Junction: Relationship to Nuclear Domains. Soc. Neurosci. Abstr. vol. 17:946.

Merlie, J.P. and Sanes, J.R., (1985), Concentration of Acetylcholine Receptor mRNA in Synaptic Regions of Adult Muscle Fibres. Nature 317:66.

Mintz, B. and Baker, W.W., (1967), Normal Mammalian Muscle Differentiation and Gene Control of Isocitrate Dehydrogenase Synthesis. Proc. Natl. Acad. Sci.USA 58:592.

Pavlath, G.K., Rich, K., Webster, S.G. and Blau, H.M., (1989), Localization of Muscle Gene Products in Nuclear Domains. Nature 337:570.

Ralston, E. and Hall, Z.W., (1989), Transfer of a Protein Encoded by a Single Nucleus to Nearby Nuclei in Multinucleated Myotubes. Science 244:1066.

Rieger, F., Koenig, J., and Vigny, M., (1980), Spontaneous contractile activity and the presence of the 16S AChE form in rat muscle cells in culture: Reversible suppressive action of tetrodotoxin. Dev. Biol. 76: 358-365.

Rossi, S.G. and Rotundo, R.L, (1991), Cell Surface Acetylcholinesterase on Quail/Mouse Mosaic Muscle Fibers is Preferentially Localized Over the Quail Nuclei. J. Cell Biology 115: 30a.

Rossi, S.G. and Rotundo, R.L., Skeletal Muscle Acetylcholinesterase is Selectively Targeted and Localized to Clusters on the Region of Membrane Overlying the Nucleus of Origin. (1992, in preparation).

Rotundo, R.L., (1990), Nucleus-specific Translation and Assembly of Acetylcholinesterase in Multinucleated Muscle Cells. J. Cell Biol. 110: 715.

Rotundo, R.L., (1987), Biogenesis and Regulation of Acetylcholinesterase. In: *The Vertebrate Neuromuscular Junction,* edited by M.M. Salpeter. Alan R. Liss, New York.

Rotundo, R.L., Fernandez-Valle, C., Gomez, A., Barton, F., and Leff, P., (1991), Regulation of Acetylcholinesterase Expression in Electrically Excitable Cells, In: Cholinesterases, Structure, Function, Mechanism, Genetics, and Cell Biology. J. Massoulié, et al.(Eds) ACS Publications.

Rotundo, R.L., Gomez, A.M., Fernandez-Valle, C. and Randall, W.R., (1988), Allelic Variants of Acetylcholinesterase: Genetic Evidence that All Acetylcholinesterase Forms in Avian Nerves and Muscles Are Encoded by A Single Gene, 1988, Proc. Natl. Acad. Sci. USA 85:7805.

Salpeter, M.M., (1967), Electron Microscope Radioautography as a Quantitative Tool in Enzyme Cytochemistry. I. The Distribution of Acetylcholinesterase at Motor Endplates of a Vertebrate Twitch Muscle. J. Cell Biology 32: 379-389.

Toutant, J-P. and Massoulié, J., (1987), Acetylcholinesterase. In: Mammalian Ectoenzymes, edited by A. Kenny and A.S. Turner. Elsevier Science Publishers, Amsterdam.

TOWARDS A FUNCTIONAL ANALYSIS OF CHOLINESTERASES IN NEURO-GENESIS: HISTOLOGICAL, MOLECULAR, AND REGULATORY FEATURES OF BCHE FROM CHICKEN BRAIN

Paul G. Layer

Technische Hochschule Darmstadt
Zoologisches Institut
Schnittspahnstrasse 3
6100 Darmstadt, Germany

INTRODUCTION

A Neurogenetic Period of Sequential Cholinesterase Expression

Along with the differentiation of synapses, the enzyme acetylcholinesterase (AChE) and to some extent butyrylcholinesterase (BChE) appear as part of the cholinergic apparatus. As an example, both AChE and BChE are found in particular synaptic sublaminae of the so-called inner plexiform layer of the embryonic chicken retina. It is noteworthy, that not in all subbands BChE and AChE are colocalized, but there is at least one subband that only presents BChE (Shen et al., 1956; Layer, 1983). The function of AChE at the adult synapse is to terminate the signal of the neurotransmitter ACh. Whether BChE at synapses has a similar role is unclear. However, there is a much earlier period of cholinesterase expression which correlates with the time when mitotic neuroblasts of the early neuroepithelium leave the proliferative cycle and start to differentiate (Fig. 1).

In all parts of the early brain, a distinct mantle layer of AChE-positive cells will develop. By using biochemical techniques, but also double-labelling of sectioned material, we have correlated the time of expression of both types of cholinesterases with the end of cell division. If the survival time after a given radioactive thymidine label is short, AChE-positive and mitotic cells are cleanly separated into an outer mantle and a ventricular mitotic zone, respectively. If the lag time after the radioactive pulse is longer, then thymidine labelled cells are found within the outer mantle, thereby already expressing AChE, e.g. they are double-labelled. Similar measurements are possible by homogenizing whole brain parts after their thymidine pulsing and relate it to their quantitative expression of AChE. These studies led us and others to the conclusion that AChE is a postmitotic neuronal marker that is reliably expressed about 10 hours after the last cell division (Layer, 1990). This is the time when cells have reached the outer mantle surface of the neuroepithelium.

Multidisciplinary Approaches to Cholinesterase Functions, Edited by
A. Shafferman and B. Velan, Plenum Press, New York, 1992

223

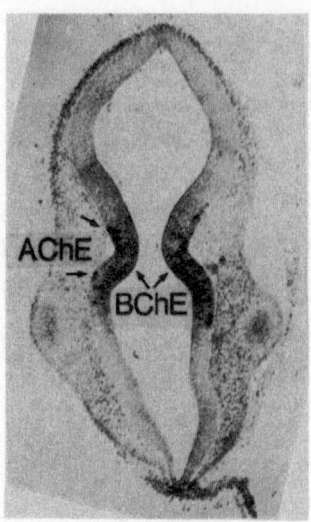

Figure 1. Cholinesterases in the early neural tube of a chicken embryo (head at stage HH14 is shown). BChE labels areas of final mitotic activity on the ventricular side of the neural tube. Located within the BChE-positive field, AChE producing postmitotic cells have reached the outer surface. Temporally, BChE precedes AChE in the tissue; it regulates AChE expression in vitro (see text).

This fact allows to reconstruct neuronal differentiation patterns by following the appearance of AChE-producing cells over the entire early brain surface. As the very start of brain differentiation, we located isolated cells at at least three distinct spots of the neural tube that express high levels of AChE. Thus following a polycentric scheme, expression starts almost simultaneously at the di-mesencephalic border and in a precise sequence of rhombomeres of the hindbrain (Puelles et al., 1987; Layer et al., 1988a; Layer and Alber, 1990). A correlation between AChE expression and expression of certain segmentation genes in the hindbrain seems indicated.

The process of subdivision of the neuroepithelium into an inner mitotic cell mass and an outer AChE-positive mantle seems to represent a general step of primary lamination. For in-vitro-cultured retinae (retinospheroids, see below, Fig. 2), we have shown recently that along the border between these two cell masses, a primary and transient fibrous network is established that corresponds to the socalled transient fiber layer of Chievitz (Willbold and Layer, 1992a). We are presently investigating whether the expression of AChE may function as guiding posts for outgrowing neurites (see below) and possibly as a repulsive signal for the immigration of mitotic cells.

By applying a double-labelling technique that delineates AChE-positive cells and neurites with an antibody to the fasciculation antigen G4 (Rathjen

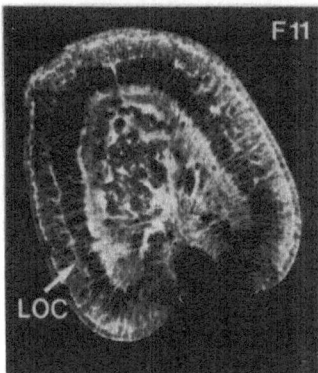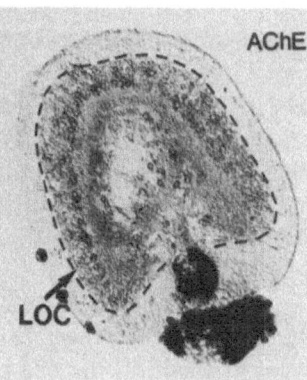

Figure 2. During neurogenesis, AChE-positive cells are strictly separated from mitotic cells. In *retinospheroids*, the two cell compartments are separated by a transient fiber layer of Chievitz (LOC). *Retinospheroids* are raised from dissociated cells of the chicken eye margin (Willbold and Layer, 1992b). (Left) staining of fibrous material by the F11 antibody, (right) AChE-positive inner zone; dashed line indicates LOC.

et al., 1987), we then learned that these cells soon after the onset of AChE expression will send out their axons (Fig. 3). In general, these are primary, long projecting systems of the CNS (Weikert et al., 1990), but also include peripheral systems like the motoraxons of the trunk (Layer et al., 1988b; Fig. 3) or cranial nerves (Layer and Kaulich, 1991).

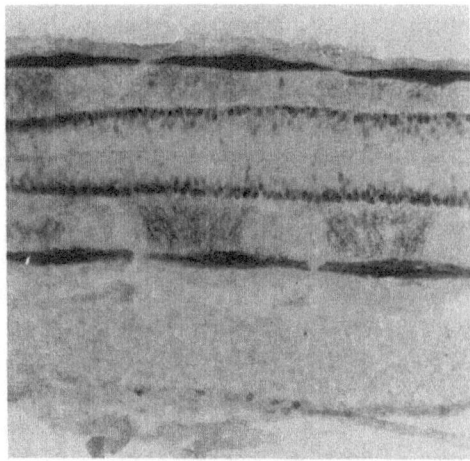

Figure 3. Early neurite tracts originate from, and (at least often) project to AChE-producing cells. Section shows segments of a chicken trunk at stage HH18. G4-stained motoraxons from AChE-positive motoneurones cross the rostral half of the sclerotome to reach their AChE-positive myotome target.

Retinospheroids Cultured From Cells of the Eye Margin Reveal
That BChE Is a Transmitotic Marker of Stem Cells in CNS

In areas that surround the newly appearing AChE-producing cells, I have detected a transient expression of BChE, typically represented by somewhat diffuse and peculiar tissue patches (Layer, 1983; Layer and Kaulich, 1991). Quantitative data showed, that BChE is elevated transiently around the time when cells leave the mitotic cycle (Vollmer and Layer, 1986; Layer and Sporns, 1987).

Normally, the levels of BChE expressed are low and transient. Therefore, they are not easily detectable. However, a special case of high and more permanent BChE expression is found at the eye periphery. In contrast to the patches in the brain, the BChE activity remains high there for a long time. Like in other parts of the brain, we find AChE-positive cells in its close vicinity in specific layers of the developping retina (Willbold and Layer, 1992b).

In the peripheral part of the eye resides a cell population that exhibits a regenerative capacity. In vivo, these cells are in a quiet proliferative state. If we take that tissue, dissociate it into single cells and take the cells into rotary culture, then these cells will start to proliferate heavily and finally reorganize into little spheres (*retinospheroids*). These spheroids resemble closely a normal retina, in particular they are fully laminated (Layer and Willbold, 1989). Moreover, different cell types are derived from multipotent precursors (Layer et al., 1990). The shape of the proliferation curve indicates strongly that the occuring proliferation is not a mere succession of what would have happened in vivo, but must include pronounced cell self renewal, e.g. stem cell activation takes place in the culture dish. At the same time, the BChE activity is down-regulated, and AChE is up-regulated (Willbold and Layer, 1992b).

This study indicates that here we may have an example where the normal transition of stem cells has been blocked, with a simultaneous block of their change from BChE to AChE production. I therefore suggest that cells with high BChE levels are involved in the regeneration process of the retina. These observations are highly reminiscent with increased BChE levels in certain tumor cells (Zakut et al., Lockridge, this Volume), but also in glia cells and in particular in Schwann cells (Dubovy and Haninec, 1990).

These developmental data indicate that cholinesterases may be linked a) with each other, b) with neuronal cell division, c) with neuronal differentiation and d) with axonal outgrowth. Retinospheroids are excellent in vitro systems that allow to test whether inhibition of cholinesterases will affect any of the histogenetic processes within spheroids.

BChE Activity Regulates AChE Expression *In Vitro*

The spatiotemporal dualism of cholinesterase expression is exciting, since it suggests that the expression of both cholinesterases must be somehow linked. Using retinospheroids, we found that inhibition of BChE activity by the addition of iso-OMPA or Ethopropazine to the culture medium leads to a decreased expression of AChE within the growing spheroids. Even more pronounced is the effect on AChE that is released into the media. More clearly, if the BChE that is supplied with the serum is irreversibly blocked by DFP or iso-OMPA, the same effect is seen. Alternatively, in cultures that are raised in artificial N2-medium (that is free of BChE), AChE expression is lower. Thus BChE can regulate AChE expression. If we apply this result to the developmental observations (AChE-positive cells differentiate within BChE-positive patches, see above and Fig. 1), then this could be one of the important functions of BChE that BChE could regulate the degree and refine the areas of neuronal differentiation by regulating AChE activity. It also indicates that the phylogenetic older enzyme can ontogenetically control the more specific younger enzyme AChE. The molecular mechanism of this regulation remains to be elucidated (Layer et al., 1992). Therefore, for understanding the functioning of cholinesterases during early neurogenesis, BChE may be of predominant interest.

A Secondary Function of Cholinesterases Regulates Neurite Growth *In Vitro,* However, Not Via Their Enzymatic Activity

Histologically, BChE appears not only in patches of the CNS, but rather is strongly expressed in areas of the neural periphery through which axons will be growing. Two examples: BChE precedes and then closely surrounds motoraxons that grow through the rostral sclerotome from their AChE-positive motorneurons to their AChE-positive muscle targets (Layer et al., 1988b; Fig. 3). Similarly, cranial nerves in the head are outlined and announced by high levels of BChE (Layer and Kaulich, 1991). We therefore were interested whether the cholinesterases could have a function in neurite guidance.

Neurite guidance can be conveyed by a number of cell adhesion molecules (CAMs), including the IgG-like CAMs and the Cadherins (Takeichi, 1988). In particular, fasciculation molecules such as L1, G4 and TAG-1 (Rathjen, 1991; Jessel, 1988) are crucial for the proper outgrowth of neurites and for keeping them together within organized fiber bundles. However, our knowledge about CAMs is incomplete, as more CAMs are still anticipated. A widespread feature of CAMs (Kruse et al., 1984; Tucker et al., 1988) is the HNK-1 sugar epitope (Abo and Balch, 1981) that may be functionally significant (Cole and Schachner, 1987). Noticeably, AChE from *Torpedo* (Bon et al., 1987) and BChE from chicken serum and brain (Treskatis et al., 1992, unpublished data) bear the HNK-1 epitope. Moreover, neurotactin and glutactin are cell adhesion molecules from *Drosophila* sharing sequence homologies with cholinesterases (Barthalay et al., 1990; De la Escalera et al., 1990; Krejci et al., 1991). All these data indicate that cholinesterases may function as cell adhesion molecules, e.g. by regulating neurite growth. Can both AChE and BChE perform such a function? If so, is the active center directly involved, or are there secondary, noncholinergic sites on these molecules?

We have provided the first direct evidence showing that neurite growth in vitro from various neuronal tissues of the chick embryo can be modified by some, but not all anticholinesterase agents. By measuring the neuritic material produced in tectal cells from individual microwells with an ELISA assay that detects the neurite-specific G4 antigen (Rathjen et al., 1987; see Fig. 4), the effect of anticholinesterases on neurite growth is directly compared with their cholinesterase inhibitory action. BW 284C51 and Ethopropazine, inhibiting AChE and BChE, respectively, strongly decrease neurite growth in a dose-dependent manner. However, Echothiophate that inhibits both cholinesterases, does not change neuritic growth. Moreover, morphological changes, such as defasciculation, and disorientation of neurite bundles are induced by iso-OMPA, BW 284C51 and Bambuterol in retinal explants grown on striped laminin carpets (Walter et al., 1987).

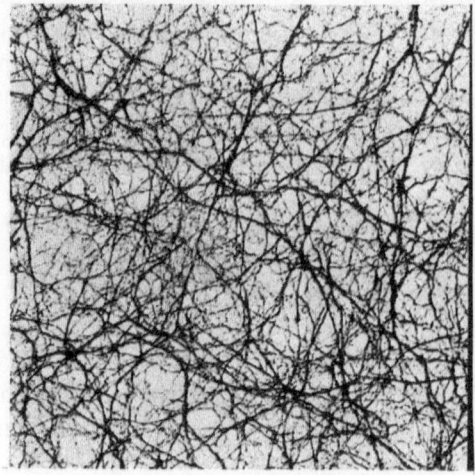

Figure 4. Neurite growth of dissociated tectal cells grown in microplates is detected by the G4 antibody (Rathjen et al., 1987). After homogenization of the tissue, an ELISA for the G4 antigen gives a measure for the extent of neurite growth on the plate. Effects of various organophosphates on neurite growth can thus be quantitatively determined (see text).

Specific Forms of Cholinesterases, Their Regulation of Glycosylation and Their Membrane Association Could Contribute to Their Functions

Histologically, we found a wide overlap of AChE and HNK-1 expression during establishment of cranial nerves. Noticeably, the HNK-1 epitope became most strongly expressed only in the target areas of the growing axons (Layer and Kaulich, 1991). Thus, a developmental regulation of cholinesterase glycosylation, but also the complexity of cholinesterase molecules (Massoulié and Bon, 1982), and their various modes of cellular localization may turn out to be functionally significant. In general, the small forms of both types of cholinesterases prevail during the early period of neurogenesis. Interestingly in the brain, the neurogenetic and the synaptogenetic periods are accompanied by an interconversion of G2 to G4 forms of AChE with a stable ratio of BChE forms, whereas in the retina the ratio of G2/G4-AChE is almost stable, but BChE quickly changes from G1 to G4-forms (Layer et al., 1987).

Applying a new 4-step isolation procedure, we have purified butyrylcholinesterase from chicken serum to homogeneity with more than 250 Units/mg specific activity. The serum enzyme was used for producing monoclonal antibodies. These BChE-specific antibodies also recognize BChE from brain, and thus enabled us to isolate the enzymes from embryonic and adult brain, that occur only in minute amounts. Their catalytic and inhibition properties are similar to those from serum.

However on SDS-Page, the serum enzyme is represented by a double-band of 79/82 kDa, while the brain enzyme has a size of 74 kDa. Limited digestion by V8-protease of the serum and brain preparations leads to similar peptide patterns. Enzymatic deglycosylation shows that their core proteins consist of 59 kDa subunits and that the different molecular weights are due to different glycosylation patterns. Furthermore, the membrane-bound brain BChE can be solubilized by pronase or protease K, but not by phosphoinositol-specific phospholipase C (PIPLC). The differently sized glycoparts of brain and serum BChE may indicate that they subserve different functions (Treskatis et al., 1992). The functional significance of these observations remains open.

CONCLUSIONS

- BChE is elevated in stem cells of the neuroepithelium at a labile state of final proliferation.
- BChE can regulate and refine patterns of cellular AChE expression, that in turn is an early sign of postmitotic neuronal differentiation.
- Both cholinesterases can regulate axonal growth, most likely by adhesive mechanisms
- This secondary function is not due to the enzyme activity per se, but rather, due to a secondary site(s) on the cholinesterase molecule.

ACKNOWLEDGEMENTS

I wish to thank my colleagues R. Alber, F. Bonhoeffer, C. Ebert, S. Kaulich, F. Rathjen, S. Rommel, S. Treskatis, G. Vollmer, T. Weikert, E. Willbold for their help and discussions. I wish to apologize that the list of references does not include all the relevant work of many other scientists. I refer to a recent issue of *Mol. & Cell. Neurobiol.* (1991, Vol. 11) that contains reviews on all topics of cholinesterase research.

REFERENCES

Abo, T. and Balch, C., 1981, A differentiation antigen of human NK and K cells identified by a monoclonal antibody (HNK-1). *J Immunol* 127: 1024-1029.

Barthalay, Y., Hipeau-Jacquotte, R., de la Escalera, S., Jimenez, F. and Piovant, M., 1990, *Drosophila* neurotactin mediates heterophilic cell adhesion. *EMBO J* 9: 3603-3609.

Bon, S., Méflah, K., Musset, F., Grassi, J. and Massoulié, J., 1987, An immunoglobulin M monoclonal antibody, recognizing a subset of acetycholinesterase molecules from electric organ of Electrophorus and Torpedo, belongs to the HNK-1 anti-carbohydrat family. *J Neurochem* 49: 1720-1731.

Cole, G.J. and Schachner, M., 1987, Localization of the L2 monoclonal antibody binding site on chicken neural cell adhesion molecule NCAM) and evidence for its role in NCAM-mediated cell adhesion. *Neurosci Letts* 78, 227-232: .

De la Escalera, S., Bockamp, E.O., Moya, F., Piovant, M. and Jiménez, F., 1990, Characterization and gene cloning of neurotactin, a *Drosophila* transmembrane protein related to cholinesterases. *EMBO J* 9: 3593-3601.

Dubovy, P. and Haninec, P., 1990, Non-specific cholinesterase activity of the developing peripheral nerves and its possible function in cells in intimate contact with growing axons of chick embryo. *Int J Dev Neurosci* 8: 589-602.

Jessel, T.M., 1988, Adhesion molecules and the hierarchy of neural development. *Neuron* 1: 3-13.

Krejci, E., Duval, N., Chatonnet, A., Vincens, P. and Massoulié, J., 1991, Cholinesterase-like domains in enzymes and structural proteins: functional and evolutionary relationships and identification of a catalytically essential aspartic acid. *Proc Natl Acad Sci (USA)* 88: 6647-6651.

Kruse, J., Mailhammer, R., Wernecke, H., Faissner, A., Sommer, I., Goridis, C. and Schachner, M., 1984, Neural cell adhesion molecules and myelin-associated glycoprotein share a common moiety recognized by monoclonal antibodies to L2 and HNK-1. *Nature* 311: 153-155.

Layer, P.G., 1983, Comparative localization of acetylcholinesterase and pseudocholinesterase during morphogenesis of the chick brain. *Proc Natl Acad Sci (USA)* 80: 6413-6417.

Layer, P.G. and Sporns, O., 1987, Spatiotemporal relationship of embryonic cholinesterases with cell proliferation in chicken brain and eye. *Proc Natl Acad Sci (USA)* 84: 284-288.

Layer, P.G., Alber, R. and Sporns, O., 1987, Quantitative development and molecular forms of acetyl- and butyrylcholinesterase during morphogenesis and synaptogenesis of chick brain and retina. *J Neurochem* 49: 175-182.

Layer, P.G., Rommel, S., Bülthoff, H. and Hengstenberg, R., 1988a, Independent spatial waves of biochemical differentiation along the surface of chicken brain as revealed by the sequential expression of acetylcholinesterase. Cell Tiss Res 251: 587-595.

Layer, P.G., Alber, R. and Rathjen, F.G., 1988b, Sequential activation of butyrylcholinesterase in rostral half somites and acetylcholinesterase in motoneurones and myotomes preceding growth of motor axons. *Development* 102: 387-396.

Layer, P.G., 1990, Cholinesterases preceding major tracts in vertebrate neurogenesis. *BioEssays* 12: 415-420.

Layer, P.G. and Alber, R., 1990, Patterning of early chick brain vesicles as revealed by peanut agglutinin and cholinesterases. *Development* 109: 613-624.

Layer, P.G., Alber, R., Mansky, P., Vollmer, G. and Willbold, E., 1990, Regeneration of a chimeric retina from single cells in vitro: cell-lineage-dependent formation of radial cell columns by segregated chick and quail cells. *Cell Tissue Res* 259: 187-198.

Layer, P.G. and Kaulich, S., 1991, Cranial nerve growth in birds is preceded by cholinesterase expression during neural crest cell migration and the formation of an HNK-1 scaffold. *Cell Tissue Res*, 265: 393-407.

Layer, P.G., Weikert, T. and Willbold, E., 1992, Chicken retinospheroids as developmental and toxicological in vitro models: acetylcholinesterase is regulated by its own and by butyrylcholinesterase activity. *Cell Tissue Res,* in press.

Massoulié, J. and Bon, S., 1982, The molecular forms of cholinesterase and acetylcholinesterase in vertebrates. *Ann Rev Neurosci* 5: 57-106.

Puelles, L., Amat, J.A. and Martinez-de-la-Torre, M., 1987, Segment-related, mosaic neurogenetic pattern in the forebrain and mesencephalon of early chicken embryos 1. Topography of AChE-positive neuroblasts up to stage HH 18. *J Comp Neurol* 266: 247-268.

Rathjen, F.G., Wolff, J.M., Frank, R., Bonhoeffer, F. and Rutishauser, U., 1987, Membrane glycoproteins involved in neurite fasciculation. *J Cell Biol* 104: 343-353.

Rathjen, F.G., 1991, Neural cell contact and axonal growth. *Curr Opin Cell Biol* 3: 992-1000.

Shen, S.C., Greenfield, P. and Boell, E.J., 1956, Localization of acetylcholinesterase in chick retina during histogenesis. *J Comp Neurol* 106: 433-461.

Takeichi, M., 1988, The Cadherins: cell-cell adhesion molecules contolling animal morphogenesis. *Development* 102: 639-655.

Treskatis, S., Ebert, C. and Layer, P.G., 1992, Butyrylcholinesterase from chicken brain is smaller than that from serum: its purification, glycosylation, and membrane association. *J Neurochem,* in press.

Tucker, G.C., Delarue, M., Zada, S., Boucaut, J.C. and Thiery, J.P., 1988, Expression of HNK-1/NC-1 epitope in early vertebrate neurogenesis. *Cell Tissue Res* 251: 457-465.

Vollmer, G. and Layer, P.G., 1986, An in vitro model of proliferation and differentiation of the chick retina: Coaggregates of retinal and pigment epithelial cells. *J Neurosci* 6: 1885-1896.

Walter, J., Kern-Veits, B., Huf, J., Stolze, B. and Bonhoeffer, F., 1987, Recognition of position-specific properties of tectal cell membranes by retinal axons in vitro. *Development* 101, 685-696.

Weikert, T., Rathjen, F.G. and Layer, P.G., 1990, Developmental maps of acetylcholinesterase and G4-antigen of the early chicken brain: Long distance tracts originate from AChE-producing cell bodies. *J Neurobiol* 21, 482-498.

Willbold, E. and Layer, P.G., 1992a, The fiber layer of Chievitz secludes AChE-positive cells from mitotic cells in chicken retinospheroids. *Cell Tissue Res* in press.

Willbold, E. and Layer, P.G., 1992b, A hidden retinal regenerative capacity from the chick ciliary margin is reactivated in vitro, that is accompanied by down-regulation of butyrylcholinesterase. *Eur J Neurosci* 4: 210-220.

ACETYLCHOLINESTERASE AS A MODULATORY NEUROPROTEIN

AND ITS INFLUENCE ON MOTOR CONTROL

Susan A. Greenfield

University Department of Pharmacology
Mansfield Rd.,
Oxford OX1 3QT
U.K.

INTRODUCTION

For over thirty years it has been speculated that acetylcholinesterase (AChE) may have novel, non-cholinergic functions (Karczmar 1969; Silver 1974). Only recently however, have we been able to gain some insight into one particular example, in a specific region of the brain: the substantia nigra.

The substantia nigra has been the focus of diverse neuroscientific investigations since its degeneration results in the movement disorder of Parkinson's disease. A key feature of the region is that it is rich in diverse neurochemicals, indeed it has been dubbed the 'treasure trove of the pharmacologist': for our purposes here, the very high levels of nigral AChE (Silver 1974) are of particular interest. In itself of course this fact is not particularly remarkable, since the classical substrate for AChE, acetylcholine, is a widespread if not the prototype neurotransmitter. However although there is evidence for a cholinergic input to the substantia nigra (Henderson and Greenfield, 1987; Beninato and Spencer 1987, 1988; Gould et al., 1989; Martinez-Murillo et el., 1989; Bolam and Henderson 1991), this input is sparse (Martinez-Murillo et al., 1989) and appears functionally to be relatively inactive (Scarnatti et al., 1986). In any event it was the disparity between very high levels of nigral AChE compared to only small amounts of choline acetyltransferase and acetylcholine, which originally prompted the hypothesis that AChE might have a second, non-cholinergic function.

In the substantia nigra, AChE is localized in dopamine (DA)-containing nigrostriatal neurons (Butcher et al., 1975), which are the specific cells lost in Parkinsonism. Although AChE can also be visualized in other, non-dopaminergic nigral neurons, ultrastructual studies suggest that the protein might have a special role in relation to the nigrostriatal cells. Only in these dopaminergic neurons is AChE associated with the Golgi apparatus, and with the plasma membrane (Henderson and Greenfield 1984). It is now well established that prior to 'externalization' AChE passes through the Golgi apparatus (Rotundo 1984), where of course it can become attached to the outer surface of the plasma membrane in its familiar ectoenzyme position. However it is perhaps less widely known that 'externalization' can also refer to the secretion of AChE into the extracellular space. This secreted AChE is a specific soluble form of the protein that is electrophoretically distinct (Chubb and Smith 1975), and indeed a different molecular form

Multidisciplinary Approaches to Cholinesterase Functions, Edited by
A. Shafferman and B. Velan, Plenum Press, New York, 1992

(Chattonet and Lockridge 1989), from the classical membrane-bound variety. Nigrostriatal neurons are characterized by large amounts of RER (Domesick et al., 1983; Henderson and Greenfield 1984), which would enable them to fulfill the requirement for the higher rates of turnover of the soluble AChE (Rotundo 1984). Hence it is possible that dopaminergic nigrostriatal neurons, specifically, secrete a soluble form of AChE, and indeed that the phenomena underlies the novel non-cholinergic function postulated earlier.

This idea, prompted by anatomical and histological observations, was readily confirmed by experiments in vivo where selective destruction of the nigrostriatal system, with the neurotoxin 6-hydroxydopamine, led to a marked reduction of an otherwise robust secretion of AChE from the substantia nigra (Greenfield et al., 1983). This secretion of AChE is a truly physiological phenomenon in that it can be evoked by depolarizing concentrations of potassium ions, and is calcium dependent (Greenfield et al., 1982). However, stimulation of nigral cholinergic receptors with carbachol is inefficacious (Burgun et al., 1985) whilst stimulus-evoked secretion of AChE from nigrostriatal cells is resistant to blockade of cholinergic receptors (Greenfield and Smith 1979): the secretion of AChE appears then to be completely dissociable from cholinergic transmission, but rather closely connected with nigral dopaminergic systems.

We are faced however with a further apparent heresy: within the substantia nigra, there are no dopaminergic axon terminals or collaterals. It seems then that secretion of AChE in this region, as already demonstrated for DA itself (Neioullon et al., 1977), must originate from nigrostriatal cell dendrites. The dendritic release of DA and AChE in the substantia nigra has already been reviewed extensively elswhere (Greenfield, 1985, 1991). In brief, it appears that the phenomena might refelect a form of neuronal signalling that is distinct from the classical series of events at the axon terminal synapse. Within the substantia nigra at least, dendritic release is non-quantal, slow and promiscuous regarding potential targets. Furthemore, the release process itself is resistant to blockade of somatic discharge and subsequent impulse flow by tetrodotoxin (TTX), and must therefore be independent of any overall changes in cell excitibility (Greenfield 1985, 1991). Within the substantia nigra therefore, the dendritic release of AChE can be viewed as a highly novel phenomenon involving a familiar protein behaving in an unfamiliar way in an unusual location: what could be the significance?

This question can be tackled by breaking it down into two broad sub-questions relating respectively to secretion and any subsequent action of AChE. Each of these sub-questions can be asked in relation to cellular events or indeed at the level of the complete system. In this review therefore we will be looking at what progress has been made, mostly in the last five years, towards understanding the non-cholinergic function of nigral AChE within this framework of four sub-questions.

1. SECRETION OF AChE: SYSTEMS LEVEL

Perhaps one of the most obvious issue to explore is whether or not the dendritic secretion of AChE within the substantia nigra reflects in any way the physiology of the system from which it occurs. Since the substantia nigra is so closely related to the generation of movement, a good place to start would be by exploring whether the secretion is indeed influenced by locomotor behaviour. However this issue was more problematic than might be initially envisaged due to the relative spatio-temporal insensitivity of most established methods for monitoring secretion of AChE from the brain in vivo. The problem was that the sensitivity of existing assays was inadequate for detecting 'on-line' behavioural events, in that it was a million times longer in duration (10 minutes) than the average action potential (1 millisecond). In order to explore the functional relevance of AChE secretion we therefore developed a chemiluminescent system (Israel and Lesbatts 1981; Birman 1985) for monitoring AChE secretion from the substantia nigra 'on-line' (Llinas and Greenfield 1987) in the freely moving guinea pig (Taylor et al., 1989; Taylor et al., 1990; Jones et al., 1990).

The immediate issue that we were able to resolve with this novel technique was that secretion of AChE within the substantia nigra had a physiological relevance: when the animal was awake and freely moving, the levels of AChE detected in nigral perfusates were far greater and more pulsatile than during anaesthesia (Taylor et al., 1990). However it was very hard to quantify or qualify the salient movements, within the animals' normal behavioural repertoire, which evoked the release of AChE. This problem was circumvented by improving the means of monitoring behaviour (Jones et al., 1990) and by ensuring the experimental paradigm was more rigorous, such that the animal moved in a particular way for a particular period of time. By placing the guinea pig on a slowly moving motorised tread- mill, we were able to relate any changes in secretion of AChE to controlled periods of locomotion. These evoked episodes of locomotion did indeed appear to cause an enhanced release of AChE within one or two minutes (Fig 1): interestingly enough, the elevated levels of released protein were usually shorter than the duration of enforced movement, and became less apparent as the episodes were repeated (Jones et al.,1991a). Hence it seemed that release of AChE was not causal to, but a corollary of, movement: furthermore the secreted protein was reflecting not merely movement per se, but some more particular abstracted aspect of the on-going activity. Corollary discharge, for example that occurring when we move our eyes, is important for proprioception (Evarts, 1985). Since AChE is not causal to the generation of movement it is conceivable that it plays a part in proprioceptive feedback regarding that movement. In this way the aspect of the 'motor' event with which we are concerned here would be effectively sensory.

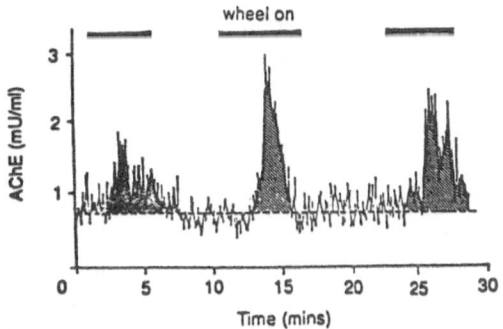

Fig. 1 A typical trace of AChE release obtained during three 5-min periods of a guinea pig walking in the motorized tread wheel, separated by two 5-min stationary periods. As seen, release occurs in a pulsatile fashion and is composed of 'peaks' of AChE (see hatched areas). Peaks of AChE release occur most often while the animal is moving. From (Jones et al., 1991a).

The role of the substantia nigra in motor control must indeed be highly sophisticated and abstracted, since a range of sensory stimuli contribute to the net excitability of nigrostriatal cells (DeLong et al.,1983; Delong et al., 1986; Schultz 1989) including visual (Steinfels et al., 1983a), olfactory (Chiodo et al., 1980), auditory (Steinfels et al., 1983b) and somatosensory (Chiodo et al., 1979) stimulation. Indeed, dendritic release of DA can be enhanced by presentaion of a flashing light (Cheramy et al., 1978). We have therefore explored whether the levels of AChE in the extracellular space of the substantia nigra of the unanaesthetized guinea pig were sensitive to the visual stimulation of a flashing light: this 'sensory' treatment does indeed evoke secretion of AChE (Jones et al., 1991b).

We can see then that the dendritic secretion of AChE in the substantia nigra probably has a physiological significance in that it reflects both motor (proprioceptive) and sensory events. On the other hand, we saw earlier that the phenomenon occurred independent of somatic discharge. It follows therefore that the sensori-motor feedback signals arriving into the substantia nigra have effects more subtle than a mere blanket excitation of the dopaminergic neurons. How could such signals evoke dendritic secretion of AChE independent of somatic discharge?

2. SECRETION OF AChE: CELLULAR LEVEL

The dendritic secretion of AChE must be evoked by some type of trigger: electrical, due to impulse propagation from soma to dendrite, or chemical ie synaptic. Since the phenomenon is TTX resistant (Greenfield 1985, 1991), we can discount the former possibility and concentrate on how a sensory 'proprioceptive' afferent signal might evoke dendritic secretion of AChE inependent of somatic events.

In order to investigate the plausibility of this hypothetical dendritic mechanism, we have recorded intracellularly from nigrostriatal neurons. In brief, we discovered that these neurons generate a TTX-resistant calcium conductance in the distal dendrites, which would normally be propagated very poorly up the long dendrite to the soma (Llinas et al., 1984; Nedergaard and Greenfield 1992). Hence this conductance, termed 'High Threshold', would provide the appropropriate means for dendritic extrusion of AChE independent of somatic excitibility. We have tested this hypothesis by seeing how substances that do modify AChE secretion (eg 5-HT, amphetamine) or do not (eg GABA) respectively might modify the High Threshold calcium conductance. To date we have found a close correspondence: 5-HT (Nedergaard et al., 1988) and amphetamine (Nedergaard et al., 1989) both enhance the dendritic calcium conductance, whereas GABA does not (Nedergaard et al., 1989). This observation suggests that AChE secretion does not reflect the activity of the neurons from which it is released, but rather that of the inputs to them. These inputs do indeed make contact at the more distal, electrotonically remote segments of the dendrite (Rinvik and Grofova, 1970) and as we have seen would at least in part, carry sensory-motor signals. Among the principal areas which send projections to the substantia nigra are the striatum, the subthalamus and the Raphe nuclei. We have therefore been examining the effects of stimulation of these input areas on the secretion of nigral AChE under various conditions in vivo.

Current experiments demonstrate that stimulation of the subthalamus, which sends a glutamate projection to the substantia nigra, does not evoke secretion of AChE unless the stimulation has also caused a movement, ie it is inefficacious in the anaesthetized animal. This observation implies that the subthalamo-nigral pathway itself does not evoke AChE secretion, and indeed direct application of exogenous glutamate into the substantia nigra is ineffective. Rather it seems that stimulation of the subthalamus evokes movement by a polysynaptic circuit (obviously only possible in the awake animal) and the proprioceptive feedback resulting from that movement, as in the case of the treadmill, will cause secretion of AChE. On the other hand secretion of AChE can be caused, even in the anaesthetized animal by stimulation of the striatum (Taylor and Greenfield, 1989b) or the Raphe nuclei. Indeed current experiments suggest the effects of Raphe stimulation can be perfectly mimiced by direct application of the transmitter used in that particular pathway, 5-HT. A rationale for interpreting these observations is as follows: the proprioceptive feedback for the secretion of AChE is ultimately carried by the striatal or Raphe, but not the subthalamic, projections. Hence when the two former pathways are selectively stimulated directly, we are effectively by-passing the need for a movement to have occurred at all.

In summary, the dendritic secretion of AChE within the substantia nigra is independent of cholinergic transmission but rather reflects the activity of certain selective inputs, which in turn are activated by sensory stimulation including proprioception during movement. This sequence of events probably most ususally occurs without affecting the overall excitability of the nigrostriatal neurons, and hence of DA release from axon terminals in the striatum. So what is the purpose of AChE secretion?

3. ACTION OF AChE: SYSTEMS LEVEL

Any potential action of secreted AChE in the extracellular space can be simulated by introducing an exogenous preparation into the substantia nigra by microinfusion, either unilaterally or bilaterally. If AChE is infused into one substantia nigra of the rat, then the animal will start to turn consistently in one direction (rotation): this effect is not replicated with butyrylcholinesterase. Furthermore, purified AChE with the same electrophoretic mobility to the endogenous secreted form, is far more potent than elliciting rotation than preparations of the commercial enzyme with a far higher activity towards acetylcholine (Greenfield et al., 1984). Clearly then this behavioural action of AChE is unrelated to hydrolysis of acetylcholine. We now need to explore the nature of that action.

Rotation is widely regarded as a crude but reliable indication of a disparity in the functional availability of DA in the two nigrostriatal pathways on either side of the brain, (Ungerstedt 1971). By studying the direction of rotation induced by AChE we can therefore gain some insight as to the net action of the protein. More recently we have discovered that AChE can induce rotation in either direction, ie can enhance or diminish the net excitibility of the nigrostriatal pathway. The critical factor appears to be whether the AChE is applied in the anterior or posterior part of the substantia nigra (Hawkins and Greenfield 1992a). AChE applied to anterior nigral neurons appears to cause a net inhibtion, whereas when applied to the posterior cells, there is a net excitation. In both cases, the effect is virtually permanent, perhaps because it does not induce a change in the target striatal DA receptor sensitivity (Hawkins and Greenfield 1992b). In any event there are only two possible but not mutually exclusive explanations for the site-dependency of the behavioural effects of AChE: either the two types of nigral cell use AChE in opposite biochemical ways, or the biochemical use is uniform, but the neurons have opposite roles in the physiology of the nigrostriatal system. This issue has yet to be resolved.

If AChE is applied bilaterally, then stereotyped movements are ehanced: again, this effect cannot be replicated following application of butyrylcholinesterase (Weston and Greenfield, 1985). Stereotyped movements, which are normally associated with excessive amounts of striatal DA, are part of the normal behavioural repertoire of an animal, except that they are repeated over and over again out of their normal context. One of the most common of these types of movement is chewing. Interestingly enough chewing movements occur just after large increases in endogenous AChE release evoked by either motor (Jones et al., 1991a) or sensory (Jones et al., 1991b) events. It seemed feasible therefore that because the mouth has a high representation in motor cortex, this particular behaviour might be one of the easiest to elicit with low, physiological levels of AChE: when exogenous AChE was applied, in amounts over a thousand times lower than in the experiments studying rotation, chewing behaviour did indeed occur (Jones et al., 1991b).

It would appear that secreted AChE can have non-cholinergic actions in the substantia nigra which can facilitate or attenuate movement, depending on its anterior-posterior disposition within the nucleus. In the case of facilitated movements, we have seen that proprioceptive feedback would provide activation of certain nigral afferents to enhance endogenous AChE secretion further. In this scheme, AChE secretion would reflect sensory events and subsequently act to enhance motor activity, or more concisely, mediate sensory-motor integration within the substantia nigra. The obvious question then is: what is the actual mechanism by which AChE can modify the output neurons (nigrostriatal cells) affecting movement?

4. ACTION OF AChE: CELLULAR LEVEL

In order to study the actual mechanisms by which secreted AChE might have non-cholinergic actions, substantia nigra neurons are again recorded intacellularly in vitro, and the effects of exogenous AChE observed upon the membrane properties. In brief, the protein hyperpolarizes the membrane via a net opening of potassium channels: this effect is once more not related to hydrolysis of acetylcholine since it occurs even when the AChE has been pretreated with the irreversible inhibitor Soman, and cannont be replicated using butyrylcholinesterase (Greenfield et al., 1988, 1989). Highly purified AChE is even more potent and the boiled protein is equally effective: hence it seems the action is not attributable to either a contaminant nor indeed to enzymatic action towards any other unknown substrate (Webb and Greenfield 1992). It also seems that this hyperpolarizing action of AChE is independent of DA transmission. First, the effect persists in the presence of the DA receptor blocker, sulpiride (Greenfield et al., 1989); secondly, the effects of AChE display neither tachyphylaxis nor sensitivity to tetraethylammonium, both of which are features of DA receptor activation (Webb and Greenfield 1992); thirdly, current experiments in our laboratory suggest that the effects of AChE persist when cells are pharmacologically depleted of DA with reserpine and alpha-methyl-paratyrosine.

Recently we have made some progress towards understanding the cellular, sub-cellular and ion channel target for the effect of AChE. Electrophysiological investigations of putative dopaminergic nigrostriatal neurons has revealed two distinct sub-populations distributed in the anterior and posterior parts of the substantia nigra (Nedergaard and Greenfield, 1992). Coincidentally, the location of these two types of cell correspond to the levels of substantia nigra where differential effects of AChE application were observed on behaviour (Hawkins and Greenfield 1992). Neurons in the anterior substantia nigra are more sensitive to AChE than their posterior counterparts (Webb and Greenfield 1992). However current experiments are suggesting that this response can be modified if the long 'apical' dendrites are cut off from the neuron. In these cases the action of AChE is no longer to open a potassium channel. We can deduce then, that the prime target of AChE is a potassium channel located on the distal dendrites of anterior nigral neurons. However we have already seen that the action of AChE is resistant to the presence of the fairly widespread potassium channel blocker, tetraethylammonium (Webb and Greenfield, 1992).

On the other hand, there is a special potassium channel that is plentiful in the substantia nigra (Mourre et al., 1990; Treherne and Ashford, 1991) yet insensitive to tetraethylammonium: the ATP sensitive potassium channel 'K-ATP', which is opened when intracellular ATP levels are low (Ashcroft and Ashcroft, 1990). Could AChE be working via this channel? Recent evidence suggests that it could. First, the anterior cells which are the most sensitive to AChE are also the most sensitive to K-ATP channel opening by cyanide (Murphy and Greenfield 1991), anoxia (Murphy and Greenfield 1992), or reduced glucose (Webb and Greenfield 1992); secondly, the effects of AChE are potentiated by reduced glucose (Webb and Greenfield 1992) (Fig 2); thirdly the effects are blocked by the K-ATP antagonist tolbutamide (Webb and Greenfield, 1992). Finally, it appears that the K-ATP could indeed be preferentially located on the nigral cell dendrites since in the single gene mutant, the 'weaver' mouse, there is a dramatic loss of K-ATP channels in the brain and a disproportionate loss of dopaminergic nigrostriatal cell dendrites (Triarhou and Ghetti, 1989).

The non-cholinergic action of AChE appears then to be nonenzymatic and to involve the selective opening of a potassium channel that is linked to the cells metabolism and perhaps located preferentially on the dendrites of anterior nigral neurons. We have seen that activation of this channel by AChE is independent of DA systems: how then can this cellular action of AChE be used to explain the extremely long term actions of AChE on DA related behaviours?

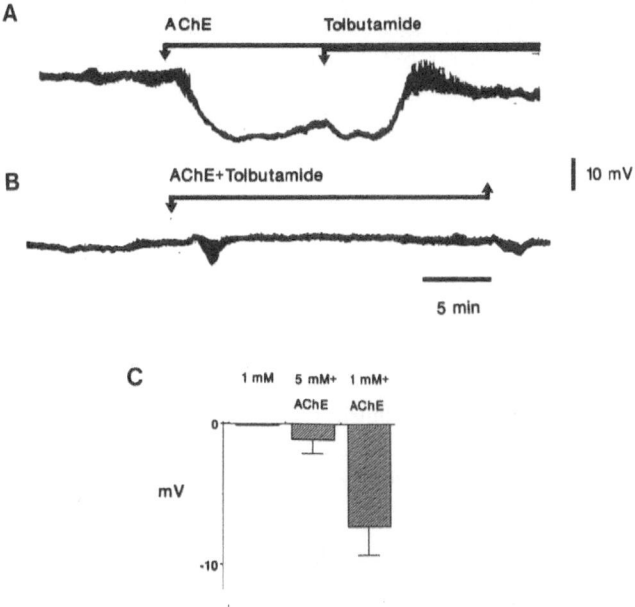

Fig. 2. Effects of AChE are mediated via the K-ATP channel. Antagonism of the effects of AChE by the K-ATP channel blocker, tolbutamide. A & B: With 30s, AChE induces a marked hyperpolarization of 15mV. Following application of tolbutamide to the perfusate, the effect is reversed within 2 min (A). When AChE and tolbutamide are applied to the perfusate simultaneously, no hyperpolarization is seen (B). From (Webb and Greenfield, 1992). C: Hyperpolarizing actions of AChE are enhanced by low glucose, which in itself has little effect.

CONCLUSIONS

An earlier attempt (see Greenfield 1985, 1991) to reconcile the electrophysiological and behavioural effects of AChE was as follows. The hyperpolarizing action of AChE would provide a prerequisite for activation of a long lasting 'Low Threshold' calcium conductance (Llinas et al., 1984). This conductance would then actually be triggered, whilst the cell was in a hyperpolarized condition, by a brief depolarization from a synaptic input. The resulting sustained entry of calcium ions would lead to a burst of action potentials and a disproportionate increase in release of the transmitter DA (Gonon 1987) from striatal axon terminals. AChE was viewed then as 'modulator' in that its function would effectively be to enhance the activity of nigrostriatal cells to synaptic input.

This theory was attractive since it explained how AChE could have a hyperpolarizing action, yet enhance synaptic activity (Last and Greenfield 1987) and cause a net behavioral excitation (Greenfield et al.,1984). However there are various anomalies that this theory fails to explain. First, the long term behavioural effects of AChE , at least a month in duration, can hardly be accounted for by the AChE molecule exerting a continued electrophysiological action within the extracellular space. Secondly, the electrophysiological effect is independent of DA systems; yet the behaviours that are modified are traditionally regarded as DA mediated and in addition we have some preliminary evidence that AChE might interact directly with the DA molecule. Thirdly, we have seen that AChE works via an opening of the K-ATP channel, yet cell metabolism was not included as a relevant factor in the original theory. Fourthly, the two types of nigral cell are not considered at all: the implications of the differential action of AChE on these cells needs therefore to be amplified somewhat here.

Anterior cells are the only type to generate a Low Threshold calcium conductance upon hyperpolarization (Nedergaard and Greenfield 1992). According to the original theory then, the net effect of AChE on these anterior cells should be one of excitation, whereas recent behavioural studies have shown the opposite, a net inhibition (Hawkins and Greenfield 1992). This argument could be countered by suggesting that the hyperpolarizations of exogenous AChE were too large such that the dendrites were functionally uncoupled from the soma (Llinas et al., 1984) and that more modest, physiological hyperpolarizations would affect only the dendrites and conceivably lead to acute excitation as suggested by extracellular recordings (Last and Greenfield, 1987). On the other hand, the behavioural and electrophysiological effects of AChE on the posterior cells are impossible to reconcile. We have seen that the electrophysiological effects on these cells is either non-existent or at best a modest hyperpolarization (Webb and Greenfield 1992). Even if these neurons were hyperpolarized by AChE, they would simply be more inhibited as they do not generate the Low Threshold conductance (Nedergaard and Greenfield 1992): but we have seen that the behavioural effect of AChE is a net excitation that is, incidentally, powerful enough to elevate levels of striatal DA (Hawkins and Greenfield 1992b).

These four issues: duration of effect, relation to DA, relation to cell metabolism and differential action on nigral sub-populations must be incorporated for any satisfactory model of a non-cholinergic role for AChE in the substantia nigra. The original theory might hold for acute situations for one sub-population, but on its own it is inadequate for the wider situations that we have reviewed here. All the four issues could be accomodated if we postulated an intacellular non-cholinergic role for AChE: it has already been suggested that AChE can be pinocytosed into cells (Jessen et al., 1978). The duration of the effect could well result from an intracellular cascade of reactions comparable to those seen in the adaptive mechanisms of aplysia. The dependence of AChE on DA could be intracellular, since the non-DA related electrophysiological effect reflects only ionic fluxes across the plasma membrane. Indeed, cell metabolism might be incorporated as a factor if the pinocytosis of AChE into the cell compromised energy sufficiently to lower intracellular ATP and hence open the K-ATP channels; Finally, the differential electrophysiological sensitivities of the two cell types to AChE could also be explained if the protein was absorbed into the neurons. AChE could be taken up equally well into both cell types, whence it exerted its diverse behavioural effects, but the responses of the two cell types to the same energy demanding process might be different. It is already established that the anterior cells possess a much more functionally active K-ATP channel, compared to the posterior cells (Murphy and Greenfield 1992); hence electrophysiologically they might well appear more 'sensitive' as they pinocytose AChE.

Although an intracellular non-cholinergic role for secreted AChE appears to be the most parsimonious and comprehensive explanation of all the observations described above, it really only provides the spring-board for further investigation. We still do not know why AChE should work differentially in two cell types, or why a uniform biochemical use of AChE would result in completely opposite behavioural effects. Nor in any case can we imagine how DA and AChE might interact within either cell to cause long term changes. Indeed the precise link between the K-ATP and secreted AChE remains to be fully explored. Nonetheless if AChE does play a non-cholinergic intracellular role, it opens up a realm of possible novel mechanisms for the modulatory actions of secretory neuroproteins. But then that would be another story, or at least another chapter.

REFERENCES

Ashcroft, S.J.H., and Ashcroft, F.M., 1990, Properties and functions of ATP-sensitive potassium channels Cellular Signalling 2: 197-214.

Beninato, M. and Spencer, R.F., 1987, A cholinergic projection to the rat substantia nigra from the pedunculopontine tegmental nucleus, Brain Res. 412:169-174.

Beninato, M. and Spencer, R.F., 1988, The cholinergic innervation of the rat substantia nigra: a light and electron microscopic immunohistochemical study, Exp. Brain Res. 72: 178-184.

Birman, S., 1985, Determination of acetylcholinesterase activity by a new chemiluminescent assay with the natural substrate, Biochem. J. 225: 825-828.

Bolam, J.P., Francis, C.M., and Henderson, Z., 1991, Cholinergic input to dopaminergic neurons in the substantia nigra: a double immunocytochemical study, Neuroscience 41: 483-494.

Burgun, C., Greenfield, S.A., Waksman, A. and Weston, J., 1985, Differential effects of cholinergic agonists on acetylcholinesterase release from the rat substantia nigra in vivo, J. Physiol. 369: 66P.

Butcher, L.L., Talbot, K. and Bilezikjian, L., 1975, Acetylcholinesterase neurons in dopamine-containing regions of the brain, J. Neural Trans. 37: 127-153.

Chatonnet, A. and Lockridge, O., 1989, Comparison of acetylcholinesterase and butyrylcholinesterase, Biochem. J. 260: 625-634.

Cheramy, A., Neioullon, A., and Glowinski, J., 1978, Gabaergic processes involved in the control of dopamine relase from nigrostriatal dopaminergic neurons in the cat, Eur. J. Pharmacol. 48: 281-295.

Chiodo, L.A., Antelman, S.M., Caggiula, A.R., and Lineberry, C.G., 1979, Reciprocal influences of activating and immobilizing stimuli on the activity of nigrostriatal dopamine neurons, Brain Res. 176: 385-390.

Chiodo, L.A., Antelman, S.M., Caggiula, A.R., and Lineberry, C.G., 1980, Sensory stimulation alters the discharge rate of dopamine (DA) neurons: evidence for two functional types of DA cells in the substantia nigra, Brain Res. 189: 544-549.

Chubb, I.W. and Smith, A.D., 1975, Isoenzymes of soluble and membrane-bound acetylcholinesterase in bovine splanchnic nerve and adrenal medulla, Proc. R. Soc. Lond. 191B: 245-261.

DeLong, M.R., Georgopoulos, A.P., and Crutcherm, M.D., 1986, Cortico-basal ganglia relations and coding of motor performance, Exp. Brain Res.' Suppl. 7 30-40.

Domesick, V.B., Stinus, L. and Paskevich, P.A., 1983, The cytology of dopaminergic and nondopaminergic neruons in the substantia nigra and ventral tegmental area of the rat-a light-microscopic and electron-microscopic study, Neuroscience 8: 743P.

Evarts, E.V., 1985, Sherrington's concept of proprioception, in: "The Motor System in Neurobiology," E. Evats, S.P. Wist, and D. Bousfield, eds., Elsevier, Amsterdam, pp 183-186.

Gonon, G.G., 1988, Nonlinear relationship between impulse flow and dopamine released by rat midbrain dopaminergic neurons as studied by in vivo electrochemistry, Neuroscience 24: 19-28.

Gould, E., Woolf, N.J. and Butcher, L.L., 1989, Cholinergic projections to the substantia nigra from the pedunculopontine and laterdorsal tegmental nuclei, Neuroscience 28: 611-623.

Greenfield, S.A., 1991, A noncholinergic action of acetylcholinesterase (AChE) in the brain: from neuronal secretion to the generation of movement, Cell. and Mol. Neurobiol. 11: 55-77.

Greenfield, S.A., 1985, The significance of dendritic release of transmitter and protein in the substantia nigra, Neurochem. Int. 7: 887-901.

Greenfield, S.A., and Smith, A.D., 1979, The influence of electrical stimulation of certain brain regions on the concentration of acetylcholinesterase in rabbit cerebrospinal fluid, Brain Res. 177: 445-459.

Greenfield, S.A., Grunewald, R.A., Foley, P., and Shaw, S.G., 1983, Origin of various enzymes released from the substantia nigra and caudate nucleus: effects of 6-hydroxydopamine lesions of the nigro-striatal pathway, J. Comp. Neurol. 214: 87-92.

Greenfield, S.A., Chubb, I.W., Grunewald, R.A., Henderson, Z., May, J.M., Portnoy, S., Weston, J., and Wright, M.C., 1984, A non-cholinergic function for acetylcholinesterase in the substantia nigra: behavioural evidence, Exp. Brain Res. 54: 513-520.

Greenfield, S.A., Nedergaard, S., Webb, C., and French, M., 1989, Pressure ejection of acetylcholinesterase within the guinea pig substantia nigra has non-classical actions on the pars compacta cells independent of selective receptor and ion channel blockade, Neuroscience 29: 21-25.

Hawkins, C.A. and Greenfield, S.A., 1992a, Non-cholinergic action of exogenous acetylcholinesterase in the rat substantia nigra I: Differential effects on motor behaviour, Behav. Brain Res., in press.

Hawkins, C.A. and Greenfield, S.A., 1992b, Non-cholinergic action of exogenous acetylcholinesterase in the rat substantia nigra II: Long-term interactions with dopamine metabolism, Behav. Brain Res. in press.

Henderson, Z., and Greenfield, S.A., 1984, Ultrastructural localization of acetylcholinesterase in substantia nigra: A comparison between rat and guinea pig, J. Comp. Neurol., 230: 278-286.

Henderson, Z., and Greenfield, S.A., 1987, Does the substantia nigra have a cholinergic innervation?, Neruosci. Lett. 73: 109-113.

Israel, M., and Lesbats, B., 1981, Chemiluminescent determination of acetylcholine and continuous detection of its release from Torpedo electric organ synapses and synaptosomes, Neurochem, Int. 3: 81-90.

Jessen, K.R., Chubb, I.W., and Smith, A.D., 1978, Intracellular localization of acetylcholinesterase in nerve terminals and capillaries of the rat superior cervical ganglion. J. Neurocytol. 7: 145-154.

Jones, S.A., Annetts, C., Preston, M., and Greenfield, S.A., 1990, An improved method for correlating behaviour with acetylcholinesterae release from the brain of the awake guinea pig 'on-line', J. Physiol. 430: 62P.

Jones, S.A., and Greenfield, S.A., 1991a, Behavioural correlates of the release and subsequent action of acetylcholinesterase secreted in the substantia nigra, Eur. J. Neurosci. 3: 292-295.

Jones, S.A., Ellis, J.R.C., Klegeris, A., and Greenfield, S.A., 1991b, The relationship between visual stimulation, behaviour and continuous release of protein in the substantia nigra, Brain Res. 560: 163-166.

Karczmar, A.G., 1969, Is the central cholinergic nervous system overexploited?, Fed. Proc. 28: 147-157.

Last, A.T.J., and Greenfield, S.A., 1987, Acetylcholinesterase has a non-cholinergic neuromodulatory action on the guinea-pig substantia nigra, Exp. Brain Res. 67: 445-448.

Llinas, R.R., and Greenfield, S.A., 1987, On-line visualization of dendritic release of acetylcholinesterase from mammalian substantia nigra neurons, Proc. Natl. Acad. Sci., USA 83: 3047-3050.

Llinas, R.R., Greenfield, S.A., and Jahnsen, H.J., 1984, Electrophysiology of pars compacta cells in the in vitro substantia nigra - a possible mechanism for dendritic release, Brain Res. 294: 127-132.

Martinez-Murillo, R., Villalba, R., Montero-Caballero, M.J., and Rodrigo. J., 1989, Cholinergic somata and terminals in the rat substantia nigra: An immunocytochemical study with optical and electron microscopic techniques, J. Comp. Neurol. 281: 397-415.

Mourre, C., Widmann, C., and Lazdunski, M., 1990b, Sulfonylurea binding sites associated with ATP-regulated K^+ channels in the central nervous system: autoradiographic analysis of their distribution and ontogenesis, and of their localization in mutant mice cerebellum, Brain Res. 519: 29-43.

Murphy, K.P.S.J., Greenfield, S.A., 1992, Neuronal selectivity of ATP-sensitive potassium channels in the guinea-pig substantia nigra revealed by differential responses to anoxia, J. Physiol. in press.

Murphy, K.P.S.J., and Greenfield, S.A., 1991, ATP-sensitive potassium channels counteract anoxia in neurons of the substantia nigra, Exp. Brain Res. 84: 355-358.

Nedergaard, S., Bolam, J.P., and Greenfield, S.A., 1988, Facilitation of a dendritic calcium conductance by 5-HT in the substantia nigra, Nature 333: 174-177.

Nedergaard, S., and Greenfield, S.A., 1992, Sub-populations of pars compacta neurons in the substantia nigra: the significance of qualitatively and quantitatively distinct conductances, Neuroscience in press.

Nieoullon, A., Cheramy, A., and Glowinski, J., 1977, Release of dopamine in vivo from cat substantia nigra, Nature 266: 375-377.

Rinvik, L.E., and Grofova, I., 1970, Observations on the fine structure of the substantia nigra in the cat, Exp. Brain Res. 11: 229-248.

Scarnati, E., Proia, E., Campana, E., and Pacitti, C., 1986, A microiontophoretic study on the nature of the putative synaptic neurotransmitter involved in the pedunculopontine-substantia nigra pars compacta excitatory pathway of the rat, Exp. Brain Res. 62: 470-478.

Schultz, W., 1989, Neurophysiology of basal ganglia, Handbook Exp. Pharmacol. 88: 1-43.

Silver, A., 1974, The Biology of the Cholinesterases, Elsevier, Amsterdam, pp 117-303, 428-431.

Steinfels, G.F., Heym, J., Strecker, R.E., and Jacobs, B.L., 1983a, Behavioural correlates of dopaminergic unit activity in freely moving cat, Brain Res. 258: 217-228.

Taylor, S.J., Haggblad, J., and Greenfield, S.A., 1989, Measurement of cholinesterase activity released from the brain "on-line" and in vivo, Neurochem, Int. 15: 199-205.

Taylor, S.J., and Greenfield, S.A., 1989, Pulsatile release of acetylcholinesterase following electrical stimulation of the striatum, Brain Res. 505: 153-156.

Taylor, S.J., Jones, S.A., Haggblad, J., and Greenfield, S.A., 1990, 'On-line' measurement of acetylcholinesterase release from the substantia nigra of the freely-moving guinea-pig, Neuroscience 37: 71-76.

Treherne, J.M., and Ashford, M.L.J., 1991, The regional distribution of sulphonylurea binding sites in rat brain, Neuroscience 40: 523-532.

Triarhou, L.C., and Ghetti, B., 1989, The dendritic dopamine projection of the substantia nigra: phenotypic denominator of weaver gene action in hetero- and homozygosity, Brain Res. 501: 373-381.

Ungerstedt, U., 1971, Striatal dopamine release after amphetamine or nerve degeneration revealed by rotational behaviour, Acta Physiol. Scand. Suppl. 367: 49-68.

Webb, C.P., and Greenfield, S.A., 1992, Non-cholinergic effects of acetylcholinesterase in the substantia nigra: a possible role for an ATP-sensitive potassium channel, Exp. Brain Res. in press.

Weston, J., and Greenfield, S.A., 1985, Application of acetylcholinesterase to the substantia nigra induces stereotypy in rats, Behav. Brain Res. 18: 71-74.

DIFFERENT INFLUENCE OF INHIBITORS ON ACETYLCHOLINESTERASE MOLECULAR FORMS G1 AND G4 ISOLATED FROM ALZHEIMER'S DISEASE AND CONTROL BRAINS

Albert Enz,[1] Alain Chappuis,[1] and Alphonse Probst[2]

[1]Preclinical Research SANDOZ Pharma Ltd.
[2]Department of Pathology
Division of Neuropathology, University of Basle
CH 4002 Basle/Switzerland

INTRODUCTION

Alzheimer's disease (AD) is a degenerative disorder of the CNS with severe consequences: up to 20% of all individuals beyond 80 years of age are affected (Bartus et al., 1982). In the developed countries AD is the fourth major cause of death after heart disease, cancer and stroke. The disease represents a major public problem, both because of the number of persons affected and because of the severity of the accompanying disabilities.

Currently, no pharmacologic agents are known to be effective in treating this disease. However, a decreased number of cholinergic neurons in basal forebrain nuclei and decreased levels of choline acetyltransferase in AD patients raises the possibility that at least some of the impairments observed in affected patients might result from loss of cholinergic activity.

The likely importance of a cholinergic deficit in AD is also supported by psychopharmacological studies. Anticholinergic drugs impair memory in persons not suffering from dementia (Drachman, 1977) and cause symptoms which are similar to those found in early stages of AD (Drachman and Leavitt, 1974).

Consequently, drugs that enhance cholinergic activity in human have been proposed for the treatment of AD. One approach is inhibition of acetylcholinesterase (AChE), the degrading enzyme of acetylcholine (ACh); however this leads indirectly to muscarinic and nicotinic effects. Furthermore physostigmine, the most readily available drug of this type is very short-acting (minutes). Nevertheless it has been shown to cause some memory improvement in AD patients (Davis and Mohs, 1982). The extent of improvement was closely related to the degree of AChE inhibition, measured in the cerebrospinal fluid and thus to the amount of physostigmine reaching the CNS (Thal et al., 1983). However, the bioavailability of orally administered physostigmine is unpredictable, rendering the handling of this drug difficult in the clinic.

Multidisciplinary Approaches to Cholinesterase Functions, Edited by
A. Shafferman and B. Velan, Plenum Press, New York, 1992

Another acetylcholinesterase inhibitor (AChE-I), Tacrine (1,2,3,4-tetrahydro-9-9-aminoacridine, THA), has produced a significant degree of improvement in AD patients when given over an average period of 12.6 months (Summers et al., 1986, 1989). However, the findings have not been generally accepted and several larger studies with tacrine in AD patients are now ongoing.

SDZ ENA 713 ((+)(S)-N-Ethyl-3-[(1-dimethyl-amino)ethyl]-N-methyl-phenyl-carbamate hydrogentartrate) is a novel brain selective AChE-I currently under clinical investigation in AD patients (Enz et al. 1991).

The existence of acetylcholinesterase in different molecular forms is well established (Massoulie and Bon 1982). The polymorphic enzyme can be classified based on solubility characteristics and sedimentation velocities determined by centrifugation in sucrose density gradients. These forms are divided into two classes: 1. globular forms consisting of monomer (G1), dimer, (G2), and tetramer (G4) structures with 1, 2 or 4 catalytic subunits, respectively and 2. asymmetric forms containing a collagen-like tail, possessing three possible configurations consisting of 4 (A4), 8 (A8) or 12 (A12) catalytic sub-units (Massoulie and Bon 1982). Although the significance of the different molecular forms is not yet clear, their tissue distribution is thought to reflect specific physiological functions (Massoulie and Bon, 1982). In human brain, the total AChE levels and the distribution of the molecular forms varies from region to region. The most abundant form found in the brain is the tetrameric G4. This form is functionally important for the degradation of ACh at the cholinergic synapses, but other proposed physiological functions of AChE within the CNS are probably mediated through G4 AChE (Greenfield, 1984, Greenfield et al., 1984). In smaller amounts the monomeric G1 is present in human brain.

During aging and more dramatically in AD the G4 form is decreased in neocortex and hippocampus, whereas no change or a smaller decrease of the G1 form is found. This results in an increase of the enzyme activity ratio G1/G4 which is well documented in the literature (Atack et al., 1983, Siek et al 1990). In the experiments described below, the influence of various acetylcholinesterase inhibitors (AChE-I) on the G1 and G4 forms of AChE isolated from post-mortem brain tissue from patients with AD was studied.

METHODS

The human brain tissue samples were obtained from the Institute of Pathology of the University of Basle. Tissue samples from 6 age-matched controls (mean age 84 ± 7 years, postmortem delay 9 ± 2.5 hrs) and 9 AD cases (mean age 86 ± 6 years, postmortem delay 12.5 ± 9 hrs) including temporal cortex, hippocampus, striatum and cerebellum, were homogenized in 1 M NaCl/50 mM Tris-HCl pH 7.4. Centrifugation at 100,000g for 1 hour yielded a supernatant ("salt-soluble" fraction) and a pellet which was further extracted with 1% Triton-X100. The resulting supernatant after centrifugation at 100,000g for 1 hour was used ("detergent-soluble" AChE fraction) for further processing. To the detergent-soluble fractions the following sedimentation markers were added: alcoholdehydrogenase (4.8S), catalase (11.4S) and ß-galactosidase (16.0S). The fractions were subsequently layered on continuous sucrose density gradients (5-40%). After centrifugation at 290,000g for 17 hours the gradient was fractionated by piercing the bottom of the tube and collecting fractions of 0.3 ml for enzyme assays. AChE activity in the different fractions was determined according the colorimetric method of Ellman (1961) using an automatic Micro-plate reader (Molecular Devices, UVmax). All enzyme determinations were done at 25°C and pH

7.4. Inhibition experiments were performed after 15 minutes of enzyme preincubation with the corresponding inhibitors. The total enzyme activity of each molecular form was calculated from the integrated peak using the program Fig.P (Biosoft, Cambridge, UK). The individual forms were identified on the basis of their sedimentation coefficients (Massoulie and Bon, 1982). Fractions with sedimentation coefficients of 3.5-4S and 9-10S respectively were pooled and taken as G1 and G4 respectively. Protein determination was performed using the BioRad Assay.

RESULTS AND DISCUSSION

The AChE activity determined in human brain varied among the different brain regions (cortex, striatum, hippocampus and cerebellum). As shown in Table 1 no difference in AChE activity between the control and the AD group was found in the salt-soluble fraction. However, the detergent-soluble enzyme activity was reduced in all brain regions from AD cases. The most pronounced reduction in AChE activity was found in temporal cortex (-80%) followed by the hippocampus and striatum (70% decrease). In the cerebellum of AD brains, the AChE activity was only slightly reduced in comparison to control tissue. Kinetic experiments revealed that the affinity of the enzyme for its substrate was similar in the control and AD group, as indicated by the K_m values given in Table 1. As a consequence of the loss of enzyme, the apparent V_{max} values were decreased in the regions of AD brains.

Table 1. AChE activities in Human Brain Regions of Controls and AD Subjects.

Brain Region	AChE Activity Salt-soluble Fraction	AChE Activity Detergent-soluble Fraction	Detergent-soluble Fraction K_m [mM]	Detergent-soluble Fraction app. Vmax
Controls				
Cortex	0.35±0.05	1.33±0.17	0.149	1.7
Hippocampus	0.44±0.03	2.67±0.39	0.151	3.4
Striatum	2.07±0.25	36.82±5.37	0.168	50.8
Cerebellum	1.94±0.26	17.14±1.29	0.167	23.2
Alzheimer				
Cortex	0.33±0.07	0.26±0.07[*]	0.139	0.4
Hippocampus	0.40±0.05	0.92±0.28[*]	0.145	0.9
Striatum	1.73±0.36	12.46±1.64[*]	0.141	13.7
Cerebellum	2.19±0.18	13.90±2.86	0.151	18.2

AChE activity was determined in aliquots of the supernatant obtained after centrifugation (100,000g, 1hr) of the high salt (1M NaCl) extraction (salt-soluble fraction) and that of re-extracted pellet (1% Triton, 100,000g, 1 hr, detergent-soluble fraction). Values represent means μmol x min^{-1} x g^{-1} fresh brain tissue ±SD n=6 (controls), n=9 (AD), K_m : Michaelis constant mM, app. V_{max} : relative maximal velocity: μmol x min^{-1} x g^{-1}. [*] 2p< 0.05 t-test.

Following the fractionation of the detergent-soluble AChE on a sucrose density gradient, most of the AChE activity from control brains was found in the fraction corresponding to the G4 form. In agreement with the results of other groups we found a marked decrease of AChE activity of the G4 form in the cortex and the hippocampus of the AD brains, whereas no difference in the G1 activity was observed (Table 2).

The inhibitory potency toward AChE of some drugs under clinical investigation for the symptomatic treatment of AD has been further investigated in vitro. In crude detergent-soluble enzyme preparations the following order of inhibitory potency was found: physostigmine (IC_{50}:0.05 μM) > heptylphysostigmine (IC_{50}: 0.07 μM) > tacrine (IC_{50}:0.1 μM) > SDZ ENA 713 (IC_{50}: 10 μM).

In all in vitro experiments the influence of the investigated inhibitors on the AChE's extracted from control and AD brain tissues, was similar.
compared to the G4 form.

Table 2. Activity of AChE Molecular Forms in Cortex and Hippocampus of Control and AD Brains following Sucrose Density Gradient Separation .

Brain region	AChE G1 Form	AChE G4 Form	Ratio G4/G1
Cortex			
Controls	399 ± 27	751 ± 39	1.88
Alzheimer	408 ± 22	157 ± 41[*]	0.38
Hippocampus			
Controls	357 ± 37	1824 ± 187	5.12
Alzheimer	306 ± 46	661 ± 65[*]	2.16

Aliquots of detergent-soluble extracts were loaded on sucrose density gradients (5-40%) and centrifuged as described in methods. The molecular nature of each peak was identified using its sedimentation velocity (internal markers:β-galactosidase (16S), catalase (11.4S) and alcoholdehydrogenase (4.8S)). The relative abundance was estimated from the integrated peak heights. Values shown are means ±SD in $nmol \cdot min^{-1} \cdot g^{-1}$ fresh brain tissue. * $2p < 0.05$ t-test.

Table 3. Inhibition of AChE Forms in Human Cortex and Striatum by different Compounds

Inhibitor	G4 Form IC_{50} [M]	G1 Form IC_{50} [M]	Ratio IC_{50} G4/G1
Cortex			
SDZ ENA 713	$1.33 \pm 0.17 \cdot 10^{-5}$	$3.15 \pm 0.42 \cdot 10^{-6}$	4.2
Heptylphysostigmine	$4.71 \pm 0.55 \cdot 10^{-8}$	$2.27 \pm 0.37 \cdot 10^{-8}$	2.1
Physostigmine	$3.86 \pm 0.34 \cdot 10^{-8}$	$3.60 \pm 0.41 \cdot 10^{-8}$	1.1
Tacrine	$9.69 \pm 0.68 \cdot 10^{-7}$	$1.02 \pm 0.13 \cdot 10^{-7}$	0.9
Striatum			
SDZ ENA 713	$1.62 \pm 0.27 \cdot 10^{-5}$	$2.76 \pm 0.36 \cdot 10^{-6}$	5.9
Heptylphysostigmine	$3.97 \pm 1.13 \cdot 10^{-8}$	$1.07 \pm 0.30 \cdot 10^{-8}$	3.8
Physostigmine	$4.08 \pm 0.23 \cdot 10^{-8}$	$2.82 \pm 0.27 \cdot 10^{-8}$	1.4
Tacrine	$1.08 \pm 0.12 \cdot 10^{-7}$	$9.13 \pm 0.57 \cdot 10^{-8}$	1.2

Aliquots of the pooled fractions containing G1 and G4 form from controls and AD brains were preincubated with the different inhibitors and the remaining AChE activity subsequently determined (substrate concentration 0.5 mM acetylthiocholine, pH 7.4, 25°C). The IC_{50}'s were calculated by linear regression of the log concentration vs % inhibition (range 20-80% inhibition). Values represent mean ± SD from at least 7 individual inhibition experiments.

While physostigmine and tacrine inhibited equally well the G1 and G4 forms, as expressed by the near unity of the ratio of IC_{50}'s G4/G1 in Table 3, a clear difference was found for SDZ ENA 713 and heptylphysostigmine, as shown in Table 3: in cortex and striatum respectively, SDZ ENA 713 is 4 respectively 6 times and heptylphysostigmine 2 respectively 4 times more potent in inhibiting the G1 form as compared to the G4 form.

Figure 1 illustrates the in vitro inhibition by SDZ ENA 713 of G1 and G4 enzymes from human cortex; with the IC_{50}'s being 3.5 and 12.6 μM respectively. The apparent K_i for this drug determined using enzymes separated from human striatal tissue, were 2 μM for the G1 and 11 μM for the G4 enzyme as shown in Figure 2.

The equal potency of tacrine regarding the inhibition of the G1 and G4 enzyme is evidenced by the similar K_i values, 0.43 and 0.47 μM respectively, obtained from kinetic experiments with human striatal enzymes (Figure 3).

There are several implications of a preferential inhibition of the G1 form. The membrane-bound G4 form is probably located presynaptically at cholinergic nerve endings and may be directly involved in the regulation of ACh transmission (Marquis and Fishman, 1985). It seems therefore that the loss of G4 represents a selective depletion of the membrane pool, possibly reflecting the state of degeneration of cholinergic terminals in AD. On the other hand, the activity of the G1 form, reflecting neurotransmitter release unrelated ACh degradation remains unchanged. A preferential inhibition of this enzyme could be beneficial in situations of cholinergic hypofunction. It has been reported that AChE is associated with senile plaques and tangles (Perry et al, 1980) and that the enzyme is accumulated in both plaques and tangles of AD brains (Mesulam and Moran, 1987, Mesulam and Geula, 1990). Based on histochemical investigations on AD brains, it has been suggested that these enzymes have different properties regarding interactions with substrates and inhibitors (Schätz et at. 1990). Recently, (Nakamura et al. 1990) confirmed an accumulation of AChE activities in fractions enriched with plaques and tangles of AD brains. After digestion with

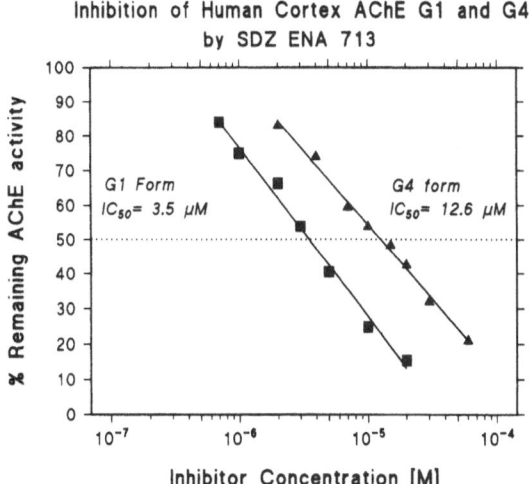

Figure 1. Example of inhibition curves obtained with SDZ ENA 713 on G1 and G4 AChE separated from human neocortex.

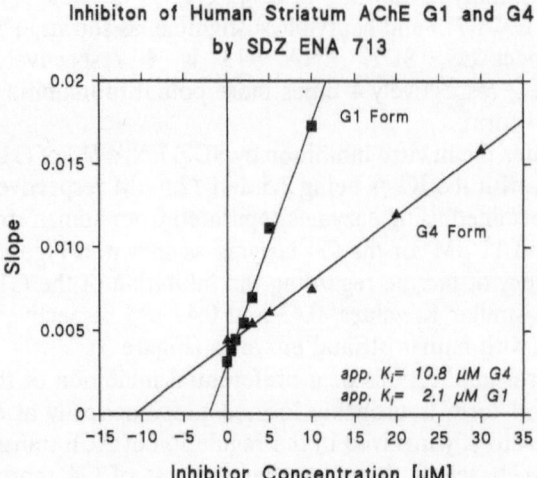

Figure 2. Inhibition of G1 and G4 AChE from striatal tissue by SDZ ENA 713. Plot of the slopes vs inhibitor concentration. The slopes were obtained from double-reciprocal plots deriving from inhibition experiments with substrate concentrations of 0, 0.1, 0.125, 0.166, 0.26 and 0.5 mM acetylthiocholine, pH 7.4, 25°C, after 15 min. of preincubation. The initial slopes were calculated by non-linear regression analysis (Enzfitter, Elsevier-Biosoft Cambridge UK) and the secondary plot (this fig.) by linear regression of the slopes vs inhibitor concentration.

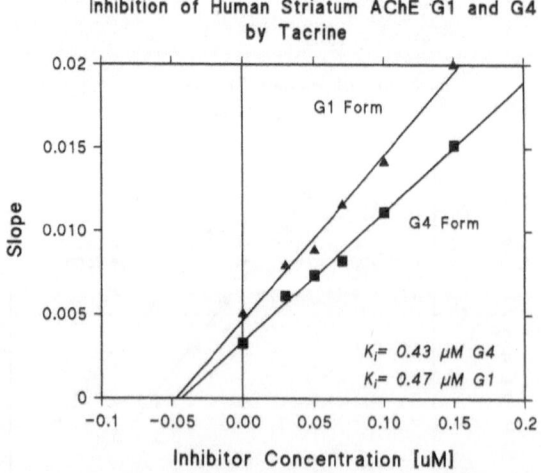

Figure 3. Inhibition of G1 and G4 AChE from human striatal tissue by tacrine. Secondary plot as indicated in legend for Figure 2.

collagenase and trypsin he found only the G4 enzyme in the supernatant. From these results the authors concluded that the AChE in plaques and tangles might be of the asymmetric type, anchored to the senile plaques through a collagen tail. Only digestion with collagenase and protease resulted in a liberation of the G4 enzyme. We found in similarly prepared fractions of AD cortex both G4 and G1 enzyme equally distributed after detergent extraction. We are currently studying the AChE accumulated in plaques and tangles, to define the forms which are present in these pathological structures.

REFERENCES

Atack, J.R., Perry, E.K., Bonham, J.R., Perry, R.H., Tomlinson, B.E., Blessed, G., and Fairbairn, A., Molecular forms of acetylcholinesterase in senile dementia of Alzheimer's type: selective loss of the intermediate (10S) form, *Neurosci.Lett.* 40:199 (1983).

Bartus, R.T., Dean, R.L., Beer,B., and Lippa, A.S., The cholinergic hypothesis of geriatric memory dysfunction, *Science* 217:408 (1982).

Davis, K.L., and Mohs, R.C., Enhancement of memory processes in Alzheimer's disease with multiple-dose intra-venous physostigmine, *Am.J.Psychiatry* 139:1421 (1982).

Drachman, D.A.; Memory and cognitive function in man: does the cholinergic system have a specific role?, *Neurology* 27:783 (1977).

Drachman, D.A., and Leavitt, J., Human memory and the cholinergic system, *Arch.Neurol.* 30:113 (1974).

Ellman, G.L., Courtney, K.D., Andres, V., and Featherstone, R.M., A new and rapid colorimetric determination of acetylcholinesterase activity, *Biochem.Pharmacol.* 7:88 (1961).

Enz, A., Boddeke, E., Gray, J., and Spiegel, R., Pharmacological and clinical-pharmacological properties of SDZ ENA 713, a centrally selective acetylcholinesterase inhibitor, Ann.NY Acad.Sci. 640:272 (1991).

Greenfield, S.A., Acetylcholinesterase may have novel functions in the brain, *Trends Neurosci.* 7:365 (1984).

Greenfield, S.A., Chubb, I.W., Grunewald, R.A., Henderson, Z., May, J., Portnoy, S., Weston, J., and Wright, M.C., A non-cholinergic function for acetylcholinesterase in the substantia nigra: behavioral evidence, *Exp.Brain.Res.* 54:513 (1984).

Marquis, J.K., and Fishman, E.B., Presynaptic acetylcholinesterase, *TIPS* 56:387 (1985).

Massoulie, J., and Bon, S., The molecular forms of cholinesterase and acetylcholinesterase in vertebrates, *Ann.Rev.Neurosci.* 5:57 (1982).

Mesulam, M.M., and Geula, C., Shifting patterns of cortical cholinesterases in Alzheimer's disease:Implications for treatment, diagnosis, and pathogenesis, *Adv.Neurol.* 51:235 (1990).

Mesulam, M.M., and Moran, M.A., Cholinesterase within neurofibrillary tangles of aging and Alzheimer's disease, *Ann.Neurol.* 22:223 (1987).

Nakamura, S., Kawashima, S., Nakano,S., Tsuji, T., and Araki, W., Subcellular distribution of acetylcholinesterase in Alzheimer's disease: abnormal localization and solubilisation, *J.Neural.Transm.*[Suppl] 30:13 (1990).

Perry, R.H., Blessed, G., Perry, E.K., and Tomlinson, B.E., Histochemical observations on cholinesterase activities in the brains of elderly normal and demented (Alzheimer-typ) patients, *Age Ageing* 9:9 (1980)

Schätz, A.S., Geula, C., and Mesulam, M.M., Competitive substrate inhibition in the histochemistry of cholinesterase activity in Alzheimer's disease, Neurosci.Lett. 117:56 (1990).

Siek, G.C., Katz, L.S., Fishman, E.B., Korosi, T.S., and Marquis, J.K., Molecular forms of acetylcholines terase in subcortical areas of normal and Alzheimer disease brain, *Biol.Psychiatry* 27:573 (1990).

Summers, W.K., Majovski, L.V., Marsh, G.M., Tachiki, K., and Kling, A., Oral tetrahydroaminoacridine in long-term treatment of senile dementia, Alzheimer type, New *Eng.J.Med.* 315:1241 (1986).

Summers, W.K., Tachiki, K., and Kling, A., Tacrine in the treatment of Alzheimer's disease. A clinical update and recent pharmacological studies, *Eur.Neurol.* 29(Suppl):28 (1989).

Thal, L.J., Fuyld, P.A., Masur, D.M., and Sharpless, N.S., Oral physostigmine and lecithin improve memory in Alzheimer's disease, *Ann.Neurol.* 13:491 (1983).

ACETYLCHOLINESTERASE INHIBITION BY NOVEL CARBAMATES:
A KINETIC AND NUCLEAR MAGNETIC RESONANCE STUDY

Marta Weinstock[1] · Michal Razin[1] · Israel Ringel[1] · Zeev Tashma[2]
and Michael Chorev[2]

Departments of Pharmacology[1] and Pharmaceutical Chemistry[2],
School of Pharmacy, The Hebrew University, Jerusalem, Israel

INTRODUCTION

Alzheimer's dementia (AD) is a condition in which there is a lack of
choline acetyl transferase, the enzyme responsible for the synthesis of
acetylcholine in the cerebral cortex and hippocampus[1]. This finding led to the
trial of anticholinesterase drugs, physostigmine and tetrahydroamino- acridine
as potential treatment for AD[2,3]. Although both drugs had some beneficial effect
in these conditions, their relatively high toxicity, and the chemical
instability and short half-life of physostigmine[4], prompted the search for safer
agents with a longer duration of action.

We have synthesized a number of derivatives of m-[1-(N,N-dimethylamino)-
ethyl] phenyl carbamate which were found to inhibit AChE in rodent brain.
Several of the compounds were significantly less toxic and had a longer duration
of action than physostigmine[5]. Since our limited <u>in vivo</u> data did not enable us
to obtain a clear understanding of the influence of the alkyl substituents in
the carbamate side chain on inhibitory potency and its duration, we synthesized
additional derivatives and studied in more detail the kinetic parameters of
their inhibition of a preparation of human erythrocyte AChE.

Carbamates become attached to the esteratic site of the enzyme through
nucleophilic attack by the OH group of the serine residue on the carbonyl of the

Multidisciplinary Approaches to Cholinesterase Functions, Edited by
A. Shafferman and B. Velan, Plenum Press, New York, 1992

phenyl carbamate group[6]. This process depends both upon the degree of positive charge on the amide carbonyl and its accessibility to the nucleophilic OH group at the active site of the enzyme. The C-N bond of amides generally displays a partial double bond character, with restricted rotation in comparison to that about a pure single bond[7]. The degree of double bond character determines the extent of the rotational barrier and charge distribution in the amide group, which are dependent on the nature of its N,N-dialkyl substituents. The rate of rotation about the C-N amide bond is temperature dependent and can be assessed by dynamic nuclear magnetic resonance (NMR) spectrometry.

The inhibition of AChE by carbamates takes place through the formation of a reversible enzyme-inhibitor complex, which is followed by the establishment of an O-C bond between the esteratic site in the enzyme and the carbonyl of the carbamate[6]. The carbamoylated enzyme then undergoes relatively slow hydrolysis to restore its activity. This reaction may be described by the dissociation constant for the formation of the reversible enzyme inhibitor complex K_i, which is a measure of the reciprocal of the affinity of the carbamate for the enzyme and the overall dissociation constant K_i^*[8]. The higher the value of K_i/K_i^*, the greater the accumulation of the inhibited enzyme complex and the more likely the inhibitor to have useful activity in vivo. These parameters were assessed in the present study, and compared to the ability of the compounds to inhibit AChE in the brain after injection to mice.

METHODS

The N-monoalkyl and N,N-dialkyl substituted phenyl carbamates were synthesized from m-[1-(N,N-dimethylamino)ethyl] phenol, as described in[5]. The structures of the phenyl carbamates reported in this paper are shown in Table 1.

Measurement of Anticholinesterase Activity

AChE activity was measured by the method of Ellman et al.[9]. In vitro kinetic measurements were carried out at 37°C in 0.1M phosphate buffer, pH 8 in a reaction volume of 1ml containing 0.01M DTNB (20μl), 0.075M AtCh (25μl), 0.05U human erythrocyte enzyme (Type XIII, Sigma, Ltd.) (50μl) and one of 5 different concentrations of inhibitor (close to the value of their K_i). The K_m value for AtCh was 2.3×10^{-4}M.

Table 1. Kinetic parameters and activation energy for internal rotation of amide bond.

Compound	R1	R2	Kix10⁻⁵M ± s.c.	Ki/Ki*	ΔG‡ (Kcal/mol)
RA2	H	CH_3	0.09±0.002	10	--
RA6	H	C_2H_5	8.61±0.75	193	--
RA15	H	$n-C_3H_7$	2.41±0.17	154	--
RA5	H	$n-C_4H_9$	0.25±0.01	45	--
RA17	H	$n-C_6H_{13}$	0.15±0.01	65	--
RA13	H	$iso-C_3H_7$	51.5±1.3	163	--
RA3	H	$sec-C_4H_9$	25.5±0.5	232	--
RA12	H	$c-C_6H_{11}$	2.5±0.1	183	--
RA10	CH_3	CH_3	0.25±0.01	16	16.9
RA7	CH_3	C_2H_5	20.2±1.9	168	16.0
RA4	CH_3	$n-C_3H_7$	6.70±0.31	265	16.6
RA16	CH_3	$n-C_4H_9$	7.86±1.51	164	16.5
RA8	C_2H_5	C_2H_5	65.6±1.1	70	15.7
Physostigmine	H	CH_3	0.056±0.007	25	--

Ex Vivo Measurements

Male mice, weighing 25-30 g were injected subcutaneously (sc) with either 1ml/100gm saline or one of at least 3 doses (4-5 mice per dose) of the above carbamates or physostigmine. They were sacrificed 15, 30 or 60 min later and the brain rapidly removed minus the cerebellum, weighed, homogenized in 0.1M phosphate buffer, pH 8, containing 1% Triton, (1ml/100mg tissue) and centrifuged to remove cell debris. Aliquots of enzyme ($25\mu l$) were incubated with AtCH and AChE activity was measured as described above. The % inhibition of whole brain AChE by each dose of drug was calculated by comparison with the enzyme activity from 3 saline-treated mice injected at the same time. The dose of each drug which inhibited AChE by 50% (ED_{50}) was calculated and was used to determine the time course of enzyme inhibition in another group of 28-30 mice per drug. These were sacrificed 15, 30, 60, 90, 120, 180 & 240 min after injection and the brain processed for AChE determination as described above.

Acute Toxicity

Acute toxicity of the carbamates, as indicated by the LD_{50} (dose to kill 50% of animals) was assessed in mice after sc. injection of at least 3 doses of each drug to groups of 10 mice per dose. The therapeutic ratio was expressed as LD_{50}/ED_{50}.

Kinetic Analysis

AChE (E) reacts with carbamates (CX) by the formation of a reversible enzyme-inhibitor complex (ECX) followed by the carbamoylation of the enzyme to yield (EC) and concomitent displacement of the leaving group, (X). EC then undergoes hydrolysis to regenerate the enzyme (E) and the inactive carbamate moiety (C). Carbamates are relatively slow binding inhibitors and conform to the following scheme:[8]

$$E + CX \xrightarrow{\text{Ki}} ECX \xrightarrow{\text{k1}} EC \xrightarrow{\text{k2}} E + C \qquad (1)$$
$$+X$$

where: $Ki = \dfrac{[E][CX]}{[ECX]}$, the dissociation constant.

k1 = the rate constant for the formation of the carbamoylated enzyme

k2 = the rate constant for the decarbamoylation of the enzyme. The overall dissociation constant Ki* may be approximated to:

$$Ki^* = \dfrac{Kik2}{k1} \qquad (2)$$

NMR Measurements of the Energy Barrier for Rotation about the C–N Bond

Solutions of identical concentration of N,N-disubstituted inhibitors (30mM) were examined by proton NMR at temperatures between 20-75°C on a Varian VXR 300s spectrometer, interfaced with a Sun-3 computer and equipped with a 5 mm multinuclear probe. Measurements could not be made on N-monosubstituted inhibitors as they do not exhibit split signals for the amide substituent. The compounds were dissolved in 0.1M phosphate buffer (pH 8.0, in D_2O, pH was calculated as pD + 0.4). Spectra were measured every 2°C from 20°C to 75°C and every degree around the coalescence temperature. The energy barrier for rotation about the C–N bond (ΔG‡) for each compound was calculated from the following approximation:[7]

$$\Delta G^{\ddagger} = 19.14 \times T_{coal}(9.97 + \log T_{coal}/\delta v) \times 2.39 \times 10^{-4} \text{ Kcalmol}^{-1}.$$

RESULTS AND DISCUSSION

Kinetics of AChE Inhibition

The kinetic parameters for AChE inhibition by mono- and disubstituted phenyl carbamates are given in Table 1. This shows that the inhibition of AChE by carbamates is markedly influenced by the nature and size of the alkyl substituents on the amide group. The affinity of the monosubstituted compounds to the active site is greatest when the alkyl substituent is a methyl group or a long linear chain such as n-hexyl. This is in accordance with recent findings of the existence of an extensive hydrophobic pocket in the structure of AChE, adjacent to the substrate binding site[10]. Our data suggest that this pocket can accomodate a linear but not a branched alkyl chain. Extension of the chain from C2 to C6 results in a progressive increase in affinity which suggests the presence of stronger interactions between the alkyl group and the hydrophobic pocket. The addition of a methyl group to form the disubstituted carbamates exerts some interference with the fit of the molecule to the site and with its accomodation within the hydrophobic pocket. This prevents the increase in affinity as the size of the second alkyl is increased from n-propyl to n-butyl. The interference becomes much more pronounced when both methyls are replaced by ethyls, or when a branched-chain moiety, such as iso-propyl or sec-butyl is introduced.

In general, a reciprocal relationship was seen between the affinity of the compounds to the enzyme and the value of Ki/Ki*, which represents the stability of the carbamoylated enzyme. (Table 1). This suggests that once the acylation of the enzyme has occurred, the presence of bulky groups at the active site can impede spontaneous hydrolysis of the carbamoylated moiety, or diffusion of the carbamate away from the hydrophobic pocket.

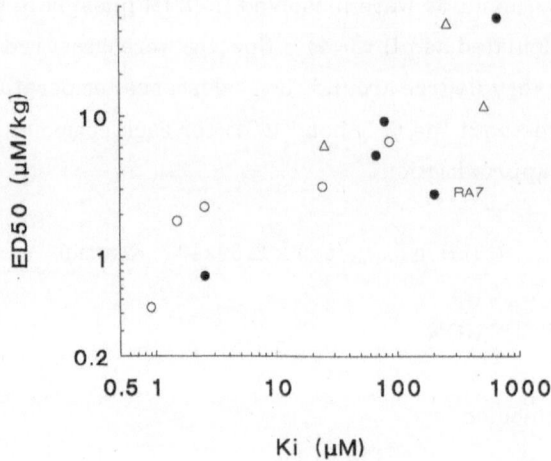

Fig 1. Relationship between the the dissociation constant for inhibition of erythrocyte enzyme and the ED_{50} for inhibition of mouse brain AChE. o N-monosubstituted compounds; ● N,N-disubstituted compounds; branched chain compounds.

Relationship Between Energy of Rotation About the C–N Bond and Affinity of Compounds to AChE.

For a given N-alkyl substituting an amide bond, the two possible configurations (syn and anti) result in different magnetic environments. The rate process of rotation about the amide bond may be observed by means of dynamic NMR. The extent of the rotational barrier for an inhibitor indicates the amount of double-bond character associated with the electrophilic carbon which participates in the inhibition of AChE.

NMR studies of the energy barrier for the rotation of N,N-disubstituted amides provided a possible explanation for the differences in the kinetic behavior of these carbamates. It was found that restriction of rotation about the amide bond favored the association of the carbamate and the enzyme. The amount of energy needed to overcome the restriction for rotation, ($\Delta G\ddagger$) was directly correlated (r=0.896) to the affinity of the inhibitor for the enzyme (1/Ki) (Table 1). This was probably due to the preferred conformational orientation of the carbamate and to the increase in the partial positive charge localized on the carbonyl C atom. The latter is the target for the nucleophilic reaction at the esteratic site, which is correlated with the extent of the double bond character at this location.

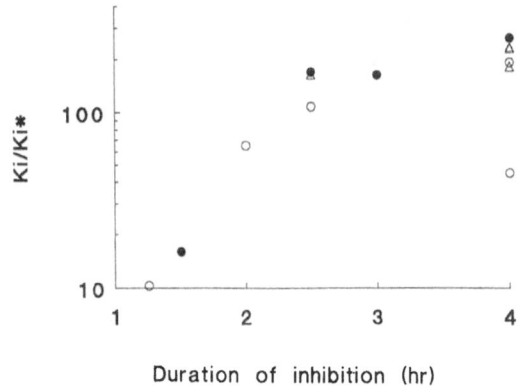

Duration of inhibition (hr)

Fig 2. Relationship between Ki/Ki*, the durability of the carbamoylated enzyme and the duration of AChE inhibition in mouse brain.

o N-monosubstituted compounds; ● N,N-disubstituted compounds; △ branched chain compounds.

Comparison of the kinetic data for enzyme inhibition by the carbamates in vitro and in vivo revealed strong correlations between the affinity to the enzyme and the dose required to inhibit brain AChE in mice by 50% (ED_{50}) (Fig 2). The ratio Ki/Ki* gave a fairly accurate prediction of the duration of action of the compound in the whole animal. The one exception was the n-butyl compound which had a long duration of action but a low Ki/Ki*. The reason for this is not clear, but it is possible that the relatively high lipid solubility of RA5 may have caused it to be taken up in fat depots in which it was protected from hydrolysis, and released slowly into the circulation.

The therapeutic ratios (TR) of the carbamates is shown in Fig 3. The branched chain derivatives were the most toxic, with similar TR to that of physostigmine (5.0), while the disubstituted analogues were the safest drugs. In each group of straight chain analogues, the lowest toxicity was found in the ethyl derivatives, and this rose with increasing chain length. RA6 & RA7 cause significantly less inhibition of AChE in peripheral tissues than in the brain[5,10]. This diminishes the liklihood of respiratory muscle paralysis and reduces the severity of other cholinergic side effects usually encountered with physostigmine at therapeutic doses.

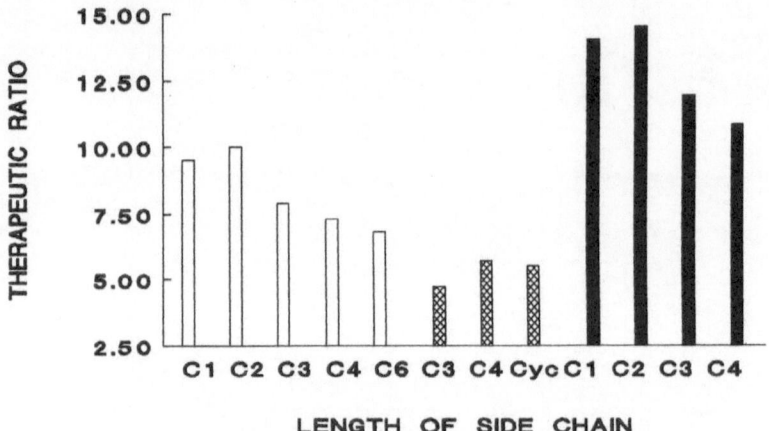

Fig 3. Therapeutic ratios of phenylcarbamates.
Open bars: N-monosubstitued compounds; filled bars: N,N-disubstituted compounds; Crossed-hatch bars: branched-chain compounds.

The results of this study demonstrate that a NMR and kinetic analysis of the interaction between carbamates and human erythrocyte AChE can provide valuable information about their potential as inhibitors of this enzyme in the whole animal. A longer duration of action can be achieved by increasing the size of the substituents on the nitrogen of the carbamate. A reduction in the incidence and intensity of cholinergic side effects may also be achieved by the use of N,N-disubstituted analogues[5,11].

REFERENCES

1. P. Davies and A.J.F. Maloney, Selective loss of central cholinergic neurones in Alzheimer's disease. Lancet, 2: 1403, 1976.

2. J. Thal, P.A. Fuld, D.M. Masur, and N.S. Sharpless, Oral physostigmine and lecithin improve memory in Alzheimer's disease. Ann. Neurol. 13: 491-496, 1983

3. W.K. Summers, L.V. Majovski, G.M. Marsh, K. Tachiki and A. Kling, Oral tetrahydroaminoacridine in long-term treatment of senile dementia, Alzheimer type. N. Engl. J. Med. 315: 1241-1245, 1986.

4. J.E. Christie, A. Shering, J. Fergusen and A.I.M. Glen, Physostigmine and arecholine: effects of intravenous infusions in Alzheimer presenile dementia. Br. J. Psychiatry, 138: 46-50, 1981.

5. M. Weinstock, M. Razin, M. Chorev, and Z. Tashma, 1986, Pharmacological activity of novel anticholinesterase agents of potential use in the treatment of Alzheimer's disease. in: "Advances in Behavioral Biology", A. Fisher, I. Hanin and C. Lachman, ed., Plenum Press, New York and London.

6. A.R. Main, Mode of action of anticholinesterases. Pharmac. Ther. 6: 579-628, 1979.

7. H. Guenther, Nuclear Magnetic Resonance Spectroscopy. Akademie Verlag, Berlin, 1980.

8. J.F. Morrison and C.T. Walsh, The behavior and significance of slow binding enzyme inhibitors. Adv. Enzym. 61: 201-299, 1988.

9. G.L. Ellman, K.D. Courtney, V. Andres, Jr. and R.M. Featherstone, A new and rapid colorimetric determination of acetylcholinesterase activity. Biochem. Pharmacol. 7: 88-95, 1961.

10. J.L Sussman, M. Harel, F. Frolow, C. Oefner, A. Goldman, L. Toker, and I. Silman, Atomic structure of acetylcholinesterase from Torpedo californica: A prototype acetylcholine-binding protein. Science, 253, 872-879, 1991.

11. E. Almalem, M. Chorev and M. Weinstock, Antagonism of morphine-induced respiratory depression by novel anticholinesterase agents. Neuropharmacology, 30: 1059-1064, 1991.

EXPERIMENTAL ACETYLCHOLINESTERASE AUTOIMMUNITY

Stephen Brimijoin[1], Vanda Lennon[2] and Zoltán Rakonczay[1,3]

[1]Department of Pharmacology
[2]Departments of Immunology and Neurology
Mayo Clinic
Rochester MN 55905, USA
[3]On leave from Central Research Laboratory
A. Szent-Györgyi Medical University
6720 Szeged, Hungary

INTRODUCTION

We are studying the immunopathology of acetylcholinesterase (AChE) with the aim of producing useful experimental models for diseases of the cholinergic nervous system. AChE is affected directly or indirectly when cholinergic neurons degenerate, as in amyotrophic lateral sclerosis (Fernandez et al., 1986), Alzheimer's disease (Atack et al., 1983; Fishman et al, 1986; Hammond and Brimijoin, 1988), Down's Syndrome (Yates et al., 1980), Parkinsonism (Perry et al., 1985), and olivopontocerebellar atrophy (Kish et al., 1988). Because of AChE's physiological importance in the cholinergic synapse, we constructed an animal model of passive AChE-autoimmunity based upon in vivo injections of monoclonal antibodies to neural AChE. The intent was not to promote the idea of an immunologic pathogenesis for any particular neural disorder, but to explore the possibilities and consequences of experimentally inducing antibody-mediated lesions in cholinergic systems. Here we present some of our current results and review recently published findings.

RESULTS AND DISCUSSION

Effects of AChE-Antibodies on the Peripheral Nervous System

AChE antibodies can be delivered systemically to adult rats by injection through the tail vein. We typically inject a defined mixture of 5 different monoclonal antibodies (ZR 2, 3, 4, 5, & 6, 300 µg each), IgG subtype 2b, raised against rat brain AChE (Rakonczay and Brimijoin., 1986). After such an injection, biologically significant amounts of antibody reach antigenic targets in muscle and sympathetic ganglia, where most of the accessible asymmetric forms and globular tetramers of AChE become complexed with IgG. Little or no AChE antibody is detected in brain or spinal cord because, except under special circumstances, the antibodies are excluded by the blood brain barrier.

The physiological effects of the AChE-antibodies suggest an immunological attack that is focussed on preganglionic sympathetic neurons, with minimal effects on other cholinergic systems. This somewhat surprising view is now supported by a large body of evidence, which can be summarized as follows.

Multidisciplinary Approaches to Cholinesterase Functions, Edited by
A. Shafferman and B. Velan, Plenum Press, New York, 1992

Adult rats given i.v. AChE antibodies pass through a brief period of sympathetic activation but soon develop a sympathetic depression, which continues indefinitely. As described elsewhere (Brimijoin and Lennon, 1990; Moser et al., 1991) the sympathetic disability is not accompanied by gross motor weakness, structural damage to the neuromuscular junction, or obvious dysfunction in the parasympathetic nervous system.

The functional effects of AChE antibodies are initially most noticeable in the eyes, where exophthalmos begins 20 min after injection, then fades. Next comes eyelid drooping (ptosis, a sign of sympathetic denervation). Ptosis is already maximal by 2 hours and continues for months, perhaps permanently, far outlasting the AChE antibody in the bloodstream (Brimijoin and Lennon, 1990; Moser et al., 1991). Because AChE stores in nerve and muscle are replaced in a few weeks, persistent ptosis must reflect cellular lesions that are not readily repaired.

There are many sites where AChE-antibodies might induce lesions that cause ptosis, among other neurologic signs (Fig. 1).

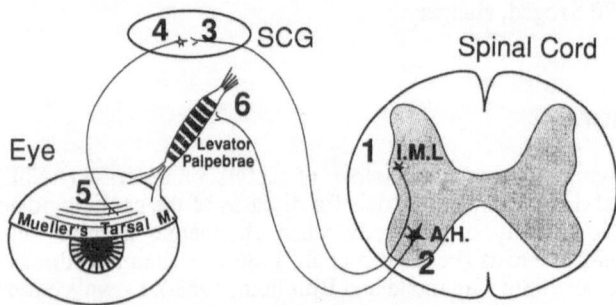

Potential Antibody Targets:

1. Intermediolateral Cell Column	➤	Preganglionic Sympathetic Neurons
2. Anterior Horn	➤	Somatic Motor Neurons
3. Superior Cervical Ganglion	➤	Preganglionic Sympathetic Terminals
4. Superior Cervical Ganglion	➤	Postganglionic Sympathetic Neurons
5. Mueller's Tarsal Muscle	➤	Sympathetic Terminals

Figure 1. AChE-rich sites and other potential targets of AChE antibodies in the somatic motor and sympathetic nervous systems.

In view of the fact that eyelid smooth muscle is under tonic adrenergic control (Isola and Bacq, 1946), sympathetic ganglia are the most likely sites of damage. Antibody-mediated damage to the superior cervical ganglion was first demonstrated by recording tension in the eyelid while electrically stimulating the sympathetic system. In control rats, lid tension increased during stimulation of the cervical preganglionic nerve and during stimulation of the ganglion itself. Eyelid contraction was also induced by dimethylphenylpiperazinium (DMPP), a selective agonist for nicotinic acetylcholine receptors of ganglia. In antibody-treated rats, tested 3 to 90 days after induction of ptosis by AChE-antibodies, preganglionic stimulation caused no eyelid response even though direct ganglionic stimulation and DMPP remained effective (Brimijoin and Lennon, 1990). These results can only be explained by assuming that the postganglionic sympathetic neurons were intact but the preganglionic terminals were disabled or destroyed.

Another sign of damage to the sympathetic nervous system is the effect of AChE-antibodies on heart rate and blood pressure. Following a latent period that lasts about 20 min after antibody-injection, most rats show transient hypertension and tachycardia (Fig. 2). By two hours, however, heart rate and blood pressure fall below normal and remain depressed for weeks (Fig. 3). Atropine does not raise blood pressure or heart rate at this stage. Hence, the cardiovascular effects of antibody do not reflect buildup of acetylcholine at muscarinic receptors. As with the control of eyelid tone, the cardiovascular effects are consistent with damage to the preganglionic system and preservation of the postganglionic system. Thus, the ganglionic agonist, DMPP, is just as effective in causing a rise in blood pressure in antibody-treated rats as it is in controls.

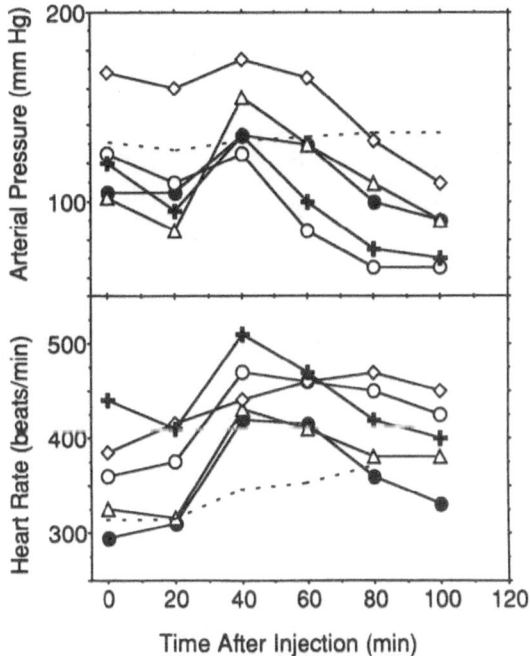

Figure 2. Acute hypertension and tachycardia after AChE-antibody. Mean arterial pressure and heart rate were recorded from a cannula in the right carotid artery. Solid lines represent individual rats receiving AChE antibodies at time zero. Dashed lines are mean values from control rats given normal IgG.

Enzyme assays and morphologic observations have confirmed the concept that damage by AChE-antibodies is selective for preganglionic sympathetic neurons. Histochemical staining at 1 to 90 days after antibody-injection reveals striking loss of AChE activity in ganglionic neuropil but not in nerve cell bodies. Adrenergic markers like dopamine-β-hydroxylase and tyrosine hydroxylase in the superior cervical ganglia are largely unaffected (Brimijoin and Lennon, 1990; Hammond and Brimijoin, unpublished data). Furthermore, norepinephrine levels in adrenergic end organs remain close to normal values (Tyce and Brimijoin, unpublished). Again, this information leads to the conclusion that adrenergic neurons must be spared.

Neuropathological studies indicate that the antibody-mediated damage to preganglionic neurons begins with lysis of the cholinergic terminals in the sympathetic ganglia. Electron micrographs of superior cervical ganglia from control rats show many

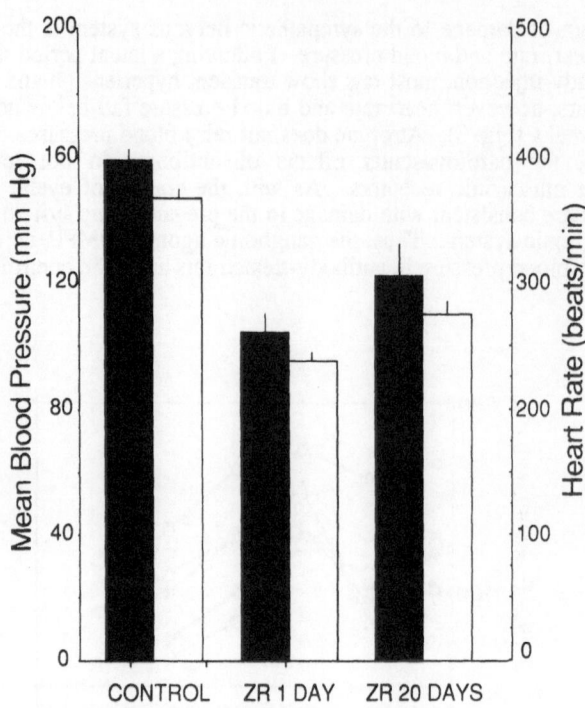

Figure 3. Chronic hypotension and bradycardia after AChE-antibody. Arterial pressure and heart rate were recorded 1 or 20 days after injection of AChE antibody or normal mouse IgG (pooled controls).

vesicle-laden terminals juxtaposed to pre- and postsynaptic densities. Ganglia examined one week after AChE-antibody have few remaining synapses (Brimijoin and Lennon, 1990). Loss of synaptic structures is paralleled by depletion of choline acetyltransferase activity (ChAT), a marker of cholinergic cytoplasm, which virtually disappears from superior cervical, stellate, thoracic, lumbar and coeliac ganglia, and adrenal gland.

The obvious inference from these findings is that AChE-antibodies globally destroy the terminals of preganglionic sympathetic neurons. Work still in progress suggests that the immunologic lesions in the peripheral fields of these neurons ultimately lead to the disappearance of the preganglionic nerve cell bodies in the spinal cord. In particular, there is a reduction in the number of cholinergic perikarya in the intermediolateral nucleus as judged by specific immunoperoxidase staining with an anti-ChAT antibody and by conventional Nissl staining (Brimijoin, unpublished). This delayed neuronal death presumably accounts for the permanent sympathetic dysfunction that follows antibody-exposure.

Experiments with purified anti-complementary factor from *Naja naja* cobra venom shed light on the mechanism of immunosympathectomy. In rats depleted of hemolytic complement by cobra venom factor, antibody-injection did not cause ptosis, and the sympathetic ganglia did not lose ChAT activity (Fig. 4). Ganglionic accumulations of IgG and complement (C3) are observed within a few hours of antibody-injection (Hammond et al., 1991). These findings imply that the neurologic lesions evoked by AChE antibodies are a direct consequence of complement activation.

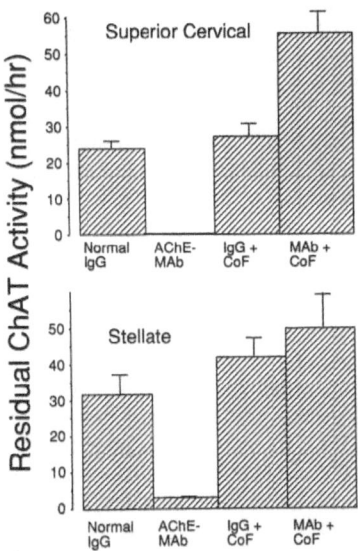

Figure 4. Inactivation of complement prevents antibody-induced loss of choline acetyltransferase. Rats were given either AChE antibody (AChE-MAb) or normal mouse IgG at zero time. Some rats also received repeated injections of cobra venom factor (CoF) to deplete complement component C3. Ganglionic ChAT activity was assayed after 5 days.

We suggest the following sequence of events in antibody-induced neural damage: 1) AChE antibodies diffuse from the circulation into ganglionic synapses; 2) Within one hour, immune complexes are formed with AChE located on the external membrane of presynaptic terminals; 3) The complement cascade is activated; 4) Attack complexes open the surface membrane and cause lysis of nerve terminals; 5) Immunologic reactions continue for many days until anti-AChE antibody is finally cleared from the circulation; 6) Prolonged damage to the terminals, with accompanying loss of trophic support from target structures, leads to reactive death of parent neurons in the spinal cord.

Effects of AChE-Antibodies on the Brain

Most immunoglobulins, including all but one of our monoclonal anti-AChE antibodies, are effectively excluded from the adult central nervous system by the blood brain barrier. The exceptional antibody, ZR-1, has been found to have a limited ability to accumulate in brain after i.v. injection (Brimijoin et al., 1990). However, ZR-1 alone does not appear to trigger neuropathologic events in AChE-expressing target neurons, either because the central accumulation is too small or because the ZR-1 epitopes are too scarce (Brimijoin, unpublished). On the other hand, when several monoclonal AChE-antibodies are injected together, directly into the caudate nucleus, they cause major damage to AChE-rich fibers in the striatum and immediately adjacent structures (Bean et al, 1991). Interesting to note, the loss of AChE-fiber staining in the injected caudate is not accompanied by a loss of AChE-positive cell bodies, nor by changes in the immunostaining of a wide variety of peptidergic neurons. These observations imply that antibody-mediated damage in the brain, as in the periphery, is highly selective.

Another line of investigation shows that AChE antibodies can induce widespread lesions of cholinergic systems throughout the brain if the antibodies are injected systemically by the intraperitoneal route, immediately after birth (Rakonczay and Brimijoin, 1991). During this period, when the blood brain barrier is immature, antibodies enter the brain, deposit themselves on AChE-rich targets, and rapidly induce lesions in AChE-containing fibers and terminals. In studies that are not yet published (Rakonczay, Brimijoin and Hammond) we have found that antibody-treatment within 24 hours of birth causes a protracted disappearance of AChE-fiber staining in most regions of the cerebral cortex. These effects, like those induced by antibodies injected into the adult brain, occur without observable changes in staining for non-cholinergic markers such as tyrosine hydroxylase and glutamic acid decarboxylase. In fact, the immunologic lesions may be confined to a subset of neuronal structures that express large amounts of AChE on their outer surface.

Applications and Implications

We have now investigated three different experimental models of cholinergic auto-immunity in the rat, all based on passive transfer by injection of murine monoclonal antibodies to brain AChE: 1) global preganglionic immunosympathectomy by systemic i.v. antibody injection in adults; 2) localized lesions of the caudate nucleus by intracerebral antibody-injection in adults; 3) general lesion of AChE rich targets throughout the developing central nervous system by i.p. injection in newborns. Besides their intrinsic interest, each of these models has potential as an investigative tool for neurobiology and neuropathology.

A unique feature of preganglionic immunosympathectomy is the probable blockade of all neural input to sympathetic ganglia and the adrenal gland. This feature contrasts with guanethidine-induced sympathectomy (Burnstock et al, 1971; Johnson and O'Brien, 1976) which targets postganglionic neurons and spares the adrenal gland. A further contrast with "conventional" sympathectomy is the preservation of cell-cell interactions between adrenergic neurons and their effector cells. The antibody-model should be useful for defining the peripheral actions of antihypertensive drugs and for studying the activity-dependent regulation of adrenergic receptors in the vasculature. Furthermore, from a clinical standpoint, the pathophysiology of preganglionic immunosympathectomy is sufficiently similar to certain human disorders, such as Shy Drager syndrome, to warrant critical analysis of the underlying mechanisms.

Localized antibody-mediated lesions after intracerebral injection have obvious utility for studies of central cholinergic circuits, provided that they can be shown to be at least as specific as that of the cholinergic neurotoxin AF64A (Eva et al., 1987; Hanin, 1990). Current information suggests that the specificity is indeed high, and it may be reasonable to plan experiments on the functional role of AChE-rich fibers and terminals in basal ganglia and cortical circuits.

Finally, the generalized cholinergic damage in the brains of antibody-treated newborns opens up a number of interesting lines of investigation. Again, more work must be done to define the extent of the damage, the degree to which it is focussed on AChE-rich neurons, and the persistence of the lesions in the maturing brain. The newborn model, once fully validated, offers additional possibilities for investigating central cholinergic function, and for learning more about the maturation of the blood-brain barrier. Its most important application could be to explore the significance of the upregulation of AChE in developing neurons and, more generally to determine whether AChE has specific roles in neuronal development.

ACKNOWLEDGEMENTS

This work was supported by the National Institute of Neurological Disorders and Stroke (grants NS 29646 and NS 15057). We thank Pamela Hammond for technical assistance.

REFERENCES

J. R. Atack, E. K. Perry, J. R. Bonham, R. H., Perry, B. E. Tomlinson, G. Blessed, and A. Fairbairn, Molecular forms of acetylcholinesterase in senile dementia of Alzheimer type: selective loss of the intermediate (10S) form, *Neurosci. Lett.* 40:199 (1983).

A. J. Bean, Z. Xu, S. Y. Chai, S. Brimijoin, and T. Hökfelt, Effect of intracerebral injection of monoclonal acetylcholinesterase antibodies on cholinergic nerve terminals in the rat central nervous system, *Neurosci. Lett.* 133:145-149 (1991).

S. Brimijoin, M. Balm, P. Hammond, and V. A. Lennon, Selective complexing of acetylcholinesterase in brain by intravenously administered monoclonal antibody, *J. Neurochem.* 54:236-241 (1990).

S. Brimijoin and V.A. Lennon, Autoimmune preganglionic sympathectomy induced by acetylcholinesterase antibodies, *Proc Natl Acad Sci USA* 87:9630-9634 (1990).

G. Burnstock, B. Evans, B. J. Gannon, J. W. Heath, and V. James, A new method of destroying adrenergic nerves in adult animals using guanethidine, *Brit. J. Pharmacol.* 43:295 (1971).

C. Eva, M. Fabrazzo, and E. Costa, Changes of cholinergic, noradrenergic and serotonergic synaptic transmission indices elicited by ethylcholine aziridinium ion (AF64A) infused intraventricularly, *J. Pharmacol Exp. Ther.* 241:191 (1987).

H. L. Fernandez, J. R. Stiles, and J. A. Donoso, Skeletal muscle acetylcholinesterase molecular forms in amyotrophic lateral sclerosis, *Muscle Nerve* 9: 399 (1986).

E. B. Fishman, G. C. Siek, R. D. MacCallum, E. D. Bird, L. Volicer, and J. K. Marquis, Distribution of the molecular forms of acetylcholinesterase in human brain: alterations in dementia of the Alzheimer type, *Ann. Neurol.* 19:246 (1986).

P. Hammond and S. Brimijoin, Acetylcholinesterase in Huntington's and Alzheimer's diseases: simultaneous enzyme assay and immunoassay of multiple brain regions, J Neurochem. 50:1111 (1988).

P. Hammond, V. A. Lennon, and S. Brimijoin, Mechanism of selective preganglionic sympathectomy by acetylcholinesterase antibodies, *Soc. Neurosci Abst.* 17:1610 (1991).

E. M. Johnson and F. O'Brien, Evaluation of the permanent sympathectomy produced by the administration of guanethidine to adult rats, J. Pharmacol. Exp. Ther. 196: 53 (1976).

I. Hanin, AF64A-induced cholinergic hypofunction, *Prog. Brain Res.* 84:289 (1990).

W. Isola and Z. M. Bacq, Innervation sympathique adrénergique de la musculature lisse des paupières, *Arch. Int. Physiol.*, 54:30 (1946).

S. J. Kish, L. Schut, J. Simmons, J. Gilbert, J. Chang, and M. Rebbetoy, Brain acetylcholinesterase is markedly reduced in dominantly inherited olivopontocerebellar atrophy, *J. Neurol. Neurosurg. Psychiat.,* 51:544 (1988).

V. Moser, S. Padilla, P. Hammond, and S. Brimijoin, Functional aspects of preganglionic sympathectomy by cholinesterase antibodies, *Soc Neurosci Abstr* 17:1610 (1991).

E. K. Perry, M. Curtis, D.J. Dick, J. M. Candy, J. R. Atack, C. A. Bloxham, G. Blessed, A. Fairbairn, B. E. Tomlinson, and R. H. Perry, Cholinergic correlates of cognitive impairment in Parkinsons's disease: comparison with Alzheimer's disease, *J. Neurol. Neurosurg. Psychiat.* 48: 413 (1985).

Z. Rakonczay and S. Brimijoin, Monoclonal antibodies to rat brain acetylcholinesterase: Comparative affinity for soluble and membrane-associated enzyme and for enzyme from different vertebrate species, *J. Neurochem.* 46: 287 (1986).

Z. Rakonczay and S. Brimijoin, Effects of acetylcholinesterase antibodies on cholinergic systems in newborn rat brain, *Soc Neurosci Abstr* 17:1610 (1991).

C. M. Yates, J. Simpson, A. F. J. Maloney, A. Gordon, and A. H. Reid, Alzheimer-like cholinergic deficiency in Down syndrome, *Lancet*, 2:979 (1980).

CLINICAL IMPLICATIONS OF CHOLINESTERASE ABERRATIONS IN SYNDROMES OF HEMOPOIETIC CELL DIVISION

Haim Zakut[1], Yaron Lapidot-Lifson[1,2,] Deborah Patinkin[2], Dalia Ginzberg[2], Gal Ehrlich [2], Fritz Eckstein[3] and Hermona Soreq[2]

[1]Dept of Obstetrics and Gynecology, The Edith Wolfson Medical Center Holon, The Sackler Faculty of Medicine, Tel-Aviv University, Israel
[2]Dept. of Biological Chemistry, The Hebrew University of Jerusalem
[3]Dept. of Chemistry, The Max Planck Institute for Experimental Medicine Gottingen, Germany

INTRODUCTION

The acetylcholine hydrolysing enzymes cholinesterases (ChEs) control the termination of intercellular communication in brain (Mesulam and Geula, 1988), muscle (Rackonzay and Brimijoin, 1988) and multiple embryonic tissues (Drews, 1975, Soreq and Zakut, 1990a). ChEs serve as the natural scavengers for a variety of drugs (i.e., Heroine derivatives), plant and animal poisons (i.e., toxic glycoalkaloids, snake venom peptides) and man-made compounds (i.e. organophosphorous (OP) nerve "gases" and insecticides). The two human genes encoding acetyl- and butyrylcholinesterase (AChE, BChE) were isolated by genetic engineering techniques (Prody et al., 1987, Soreq et al., 1990); The cloned genes served as research tools to demonstrate the expression patterns of AChE and BChE in human brain (Soreq et al., 1984, Ayalon et al., 1989), germ cells (Beeri et al., this volume), tumors of various origins (Zakut et al., 1988, 1990) and fetal chorionic villi (Zakut et al., 1991). The observation that ChEs are expressed in fetal and germ cells implies that the environmental exposure to ChE inhibitors subjects the AChE and BChE genes to evolutionary selection pressure; This pressure should potentially become more prominent in countries like Israel, where OP ChE inhibitors are intensively employed in agricultural work. Various aberrations in cholinergic signalling were found to be correlated with impaired cell proliferation control in bone-marrow cells (Burstein et al., 1980, Burstein and Harker, 1983, Lapidot-Lifson et al., 1989; Zakut et al., 1992). Therefore, we initiated a study of ChE gene expression in hemopoietic disorders and in cultured bone marrow cells.

Multidisciplinary Approaches to Cholinesterase Functions, Edited by
A. Shafferman and B. Velan, Plenum Press, New York, 1992

METHODOLOGY

ChE gene amplifications were determined by DNA blot hybridization as previously detailed (Lapidot-Lifson et al., 1989). To interfere with ChE gene expression, we employed "Antisense" phosphorothioate oligodeoxynucleotides (AS-oligos). These are short (15 mer) synthetic DNA sequences protected by the insertion of internucleotide phosphorothioate groups from nucleolytic degradation (Matsukura et al., 1987; Spitzer and Eckstein, 1988). As-oligos are actively taken up by living cells (Loke et al., 1989), where they hybridize with their complementary target mRNAs to create double-stranded DNA-mRNA hybrids (Hoke et al., 1991), interfering with translation and exposing these cellular mRNAs for nucleolytic degradation (Chiang et al., 1991). Once the mRNA is destroyed, the stable AS-oligos are released to re-hybridize with other chains of their target mRNAs. These properties make AS-oligos into long-duration drugs of potential importance (Agarwal et al., 1991), particularly when their target mRNAs are essential for vital functions such as cell division or biochemically important pathways like cholinergic signalling (Patinkin et al., 1990). The use of AS-oligos targeted against ChEmRNAs could hence demonstrate the cellular and/or developmental processes for which ChEs are essential.

RESULTS

Cholinesterase Gene and Oncogene Amplifications in Polycythemia Vera

The blood cell disorder *Polycythemia Vera* (P.Vera) is defined as a pre-cancerous hemopoietic syndrome. This is a myeloproliferative disorder classically characterized by splenomegaly and increased production of erythrocytes, granulocytes and platelets. It has a gradual onset and runs a chronic but usually slowly progressive course, with an increased incidence of leukemia (Landaw, 1986). Risks mount to 1-2% in phlebotomy-treated patients, in contrast to 10-15% in those receiving chlorambucil or irradiation (Bloomfield and Bruning, 1976). This suggests that mutagenic agents may become secondary inducers which, added to the existing state in P. Vera, lead to tumorigenesis.

Acute myelocytic leukemia (AML), when it appears secondary to P.Vera, is resistant to therapy and its progress is grave (Landaw, 1986, Bloomfield and Bruning, 1976). It is not yet clear whether predisposing factor(s) or defects in specific signalling pathway(s) contribute to the appearance of malignancy in P.Vera or what these defects may be. To examine whether the AChE and BChE genes amplify in P.Vera, peripheral blood cell DNA from a P.Vera patient prior to any treatment was subjected to slot blot hybridization with (^{32}P)-labeled probes for the AChE and BChE genes as well as to several oncogenes (i.e. c-raf-1, c-myc, c-fes/fps, v-sis and c-fos). DNA samples from 3 leukemic patients and one apparently healthy individual were tested in parallel.

Considerable amplification of both ChE genes and 3 of the examined oncogenes (c-raf1, c-myc and c-fes/fps) was observed in the P.Vera and leukemic DNAs, but not in the healthy control (Fig.1). Although gene and oncogene amplifications associated with malignancy, chemotherapy and multidrug resistance are well known phenomena (Schimke, 1990, Bishop, 1991, Schwab and Amler, 1990), the appearance of gene amplification in P.Vera cannot be classified into any of these subgroups. A careful followup of this P.Vera patient and others presenting gene amplifications will be required

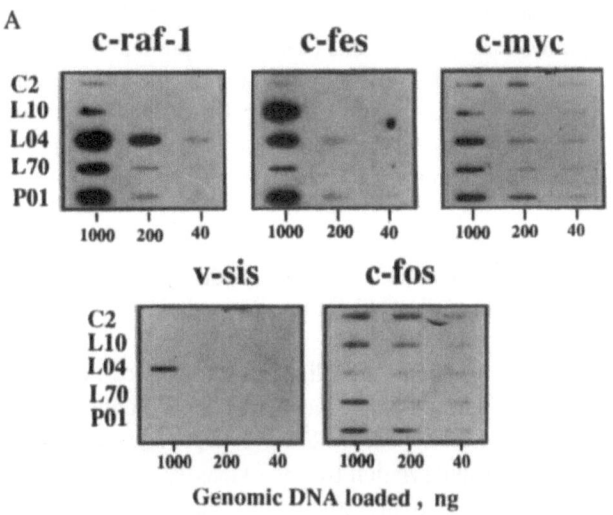

Genomic DNA loaded , ng

B. Approximate amplifications

Patients-	healthy	AML	AML	AMLM2	P. Vera
Genes	(1)	(2)	(3)	(4)	(5)
AChE	1	10	10	25	15
BChE	1	10	15	10	15
c-raf	1	5	25	15	20
c-fes	1	25	15	5	20
c-myc	1	1	5	5	5
v-sis 1	1	1	1	1	1
c-fos 1	1	1	1	1	

Figure 1. Oncogenes and cholinesterase genes amplify in leukemias and polycythemia vera.

(A). Slot blot hybridizations. Peripheral blood DNA samples from an apparently healthy individual (1), two patients with acute myelocytic leukemia (AML,2,3), one with the M2 subdefinition of AML (4) and one with polycythemia vera (5) were diluted and loaded onto Genescreen filters in the noted quantities as previously detailed (Lapidot-Lifson et al., 1989). Hybridization was with (^{32}P)-labeled cDNA probes for the c-raf1, c-fes/fps, c-myc, v-sis and c-fos oncogenes (Amersham Int.). Parallel dilution curves of the purified, cloned DNA probes served for calibration (not shown).Relative strength of the hybridization signal reflects the level of amplification.

(B). Quantification of amplification levels by densitometric analysis of autoradiography efficiencies. Amplification levels for the above detailed oncogenes and the AChE and BChE genes were determined by densitometric analysis of slot blot hybridizations and parallel calibration curves. Note that DNA from the P.Vera patient contained 15-20 copies of the AChE and BChE genes as well as amplified c-raf, c-fes and c-myc oncogenes, a pattern resembling that observed for DNA from the leukemic patients.

to determine whether the amplification of ChE genes as well as oncogenes is involved in the high vulnerability to leukemic transformation which is characteristic of this disease. The finding of amplified ChE genes in this example for pre-leukemic state supports the notion that ChEs could be functionally implicated in the maintenance of normal hemopoietic cell division. Our recent finding of parallel gene amplifications in a case of platelet deficiency associated with the auto-immune disease *lupus erythematosus* (Zakut et al., 1992) may also be linked to this phenomenon.

"Antesense" Oligodeoxynucleotide Inhibition Studies Demonstrate Linkage ıkage Between CHEs and Hemopoietic Cell Dision

We have used phosphorothioate AS-oligos to study the expression and biological significance of ChE genes in developing bone marrow cells (Patinkin et al., 1990). Both the AChE and the BChE proteins include the S/T-P-X-Z peptide motif, which makes them potential substrates for phosphorylation by cdc_2 kinases, controllers of the cell cycle (Moreno and Nurse, 1990). To investigate this putative linkage in hemopoietic development, AS-oligos were further designed to interrupt the expression of cdc_2-related kinases produced in mammalian bone marrow. These include the well known 2Hs homolog of yeast cdc_2 (Lee and Nurse, 1987) and a novel, larger cdc_2 homolog designated CHED, which was molecularly cloned and characterized in the course of our work (Lapidot-Lifson et al., 1992). Table I displays the AS-oligos employed in this study and their effects on hemopoietic cell development in bone marrow cultures.

The efficient uptake of such AS-oligos into living bone marrow cells, their ability to interact with their target mRNAs and induce their destruction and the stability of these oligos within cells were directly examined by RNA-PCR amplification. Thus, AS-CHED significantly reduced the ratio between CHEDmRNA and Actin mRNA within 1 hr of its administration to bone marrow cultures, an effect which persisted for at least 4 days (Lapidot-Lifson et al., 1992). Moreover, the examined AS-oligos induced biologically distinct effects on hemopoietic development: AS-2Hs blocked bone marrow cell proliferation in general, without altering the cell type composition in surviving colonies. In contrast, both AS-BChE and AS-CHED reduced colony counts while selectively inhibiting the development of megakaryocytes, the progenitors of blood platelets. These two AS-oligos did not interfere with other hemopoietic pathways, corroborating our previous observations (Patinkin et al., 1990). Therefore, increasing numbers of macrophages were observed in the treated cultures (Table I).

DISCUSSION

The findings derived from our DNA hybridization and AS-oligo studies reinforce the connection between cholinergic signalling and cell division control. In addition, this study demonstrates the efficacy of AS-oligos as a research tool to study the biological role(s) of ChEs. The aberrations observed in the leukemic, P.vera and lupus erythematosus patients could hence be causally related with defects in the AChE and BChE genes and/or their transcription products, particularly in cases of platelet abnormalities (see Zakut et al., 1992 for further discussion of this issue).

Table I. Effects of AS-oligos on CFU-MEG development

AS-oligo[1]	Nucleotide sequence[2]	Effects on CFU-MEG[3]
1. AS-BChE	CAC TTT GCT ATG CAT	a. Reduced No. of colonies b. Arrested megakaryocytopoiesis c. Increased macrophages
2. AS-CHED	TTT TCC CCA GTC CAT	Similar to No. 1
3. AS-2Hs	GGT ATA ATC TTC CAT	a. Reduced No. of colonies b. Unchanged composition
4. S-BChE (control)	ATG CAT AGC AAA GTC	None

1. See Lapidot-Lifson et al., 1992 for details on the targetted genes.
2. AS-oligos were synthesized as phosphorothioates; see Patinkin et al. (1990) for details on the synthesis.
3. Colony forming units (CFU) were grown in the presence of interleukin 3, conditions allowing for megakaryocyte growth.

The suggested correlation between cholinergic signalling and cell division bears worrysome implications related to the environmental exposure to natural poisons like solanidine, quinoline-derived drugs like dibucaine, warfare organophosphorous poisons or commonly used agricultural insecticides like paraoxon, the metabolite of parathion (Neville et al., 1992). The multipurpose use of ChE inhibitors in daily life implies that the risk for subacute poisoning is serious both for the occupationally exposed individuals and for others (Prody et al., 1989, Soreq and Zakut, 1990b). However, it would be rather difficult to determine whether subacute exposure to ChE inhibitors is causally related to tumorigenesis. Several studies related to this notion should be discussed to evaluate this issue.

To be tumorigenic, a poisonous substance must first be cytotoxic, so that cells exposed to it will be under selection pressure favoring changes in their mode of cell division. The cytotoxicity of subacute exposure to the commercially available organophosphorous insecticide Fenthion was recently demonstrated for cholinoceptive hippocampal neurons (Verones et al., 1990). Furthermore, a pharmacokinetic model for AChE inhibition by the organophosphorous poison diisopropylfluorophosphate (DFP) demonstrated that brain AChE levels do not return to normal under prolonged DFP exposure, suggesting modulation of AChE gene expression (Gerhart et al., 1990). Thus, interference with ChE activities appears to be cytotoxic in brain, in addition to its ability to alter cell division mechanisms in bone marrow.

Farmers represent the human group with the highest risk for occupational OP exposure. Careful epidemiological studies demonstrate that farmers are less vulnerable than average for lung and bladder cancers, perhaps due to lower exposure to nicotine, but are significantly more vulnerable than average to leukemias and brain tumors; and the excess risk to leukemias could be correlated with the extent and duration of exposure to anti-ChE organophosphorous insecticides (Blair et al., 1989). In view of our findings and those of others, demonstrating that acetylcholine analogs stimulate DNA synthesis in brain-derived cells (Ashkenazi et al., 1989), this epidemiological observation strongly supports the notion of causal relationship between ChEs and brain and hemopoietic tumorigenesis.

Another research approach emphasizing the relationship between ChE inhibition and environmental pressure is that of molecular evolution. The emergence of phenotypic novelties through progressive genetic change is considered to reflect environmental stress (Agur and Kerzberg, 1987), resulting in speciation events that are very rapid with respect to the total duration of species as unchanging entities. The abundant point mutations in the well conserved human BChE gene (Neville et al., 1992, Primo-Parmo et al., this volume) may thus represent speciation events providing resistance to ChE intoxication, while tumorigenicity induced by OP exposure could reflect another type of response to environmental poisoning.

ACKNOWLEDGEMENTS

Supported by the U.S. Army Medical Research and Development Command (Grant No.DAMD17-90-Z-0038) and by the American-Israel Binational Science Foundation (Grant No 89-00205) toH.S.& H.Z., and by the BMFT German-Israel Binational Biotechnology program (Grant No. GR01001) toH.S. and F.E.

REFERENCES

Agarwal, S., J. Temsamani and J.Y. Tang. 1991. Pharmacokinetics, biodistribution and stability of oligodeoxynucleotide phosphorothioates in mice. Proc. Natl. Acad. Sci. USA 88:7595-7599.

Agur, Z. and M. Kerzberg. 1987. The emergence of phenotypic novelties through progressive geneticchange. The American Naturalist 129: 862-865.

Ashkenazi, A., J. Ramachandran, and D.J. Capon. 1989. Acetylcholine analogue stimulates DNA synthesis in brain-derived cells via specific muscarinic receptor subtypes. Nature 340:146-150.

Ayalon, A., Zakut, H., Prody, C.A. and Soreq, H. (1990) Preferential transcription ofacetylcholinesterase over butyrylcholinesterase mRNAs in fetal human cholinergic neurons. In: Gene Expression in the nervous system. (A.M. Giuffrida-Stella, Ed.) Alan R. Liss, New York pp.191-203.

Beeri, R., A. Gnatt, Y. Lapidot-Lifson, D. Ginzberg, M. Shani, H. Zakut and H. Soreq. 1992. Testicular gene amplification and impaired BCHE transcription induced in transgenic mice by the human BCHE coding sequence. Proceedings of the Oholo Conference on cholinesterases. in press.

Bishop, J.M. 1991. Molecular themes in oncogenesis. Cell 64:235-248.

Blair, A., S.H. Zahn, K.P. Cantor, and P.A. Stewart. 1989. Estimating exposure to pesticides in epidemiologic studies of cancer. In: Wang, R.G.M., Franklin, C.A., Honeycutt, R.C., et al., eds. Biological monitoring for pesticide exposure-measurement, estimation, and risk reduction. Washington, D.C: American chemical Society, ACS symposium series. 382:38-46. Bloomfield, C.D., and R.D. Bruning. 1976. Acute leukemia as a terminal event in non leukemic hematopoietic disorders. Semin. Oncol. 3:297-317.

Burstein, S.A., J.W. Adamson, and L.A. Harker. 1980. Megakaryocytopoiesis in culture: Modulation by cholinergic mechanisms. J. Cell Physiol. 103:201-208.

Burstein, S.A., and L.A. Harker. 1983. Cholinergic control of platelet production. Clin.Haemotol. 12:3-27.

Chiang, M.Y., H. Chan, M.A. Zounes, S.M. Freier, W.F. Lima and C. Bennett. 1991. Antisense oligonucleotides inhibit intercellular adhesion molecule 1 expression by two distinct mechanisms. J. Biol. Chem. 266:18162-18171.

Drews, E. 1975. Cholinesterase in embryonic development. Prog. Histochem. Cytochem. 7:1-52.

Gerhart, J.M., G.W. Jepson, H.J. Clewell, M.E. Andersen, and R.B. Conolly. 1990. Physiologically based pharmacokinetic and pharmacodynamic model for the inhibition of acetylcholinesterase by diisopropylfluorophosphate. Toxicology and applied pharmacology 106:295-310.

Hoke, G.D., K. Draper, S.M. Freier, C. Gonzalez, V.B. Driver, M.C. Zounes and D.J. Ecker. 1991. Effects of phosphorothioate capping on antisense oligonucleotide stability, hybridization and antiviral efficacy versus herpes simplex virus infection Nuc. Acids Res. 19:5743-5748.

Landaw, S.A. 1986. Leukemia in Polycythemia Vera. Semin. Haematol. 23:256-265.

Lapidot-Lifson, Y., C.A. Prody, D. Ginsberg, D. Meytes, H. Zakut, and H. Soreq. 1989. Coamplification of human acetylcholinesterase and butyrylcholinesterase genes in blood cells; Correlation with various leukemias and abnormal megakaryocytopoiesis. Proc. Natl. Acad. Sci.USA 86:4716-4719.

Lapidot-Lifson, Y., Patinkin, D., Prody, C.A., Ehrlich, G., Seidman, S., Ben-Aziz, R., Benseler, F., Eckstein, F., Zakut, H. and Soreq, H. 1992. Cloning and antisense oligodeoxynucleotide inhibition of a human cdc2 homologue required in hematopoiesis. Proc. Natl. Acad. Sci. USA 89: 579-583.

Lee, M.G., and P. Nurse. 1987. Complementation used to clone a human homologue of the fission yeast cell cycle control gene cdc2. Nature 32:31-35.

Loke, S.L., C.A. Stein, X.H. Zhang, K. Mori, M. Nakanishi, C. Subasinghe, J.S. Cohen and L.M. Neckers. 1989. Characterization of oligonucleotide transport into living cells. Proc. Natl. Acad. Sci. USA 86:3474-3478.

Matsukura, M., Shinozuka, K., Zon, G., Mitsuya, H., Reitz, M., Cohen, J.S. and Broder S. 1987. Phosphorothioate analogues of oligodeoxynucleotides: Inhibitors of replication and cytopathic effects of human immunodeficiency virus. Proc. Natl. Acad. Sci. USA 14:7706-7710.

Mesulam, M.M., and C. Geula. 1988. Nucleus basalis (ch4) and cortical Cholinergic innervation in the human brain: Observations based on the distribution of acetylcholinesterase and choline acetyl transferase. J. Comp. Neurol. 275:216-240.

Moreno, S. and P. Nurse. 1990. Substrates for p34 cdc2: in vivo veritas? Cell 61:549-551.

Neville, L.F., A. Gnatt, Y. Loewenstein, S. Seidman, G. Ehrlich and H. Soreq. 1992. Intramolecular relationships in cholinesterases revealed by oocyte expression of site-directed and natural variants of human BCHE. The EMBO J. Vol. 11 pp 1641-9.

Patinkin, D., S. Seidman, F. Eckstein, F. Benseler, H. Zakut and H. Soreq. 1990. Manipulations of cholinesterase gene expression modulate murine megakaryocytopoiesis in vitro. Mol. Cell. Biol. 10:6046-6050.

Paulus, J.M., J. Maigen and E. Keyhani. 1981. Mouse megakaryocytes secrete acetylcholinesterase. Blood 58:1100-1106.

Prody, C. A., Zevin-Sonkin, D., Gnatt, A., Goldberg, O. and Soreq, H. (1987). Isolation and characterization of full-length cDNA clones coding for cholinesterase from fetal human tissues. Proc. Natl. Acad. Sci. USA 84:3555-3559.

Prody, C.A., P.A. Dreyfus, R. Zamir, H. Zakut, and H. Soreq. 1989. De novo amplification within a"silent" human cholinesterase gene in a family subjected to prolonged exposure toorganophosphorous insecticides. Proc. Natl. Acad. Sci. 86:690-694.

Rackonzay, Z., and S. Brimijoin. 1988. Biochemistry and pathophysiology of the molecular forms ofcholinesterases. In: Subcellular Biochemistry, vol. 12, Harris, J.R. ed (New York: Plenum Press) pp 335-378.

Schimke, R.T. 1990. The search for early genetic events in tumorigenesis: an amplification paradigm. Cancer cells 2:149-151.

Schwab, M., and C. Amler. 1990. Amplification of cellular oncogenes: a predictor of clinical outcome in human cancer. Genes, Chromosomes and Cancer 1:81-93.

Soreq,H., D. Zevin-Sonkin and N. Razon. 1984. Expression of cholinesterase gene(s) in human brain tissues: Translational evidence for multiple mRNA species. EMBO J. 3:1371-1375.

Soreq, H., R. Ben-Aziz, C. Prody, S. Seidman, A. Gnatt, L. Neville, E. Lieman-Hurwitz, E. Lev-Lehman, D. Ginzberg, Y. Lapidot-Lifson, and H. Zakut. 1990. Molecular cloning and construction of the coding region for human acetylcholinesterase reveals a G,C-rich attenuating structure. Proc. Natl. Acad. Sci.USA 87:9688-9692.

Soreq, H. and H. Zakut. 1990a. Human cholinestrerase genes: multileveled regulation. Monographs in Human Genetics Vol.13. R.E. Sparkes, Ed. Karger, Basel.

Soreq,H., and H. Zakut. 1990b. Amplification of butyrylcholinesterase and acetylcholinesterase genes in normal and tumor tissues: putative relationship to organophosphorous poisoning. Pharmaceutical Res. 7:1-7.

Spitzer, F. and F. Eckstein. 1988. Inhibition of deoxyribonucleases by phosphorothioate groups in oligodeoxyribonucleotides. Nuc. Acids Res. 18:1763-1768.

Verones, B., K. Jones, and C. Pope. 1990. The neurotoxicity of subchronic acetylcholinesterase inhibition in rat hippocampus. Toxicology and applied pharmacology 104:470-456.

Zakut, H., L. Even, S. Birkenfeld, G. Malinger, R. Zisling, and H. Soreq. 1988. Modified properties of serum cholinesterases in primary carcinomas. Cancer 61:727-737.

Zakut, H., G. Ehrlich, A. Ayalon, C.A. Prody, G. Malinger, S. Seidman, R. Kehlenbach, and H. Soreq. 1990. Acetylcholinesterase and butyrylcholinesterase genes co-amplify in primary ovarian carcinomas. J. Clin. Invest. 86:900-908.

Zakut, H., J. Lieman-Hurwitz, R. Zamir, L. Sindell, D. Ginzberg, and H. Soreq. 1991. Chorionic villi cDNA library displays expression of butyrylcholinesterase; putative genetic disposition for ecological danger. Prenatal Diagnosis 11:597-607.

Zakut, H., Lapidot-Lifson, Y., Leibson, R., Ballin, A. and Soreq, H. 1992. In vivo gene amplification in non-cancerous cells: Cholinesterase genes and oncogenes amplify in thrombocytopenia associated with Lupus Erythematosus. Mutation Research, in press.

ACETYLCHOLINESTERASE:

A PRETREATMENT DRUG FOR ORGANOPHOSPHATE TOXICITY

Bhupendra P. Doctor,[1] Dennis W. Blick,[2] Mary K. Gentry,[1]
Donald M. Maxwell,[2] Stephanie A. Miller,[3] Michael R. Murphy,[3]
and Alan D. Wolfe[1]

[1] Walter Reed Army Institute of Research
   Washington, DC 20307-5100
[2] U.S. Army Medical Research Institute of Chemical Defense
   Edgewood, MD 21010-5425
[3] Armstrong Laboratory, Brooks Air Force Base
   San Antonio, TX 78235-5000

INTRODUCTION

Accumulation of acetylcholine at cholinergic receptor sites as a result of inhibition of acetylcholinesterase (AChE; EC 3.1.1.7) by organophosphorus compounds (OP) produces effects equivalent to excessive stimulation of these receptors throughout the central and peripheral nervous system.[1,2] The cholinergic crisis thus produced is characterized by miosis, increased salivary and tracheobronchial secretions, bradycardia, muscle weakness, fasiculations, and convulsions, resulting in death by respiratory failure.[3] The multiple drug regimen used for OP poisoning consists of pretreatment with a reversible ChE inhibitor, pyridostigmine bromide, and treatment with a combination of atropine, to counteract the effect of accumulated acetylcholine, and pralidoxime chloride (2–PAM), to reactivate inhibited AChE.[4] This drug regimen is very effective in protecting experimental animals against death by OP poisoning. However, it is ineffective in protecting against performance deficits, convulsions, or permanent brain damage.[4-7] An anticonvulsant drug, diazepam, was recently included as a treatment to minimize convulsions and risk of permanent brain damage.[8,9]

The problems associated with use of these antidotes have stimulated attempts to develop a single protective drug, itself devoid of pharmacological effects, that would provide complete protection against lethality of OPs and prevent post–exposure incapacitation.[4] One approach to prevent lethality and side effects or performance deficits is the use of enzymes such as ChEs as single pretreatment drugs to sequester highly toxic OP anti–ChEs before they reach their physiological targets.[10-19] This approach focuses on cholinesterase as an anti–OP rather than OP as an irreversible anticholinesterase. Pretreatment of mice with purified ChEs successfully protected the animals against 2–4 LD_{50}

Multidisciplinary Approaches to Cholinesterase Functions, Edited by
A. Shafferman and B. Velan, Plenum Press, New York, 1992

277

of a variety of highly toxic OPs without requiring any post–treatment therapy. These studies established a quantitative correlation between the degree of protection against OP and the level of inhibition of administered enzyme.[11-18] We summarize here results on the protection of nonhuman primates, rhesus monkeys, from soman (pinacolyl methylphosphonofluoridate) toxicity as high as 5 LD_{50} by pre–treatment with fetal bovine serum (FBS) AChE. The effectiveness of the use of FBS AChE for protection against soman toxicity was evaluated because this OP is the most refractory to current therapy.[4]

METHODS

Behavioral performance was measured by two different methods: (1) using a sensitive test of cognitive function, serial probe recognition (SPR) task,[20] judged to be highly sensitive to disruption after soman doses as low as 1.5 to 2.0 $\mu g/kg$,[21] and (2) by the Primate Equilibrium Platform (PEP) task.[22] The PEP task, in which a seated monkey is trained to manipulate a joystick to maintain orientation in space, was originally developed as an animal model of pilot performance for aerospace medicine studies. PEP performance is a sensitive measure of cholinergic toxicity, and amounts of OPs not producing overt signs of intoxication cause significant performance deficits in PEP tasks.[22-24] Although FBS AChE, horse serum butyrylcholinesterase (BChE)[18,19] and human serum BChE (Ashani, personal communication) have been tested as pretreatment drugs for OP toxicity in nonhuman primates, we will describe only the results with FBS AChE in this report.

The *in vitro* titration of purified FBS AChE by soman and sarin showed that the molar concentration of soman required for 100% inhibition of enzyme was twice the molar concentration of FBS AChE.[12,17,18] This suggested that only two of the four enantiomers of the racemic mixture of these OPs inhibited FBS AChE.[25] One nmole of purified FBS AChE had 400 units of activity, and one mg of protein contained 14 nmoles of enzyme. One μg of racemic soman completely inhibited 1,100 units (2.75 nmoles) of FBS AChE under the experimental conditions; therefore, the *in vitro* stoichiometry for the reaction of soman with FBS AChE was 1:1. After i.v. administration of 30,000 units of enzyme, blood concentration of AChE was raised more than 100–fold and remained at a constant value up to 8 hr. More than 95% of the administered enzyme was accounted for in circulation. The half–life of this enzyme in nonhuman primates was essentially as long as that previously observed in rodents (40 hr).[12] Because the enzyme had a relatively long half–life and was maintained at a constant level in blood for 6–8 hr post–injection, all *in vivo* titrations were carried out during this period to avoid variations in enzyme level that would occur due to its clearance.

In the first series of experiments, two monkeys were injected i.v. with 7.5 mg (105 nmole active site) of FBS AChE to titrate the enzyme with soman *in vivo*. One half–hour later, blood was withdrawn for AChE determination. Each animal was then injected with 6.5 $\mu g/kg$ (1 LD_{50}) of soman,[26,27] followed one half–hour later by blood withdrawal and AChE determination.[28] Based on blood AChE concentration, a second i.v. injection of soman was given to each animal so that 15–20% of the administered AChE remained in circulation after inhibition with the second soman injection. A final blood sample was taken one–half hour after the second soman injection to determine the residual AChE level after both soman injections. Animals were transferred to SPR test chambers to perform SPR trials within 15–20 min after obtaining blood samples. Details of the SPR apparatus, training procedures, and testing are described elsewhere.[8,17]

A second set in this series of FBS AChE protection experiments using the same experimental design was conducted with 15 mg (84,000 units, 210 nmole active site) of FBS AChE in two additional monkeys that received cumulative doses of 16–18 $\mu g/kg$ of soman (2.5–2.7 LD_{50}) in two injections. The behavioral performance of these monkeys was also tested using the SPR task.

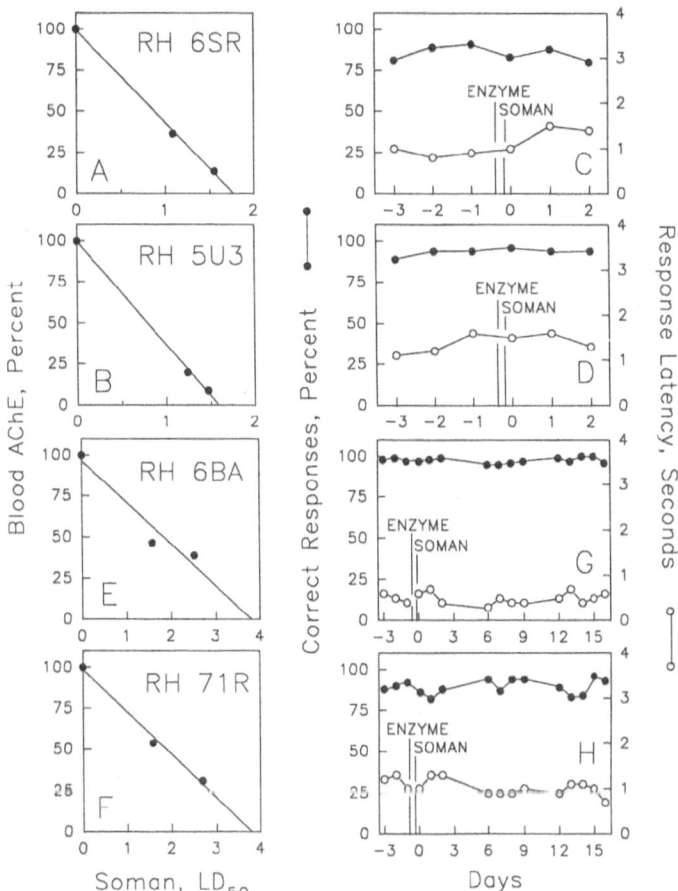

Figure 1. A, B: *In vivo* titration of blood AChE in rhesus monkeys pretreated with 105 nmoles of FBS AChE. See text for details. Initial blood AChE levels were subtracted before calculating percent residual levels. Soman dose shown is the cumulative LD_{50}. **C, D:** Percent correct responses and response latencies for rhesus monkeys. SPR scores (list length of one item) were obtained at indicated times before administration of 105 nmoles of FBS AChE and after challenge with 1.5 LD_{50} of soman, i.v., in two injections. **E, F:** *In vivo* titration of blood AChE in rhesus monkeys pretreated with 210 nmoles of FBS AChE. Monkeys were challenged with 2.5–2.7 LD_{50} of soman. **G, H:** Percent correct responses and response latencies for SPR scores (list length of six items) before injection of 210 nmoles of FBS AChE and after challenge with 2.5–2.7 LD_{50} of soman, i.v., in two injections.

The *in vivo* neutralization of soman by FBS–AChE in monkeys is shown in Figures 1A and 1B. There was a linear relationship between the progressive inhibition of *in vivo* blood AChE activity in these monkeys and the cumulative dose of soman administered. This linear relationship between *in vivo* AChE inhibition and increasing soman levels was consistent with the linear relationship observed with the *in vitro* titration of FBS–AChE by soman. Figures 1C and 1D show the behavioral performance and latency of response of the same two monkeys on the SPR task after being pretreated with FBS–AChE and injected with a cumulative dose of 9.8 $\mu g/kg$ (1.5 LD_{50}) of soman. Monkeys were tested before and after soman exposure on a single–item SPR task to compare their control performance with their performance after the combined effect of FBS–AChE and soman. The performance of both monkeys on the SPR task was unchanged after the soman challenge, indicating that no cognitive deficits were produced as a result of the soman exposure in enzyme–protected monkeys. The monkeys showed no physical signs of OP intoxication.

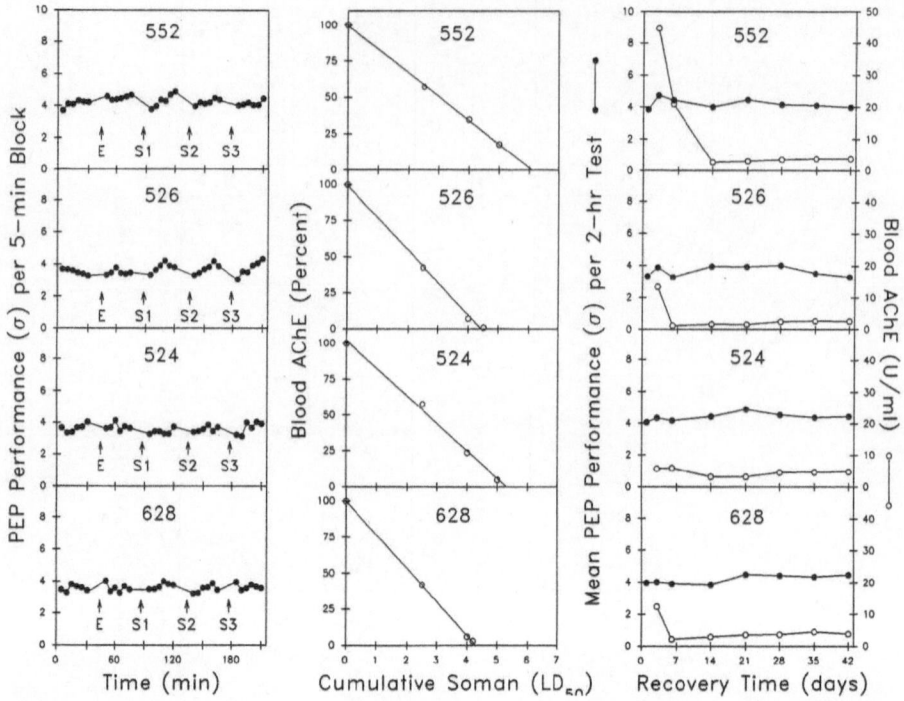

Figure 2. Left: Effects of i.v. administered, purified FBS AChE on PEP task performance by rhesus monkeys before and after challenge with approximately 2.5, 1.5, and 1.0 LD$_{50}$ of soman. See text for experimental details. **Middle:** *In vivo* titrations of blood AChE in monkeys pretreated by i.v. injection with FBS AChE. **Right:** Long-term effects on PEP task performance of i.v.–administered FBS AChE and challenge with a total of approximately 5 LD$_{50}$ of soman with residual blood AChE levels. PEP performance scores are the mean of data from 24 separate 5–min blocks comprising the 2–hr test.

In the second set of this series, animals were tested to examine the protective effect of FBS–AChE against a higher dose of soman and their ability to perform a more complicated SPR task. The results of the *in vivo* titration of FBS–AChE by soman in these monkeys are shown in Figures 1E and 1F. Figures 1G and 1H show the percentage correct responses and response latencies of both monkeys trained on an SPR task to a list length of 6 items. Again, complete protection against behavioral incapacitation by soman was obtained, and no signs of OP toxicity were apparent. In only two trials out of several thousand (4,200) did the monkeys fail to respond within the 10–sec interval.

Protection of rhesus monkeys from soman doses as high as 5 LD$_{50}$ by pretreatment with FBS AChE was also evaluated by the Primate Equilibrium Platform (PEP) task.[17] PEP tests after soman challenge were conducted with one monkey at a time. After a 30–min baseline test, (Figure 2, left panel, start to E), monkeys were given FBS AChE i.v. in an amount sufficient to neutralize, *in vitro*, 32 μg/kg (5 LD$_{50}$) of soman.[12,13,15,17–19,26,27] As serum proteins, both enzymes have long half–lives, 30–40 hrs,[12,13,17,18] and without soman challenge remained at constant levels in the blood for more than eight hrs, well beyond duration of the PEP test (3 hrs). A total of four monkeys receiving FBS AChE were tested. No other treatment or prophylactic agent was given at any time. After AChE injection, monkeys were tested for 30 min on the PEP to determine whether enzyme affected performance. Next, enzyme level in whole blood was measured, followed by the first soman challenge (16 μg/kg, 2.5 LD, i.m.) and 30 min of PEP performance. Blood AChE levels were again determined, and the second soman challenge (9.6 μg/kg, 1.5 LD$_{50}$) was administered. After another 30 min of PEP testing, blood AChE levels were measured for the third time. The result of this measurement was used to calculate a third soman challenge (up to 6.4 $\mu g/kg$, 1.0 LD_{50}) based on residual circulating enzyme activity.

Although blood AChE levels were elevated more than 100-fold, AChE injections alone caused no apparent physiological or neurological effect or deficit, as measured by PEP task performance (Figure 2, left panel, E-S1). None of the four monkeys showed any OP toxicity after soman challenges; protection was so complete that there were no fasciculations even at the site of soman injections. This level of PEP performance continued after the third soman challenge (Figure 2, left panels). Based upon this comparison, we believe that the protective ratio afforded by our AChE pretreatment can be estimated to be between 10 and 15.

Blood AChE levels measured before and after each soman challenge (Figure 2, middle panel), showed linear decline with increasing soman dose. This pattern continued until nearly all enzyme in blood was inhibited. The linear relationship between OP dose and residual AChE *in vivo* agreed with previous observations made with mice[11-13] and confirmed the hypothesis that a stoichiometric reaction is responsible for observed protection. Soman was sequestered by exogenous and endogenous AChE in blood before it reached physiologically-critical AChE. The extent of protection afforded by exogenous AChE was dependent upon concentration of AChE in blood and OP dose. The rate of circulation and the clearance of exogenous enzyme and OP influenced the extent of protection. As shown here, twice the amount of AChE would be required to double the protection.

During the period after soman challenge, blood AChE levels quickly returned to pre-FBS AChE administration levels (Figure 2, right panel). During a six-week period of post-soman testing, no monkey showed signs of delayed toxicity, convulsions, or other OP-exposure symptoms, including abnormality in PEP performance.

RESULTS AND DISCUSSION

Results presented here are in sharp contrast to findings of previous studies testing the effectiveness of the multiple drug regimen currently used against OP (soman) toxicity. In those studies, pyridostigmine-pretreated rhesus monkeys receiving soman and the appropriate regimen of 2-PAM chloride, atropine, and diazepam were partially protected against lethality and extensive convulsions but still displayed serious symptoms of OP poisoning, including convulsions, a period of unconsciousness,[22-24] and, in one study,[24] a period of delayed and extended performance debilitation. We have shown that ChE pretreatment protected against OP-induced lethality, convulsions, and other overt symptoms and performance deficits without delayed effects.

Adverse immunological reactions are among possible drawbacks to the present approach, and our current research is addressing this problem. Because large (stoichiometric) amounts of enzyme will be required to provide *in vivo* protection against OP toxicity, we are also evaluating the possibility that addition of one or more currently used drugs may multiply effectiveness of exogenous ChE. For example, oximes are exceptionally efficient nucleophilic reactivators of OP-inhibited AChE,[30] and bis-quaternary oximes, particularly HI-6, are capable of reactivating OP-inhibited AChE at a rapid rate. Therefore, it is possible that when animals are pretreated with a sufficient quantity of enzyme, followed by an appropriate dose of a reactivator such as HI-6, and then challenged with repeated doses of OP, the enzyme will be continually reactivated. If this is successful, then the potency of the OP as an inhibitor of AChE will be negated. This process should continue as long as the molar concentration of administered OP is lower than the molar concentration of circulating enzyme at any given time.

We have investigated this hypothesis using FBS AChE, HI-6, and repeated doses of sarin, which ages enzyme more slowly than soman. Extrapolation of the *in vitro* titration of FBS AChE by sarin in the presence of HI–6 (Figure 3) indicates that greater than 1:400 stoichiometry of AChE:sarin was achieved. *In vivo* experiments in mice yielded a stoichiometry of greater than 1:65 for sarin. Similar but less efficient reactivation of soman–, tabun–, and VX–inhibited FBS AChE was obtained *in vitro* using 2–PAM, TMB$_4$, and MMB$_4$ as oximes.

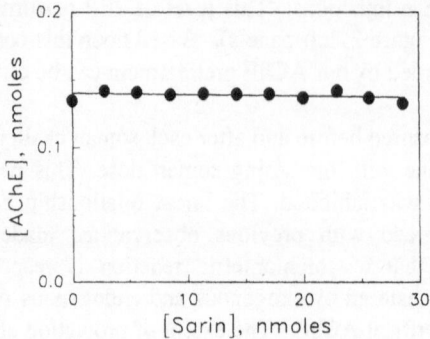

Figure 3. Reactivation of FBS AChE in the presence of HI–6 after repeated additions of sarin.

CONCLUSION

Extensive use of OPs by the pesticide industry, continued threat of use in warfare or by terrorists, and need for destruction of extensive existing stocks of these chemicals make pharmacological and environmental protection against OP toxicity an important issue. We have shown that pretreatment with ChE is the most effective approach found to date for protection against OP toxicity. The approach demonstrated here should provide a foundation for future prophylaxis and therapy for organophosphate toxicity.

REFERENCES

1. A.G. Karczmar, History of the research with anticholinesterase agents, *in*: "Anticholinesterase Agents, International Encyclopedia of Pharmacology and Therapeutics," A.G. Karczmar, ed., Pergamon Press, Oxford (1970).
2. P. Taylor, Anticholinesterase agents, *in*: "The Pharmacological Basis of Therapeutics," A.G. Gilman, L.S. Goodman, T.W. Rall, and F. Murad, eds., MacMillan, New York (1985).
3. R.W. Brimblecombe, Drugs acting on central cholinergic mechanisms and affecting respiration, *Pharmacol. Ther.* [B] 3:65 (1977).
4. M.A. Dunn and F.R. Sidell, Progress in medical defense against nerve agents, *JAMA* 262:649 (1989).
5. P. Dirnhuber, M.C. French, D.M. Green, L. Leadbeater, and J.A. Stratton, The protection of primates against soman poisoning by pretreatment with pyridostigmine, *J. Pharm. Pharmacol.* 31:295 (1979).
6. W.M. Kluwe, J.L. Chin, P. Feder, C. Olson, and R. Joiner, Efficacy of pyridostigmine pretreatment against acute soman intoxication in a primate model, *in*: "Proceedings of the Sixth Medical Chemical Defense Bioscience Review," U.S. Army Medical Research Institute of Chemical Defense, Edgewood, MD (1987).
7. C.G. McLeod, Jr., Pathology of nerve agents: Perspectives on medical management, *Fundam. Appl. Toxicol.* 5:S10 (1985).
8. C.A. Castro, T. Larsen, A.V. Finger, R.P. Solana, and S.B. McMaster, Behavioral efficacy of diazepam against nerve agent exposure in rhesus monkeys, *Pharm. Biochem. Behav.* 41:159 (1991).

9. M.R. Murphy, D.W. Blick, J.W. Fanton, S.A. Miller, S.Z. Kerenyi, F.R. Weathersby, G.C. Brown, and S.L. Hartgraves, Effects of diazepam on soman-induced lethality, convulsions, and performance deficit, Technical Report USAFSAM-TR-89-34, U.S. Air Force School of Aerospace Medicine, Brooks Air Force Base, TX (1989).

10. A.D. Wolfe, R.S. Rush, B.P. Doctor, I. Koplovitz, and D. Jones, Acetylcholinesterase prophylaxis against organophosphate toxicity, *Fundam. Appl. Toxicol.* 9:266 (1987).

11. B.P. Doctor, L. Raveh, A.D. Wolfe, D.M. Maxwell, and Y. Ashani, Enzymes as pretreatment drugs for organophosphate toxicity, *Neurosci. Biobehav. Rev.* 15:123 (1991).

12. L. Raveh, Y. Ashani, D. Levy, D. De La Hoz, A.D. Wolfe, and B.P. Doctor, (1989). Acetylcholinesterase prophylaxis against organophosphate poisoning. Quantitative correlation between protection and blood-enzyme level in mice, *Biochem. Pharmacol.* 38:529-534 (1989).

13. Y. Ashani, S. Shapira, D. Levy, A.D. Wolfe, B.P. Doctor, and L. Raveh, Butyrylcholinesterase and acetylcholinesterase prophylaxis against soman poisoning in mice, *Biochem. Pharmacol.* 41:37 (1991).

14. A.D. Wolfe, D.M. Maxwell, L. Raveh, Y. Ashani, and B.P. Doctor, *In vivo* detoxification of organophosphate in mammosets by acetylcholinesterase, in: "Proceedings of the 1991 Medical Defense Bioscience Review," U.S. Army Medical Research Institute of Chemical Defense, Edgewood, MD (1991).

15. B.P. Doctor, K. Brecht, C. Castro, D. De La Hoz, A. Finger, M.K. Gentry, G. Gold, H. Hively, R. Larrison, D. Maxwell, S. McMaster, R. Solana, A.D. Wolfe, and C. Woodard, in: "Proceedings of the 1991 Medical Defense Bioscience Review," U.S. Army Medical Research Institute of Chemical Defense, Edgewood, MD (1991).

16. D.M. Maxwell, A.D. Wolfe, Y. Ashani, and B.P. Doctor, in: "Cholinesterases: Structure, Function, Mechanism, Genetics, and Cell Biology," J. Massoulié, F. Bacou, E. Barnard, A. Chatonnet, B.P. Doctor, and D.M. Quinn, eds., American Chemical Society, Washington, DC (1991).

17. D.M. Maxwell, C.A. Castro, D.M. De La Hoz, M.K. Gentry, M.B. Gold, R.P. Solana, A.D. Wolfe, and B.P. Doctor, Protection of rhesus monkeys against soman and prevention of performance decrement by pretreatment with acetylcholinesterase, accepted for publication, *Toxicol. Appl. Pharmacol.* (1992).

18. A.D. Wolfe, D.W. Block, M.R. Murphy, S.A. Miller, M.K. Gentry, S.L. Hartgraves, and B.P. Doctor, Use of cholinesterases as pretreatment drugs for the protection of non-human primates against soman toxicity, submitted for publication, *Toxicol. Appl. Pharmacol.* (1992).

19. C.A. Broomfield, D.M. Maxwell, R.P. Solana, C.A. Castro, A.B. Finger, and D.E. Lenz, Protection by butyrylcholinesterase against organophosphorus poisoning in nonhuman primates, *J. Pharmacol. Exp. Therap.* 259:633 (1991).

20. S.F. Sands, and A.A. Wright, Primate memory: Retention of serial list items by a rhesus monkey, *Science* 209:938 (1980).

21. C. Castro, and A. Finger, The use of serial probe recognition in nonhuman primates as a method for detecting cognitive deficits following CNS challenge, *Neurotoxicology*, 125:125 (1991).

22. D.W. Blick, S.Z. Kerenyi, S. Miller, M.R. Murphy, G.C. Brown, and S.L. Hartgraves, Behavioral toxicity of anticholinesterases in primates: Chronic pyridostigmine and soman interactions, *Pharmcol. Biochem. Behav.* 38:527 (1991).

23. D.N. Farrer, M.G. Yochmowitz, J.L. Mattsson, N.E. Lof, and C.T. Bennett, Effects of benzactyzine on an equilibrium and multiple response task in rhesus monkeys, *Pharmcol. Biochem. Behav.* 16:605 (1982).

24. D.W. Blick, M.R. Murphy, J.W. Fanton, S.Z. Kerenyi, S.A. Miller, and S.L. Hartgraves, Incapacitation and performance recovery after high-dose soman: Effects of diazepam, in: "Proceedings of the 1989 Medical Chemical Defense Bioscience Review," U.S. Army Medical Research Institute of Chemical Defense, Edgewood, MD (1989).

25. H.P. Benschop, C.A.G. Konigs, J. Van Geuderen, and L.P.A. De Jong, Isolation, anticholinesterase properties, and acute toxicity in mice of the four stereoisomers of the serve agent soman, *Toxicol. Appl. Pharm.* 72:61 (1984).

26. M.G. Hamilton and P.M. Lundy, HI-6 therapy of soman and tabun poisoning in primates and rodents, *Arch. Toxicol.* 63:144 (1989).

27. N.L. Adams, J. von Bredow, and H.V. de Vera, Intramuscular lethality of GD (soman) in the rhesus monkey. Edgewood Arsenal Technical Report #EB-TR-76039, Edgewood, MD (1976).

28. G.L. Ellman, D. Courtney, V. Andres, R.M. Featherstone, A new and rapid colorimetric determination of acetylcholinesterase activity, *Biochem. Pharmacol.* 7:88 (1961).

29. D.M. Maxwell and K.M. Brecht, The role of carboxylesterase in species variation of oxime protection against soman, *Neurosci. Biobehav. Rev.* 15:135 (1991).
30. I.B. Wilson, Molecular complementarity and antidotes for alkyl phosphate poisoning, *Fed. Proc.*, 18:752 (1959).

NOTE: Figures 1 and 2 are reproduced with the permission of *Toxicology and Applied Pharmacology*, refs. 17 and 18.

RECOMMENDATIONS FOR NOMENCLATURE IN CHOLINESTERASES

Jean Massoulié, Joel L. Sussman, Bhupendra P. Doctor,
Hermona Soreq, Baruch Velan, Miroslaw Cygler, Richard Rotundo,
Avigdor Shafferman, Israel Silman, and Palmer Taylor

Information gathered on the genomic organization of cholinesterase genes, advances in biochemical characterization of the various subunits and the elucidation of the tertiary structure of the *T. californica* enzyme can serve as a basis for a unified nomenclature in cholinesterase research. Herein are some recommendations for nomenclature of catalytic and structural subunits, for designation of cholinesterase gene exons and secondary structure motifs, and for numeration of amino acid positions.

Exon Numbering

The exon containing the predominantly used CAP site will be numbered exon-1 (alternative exons for the start of transcription will be designated 1α,1β) and the following exons will be marked by consecutive numbers. Exon numbers in different genes will not necessarily refer to homologous coding blocks, since cholinesterase sequences are interrupted in different ways in genomes of various species. As opposed to mammalian AChEs, the avian and the *Torpedo* AChE as well as mammalian butyrylcholinesterase have no interruption in the first exon of the open reading frame (Taylor *et al.*, 1992). In addition, alternative splicing has not been identified in the 3' end of the avian AChE and in the butyrylcholinesterase gene. Exon numbering in several cholinesterases, for which gene organization has been elucidated are presented in Figure-1.

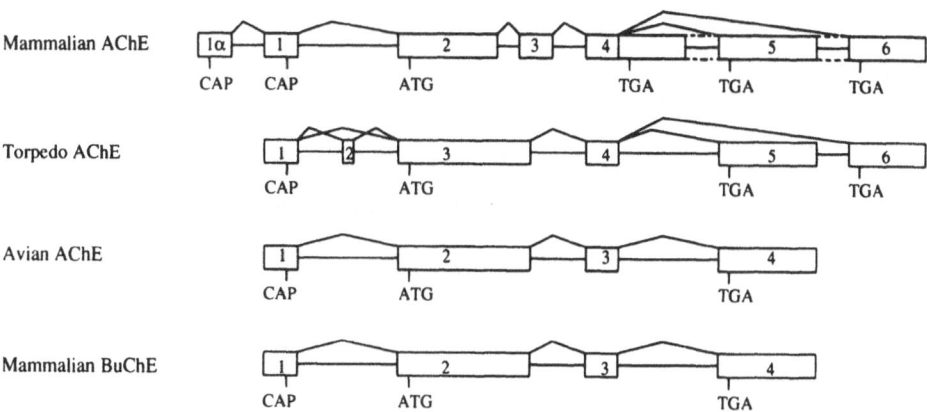

Fig. 1 Schemes of exon-intron organization in cholinesterase genes. Exon and intron distances are not drawn to scale.

Multidisciplinary Approaches to Cholinesterase Functions, Edited by
A. Shafferman and B. Velan, Plenum Press, New York, 1992

Subunit Nomenclature

The simplified system for designation of cholinesterase subunits relies mainly on coding sequences utilized for their generation. Catalytic subunits are divided into types H, T and R (the latter polypeptide remains to be identified in the purified protein) and structural subunits are designated Q and P (Figure 2). Subtypes within these subunits, once characterized, could be named T_I, T_{II} etc.

Subunit T - Catalytic subunit which can form soluble homo-oligomers or can be linked to the collagen-tail and other structural subunits.

Subunit H - Catalytic subunit containing the glycophospholipid anchor or encoding a sequence which enables this processing.

Subunit R - Catalytic subunit encoded by reading through the 3' splice site of the last invariant exon (exon 4 in mammals) in the open reading frame.

Subunit Q - Collagen-like structural subunit.

Subunit P - Lipid-linked structural subunit

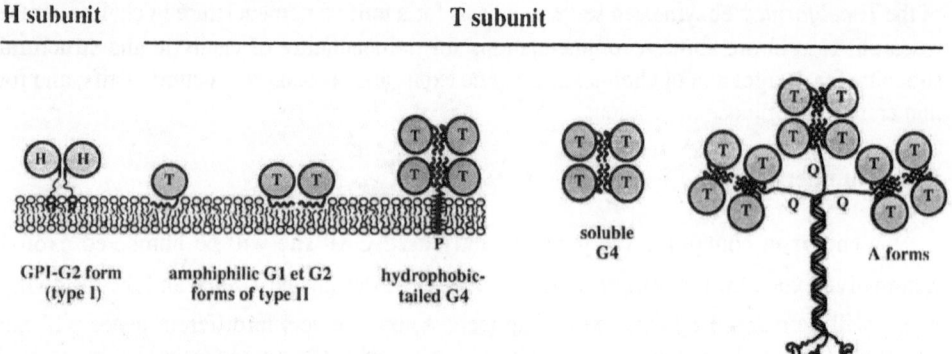

Fig. 2 Catalytic subunits and molecular forms of acetylcholinesterase (Massoulié *et al.*, 1992).

Nomemclature of Secondary Structure Motifs

Nomenclature of the structure element of the cholinesterase family fold (Fig. 3) is based on the 3-dimensional X-ray structures of *Torpedo californica* (AChE) (Sussman *et al.*, 1991) and *Geotrichum candidum* lipase (GLIP) (Schrag *et al.*, 1991). The strands of the small N-terminus β-sheet are marked by b_j where j ranges from 1 to 3. Strands of the large central β-sheet are marked by the Greek letter β with subscripts from 0 to 10. The numbering starts at 0 in order to be consistent with the nomenclature of the 'α/β hydrolase fold' motif (Ollis *et al.*, 1992). The connection (loop) between strands i and j of the β-sheet is marked as $l_{i,j}$ or $L_{i,j}$, for the small of large sheet respectively. The α-helices are referenced as $\alpha^k_{i,j}$, where the subscripts refer to the loop in which the helix is embedded, and superscript k refers to the sequential number of this helix within the loop. If there is only one helix in the connecting loop the superscript is not used. In some cases, the symbol $L^k_{i,j}$ is used, in which the superscript refers to the sequential order of a part of the $L_{i,j}$ loop contained between two

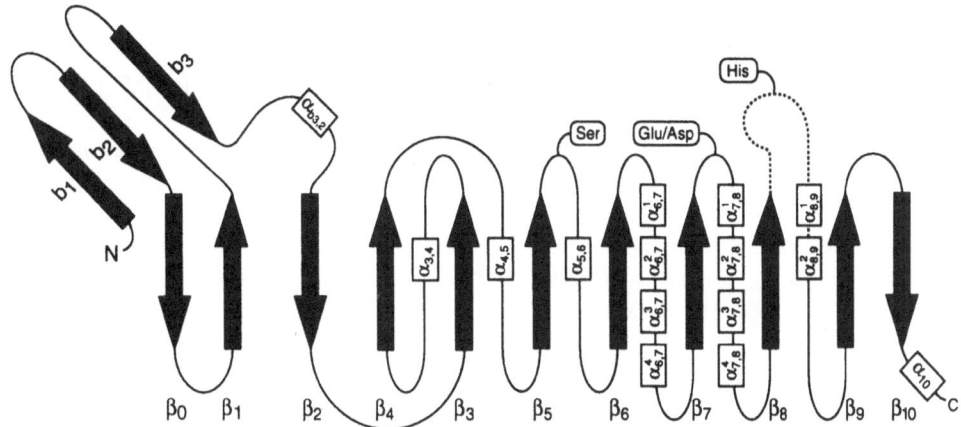

Fig. 3 Secondary structure topology diagram of cholinesterases

β-strands are indicated with black arrows and α-helices with open rectangles. The position of the 3 residues of the active site, i.e. Ser-Glu-His are marked on the loops where they have been observed in AChE and GLIP. For some members of the cholinesterase family, Asp replaces Glu as the *acid* in the triad, and therefore on this figure the *acid* is indicated as Glu/Asp.

secondary structural elements. As the β strands are likely to be well conserved in the different cholinesterase structures (Cygler *et al.*, 1992), the numbers should remain relatively fixed for them. Based on the AChE and GLIP structures, it is clear that the number and position of the α-helices may differ slightly, however this would only affect local superscripts, and therefore most changes based on the determination of new 3-D structures should only result in local modification of the secondary structure numbering. For example, there are two α-helices in each of the two loops $L_{b3.2}$ and $L_{4.5}$ of GLIP, while only one α-helix is found in each of these loops in AChE.

Numbering of Amino Acids

Numbering will begin at the first amino acid of the mature form of each individual cholinesterase catalytic polypeptide (Gentry and Doctor, 1991). If alternative signal-processing leads to generation of more than one mature form, as appears to be the case in human AChE, (Velan *et al.*, 1992) numbering will begin at the longer polypeptide. To allow alignment of specific, functional amino acids from different sources, marking of the homologous position in the sequence of the *Torpedo californica* AChE is recommended. The *Torpedo* amino-acid number will be written in italics and will appear in parentheses following the species specific number. For example the active-site serine of human acetylcholinesterase which is located at position 203 will be designated S203(*200*). For sake of convenience, this composite numeration could be used only for the first reference to the particular residue, or presented as a table (Shafferman *et al.*, 1992). Whenever point mutations are referred to, the native residue will appear before the residue number and the substituted one after it. This will apply both for natural variants and site-directed mutants (Gnatt *et al.*, 1992). Thus, the natural variant of human BChE in which aspartate 70 is modified into glycine will be D70G.

REFERENCES

Cygler, M., Schrag, J.D., Sussman, J.L., Harel, M., Silman, I., Gentry, M.K., and Doctor, B.P. 1992. This volume.

Gentry, M.K., and Doctor, B.P. (1991) in *Cholinesterases: Structure, Function, Mechanism, genetics, and Cell Biology.* (Massoulié, J., Bacou, F., Barnard, E.A., Doctor, B.P. and Quinn, D.M. eds) pp. 394-398. Am. Chem. Soc., Washington.

Gnatt, A., Loewenstein, Y. and Soreq, H. 1992. This volume.

Massoulié, J., Bon, S., Anselmet, A., Chatel, J.M., Coussen, F., Duval, N., Krejci, E., Legay, C., and Vallette, F. 1992. This volume.

Ollis, D.L.,, Cheah, E., Cygler, M., Dijkstra, B., Frolow, F., Franken, S. M., Harel, M., Remington, S.J., Silman, I., Schrag, J., Sussman, J.L., Verschueren, K. H. G., and Goldman, A., 1992, protein Engineering. 5:197.

Schrag, J. D., Li, Y., Wu, S., and Cygler, M., 1991, *Nature.* 351:761

Shafferman, A., Velan, B., Ordentlich, A., Kronman, C., Grosfeld, H., Leitner, M., Flashner, Y., Cohen , S., Barak, D. and Ariel, N. 1992. This volume.

Sussman, J.L., Harel, M., Frolow, F., Oefner, C., Goldman, A., Toker, L., and Silman, I. 1991, *Science.* 253:872

Taylor, P. 1992. This volume

Velan, B., Kronman, C., Leitner, M., Grosfeld, H., Flashner, Y., Marcus, D., Lazar, A., Keren, A., Bar-Nun, S., Cohen, S. and Shafferman, A. 1992. This volume.

INDEX